LAMINATED SPRINGS

INTERIOR OF A SPRING SHOP IN SHEFFIELD. *From a painting by W. Luker.*

Reproduced by courtesy of Messrs. Brown Bayleys Steel Works Ltd.

LAMINATED SPRINGS

(Leaf Spring Design)

BY

T.H. Sanders,

M.I. Mech. E., M.I. & S.I.

Part A—Calculation and Design
Part B—Manufacture

Wexford College Press
2007

CONTENTS

PART A :—CALCULATION AND DESIGN

PART B:—MANUFACTURE

B.S. This abbreviation stands for " British Standard."

R.C.H. These initials mean " Railway Clearing House " a British Institution supported by all the railway companies of the United Kingdom, a branch of which deals with the supervision of the standardized wagon stock of the " private owners."

EXPLANATION OF SYMBOLS USED

(except in Chapter IX.)

BM	Bending Moment of any beam.
B	Total Breadth or Width of any beam or spring = also bn.
b	Breadth or Width of one plate of a spring.
D	Test Deflection for a spring. Also the camber of a spring.
d	Unit deflection (inches per ton, or metric units) of a beam or spring.
E	Modulus of Elasticity, 13000 tons per sq. inch.
f	Unit Stress in a beam or spring under load.
G	Number of elliptic springs in a complete working Group.
H	Width of spring hoop, or axlepad bearing.
I	Moment of Inertia of any section.
i	Elastic Elongation ("allongement elastique") of the surfaces ot beams or spring plates.
J	Offset of spring plates.
K	A Constant, as used in formulae for unit deflection.
k	A Constant, as used in formula for spring weights.
L	Length of a beam or spring, between bearings, measured as "straight" or "full" length.
Ls	Length of the "short" plate of a spring.
n	Number of plates in a spring, of uniform thickness.
P	Periodicity of a spring.
Q	Difference in the length of the two arms of an asymmetrical semi-elliptic.
R	Radius of a curved beam or spring.
r	Resiliency of a spring, also "resiliency efficiency per cent."
S	Span, or full chord, between bearings, of a cambered spring.
T	Thickness of a beam or spring plate.
TT	Total Thickness of the plates of a spring, equal to the product nT if all plates are the same thickness.
t	Thickness of a spring plate, in sixteenths of an inch, $\frac{1}{4}$ in. = 4, $\frac{3}{8}$ in. = 6, and so on.
W	Load on a beam or spring. (Whether Test Load, Working Load, or Safe Load, is specially stated).
w	Weight of a spring (without the hoop or fittings).
wh	Weight of a spring hoop.
X	The result obtained from the expression Test Load (or Deflection) $\div$ Working Load (or Deflection). Can be termed "Factor of Safety," bearing in mind that the ultimate stress figure is taken as the Elastic Limit of the steel, and not as the final Breaking Stress, as usually meant when the term "Factor of Safety" is employed.
Z	Modulus of any Section.

LAMINATED SPRINGS

PART A :—CALCULATION AND DESIGN

CHAPTER I

THE SPRING AS A BEAM

THE laminated spring is a beam in the ordinary mechanical sense of the word, differing chiefly from normal weight carrying beams in the fact that the spring is a beam of relatively very high deflection, whereas the ordinary constructional type of beam, whether solid or built-up, has necessarily to be designed in such a way as to give a negligible deflection, in order to maintain the general lines of the structure. The function of a spring is to absorb shock, and this absorption is performed during the travel of the spring consequent upon the shock. Clearly, the greater this travel, the less obvious will be the displacement of the vehicle body, and as the converse holds good, if no spring were interposed, or if the spring were of negligible deflection, the vehicle body would be instantaneously aware of the receipt of the shock. It is very necessary that all types of vehicles, whether railway or road, should be efficiently spring-borne in order to permit of smooth running, not only from the passenger or freight point of view, but also from the aspect of the prolongation of the life of the vehicle itself. Many practical considerations, however, arise in the spring design as regards permissible flexibilities, such as the necessity for maintaining buffer heights in railway work, and the chassis clearances to tyres, etc., in road work, and general design therefore is subordinated in part to these points, as will be regarded hereafter.

The spring is attached and arranged in various methods, of which the most used can be conveniently dealt with under six heads. Fig. 1 illustrates examples, and shows adjacent the conventional beam diagram. In this drawing :—

A. Instances a " Semi-Elliptic " Spring. This is the most used of all forms, both in railway and road work. For locomotives this type is nearly universal, and, with the exception of American type freight stock, nearly all passenger and freight vehicles throughout the world are borne on this design. Probably 95 per cent. of the world's automobiles are also carried on this pattern. As shown, which is in its most used form, the weight is carried at each end, and the spring is supported at the centre, in other words the ends deflect. The illustration is of a reverse cambered engine spring, but most semi-elliptics loaded at the ends are either normally cambered, or straight, when working.

B. Shows the same design of " Semi-Elliptic " spring, but amended in situation, having the weight applied at the centre, and the ends supported, the deflection therefore taking place at the centre. This arrangement is used very largely in railway work for locomotive bogies, and in automobile work for transverse front axle suspension.

C. Indicates a built up arrangement of " Semi-Elliptics," known in the form shown as an "Elliptic" or "Bolster" spring. This is chiefly employed for the bogies of railway passenger stock, and very rarely used in road work. The deflection of the double spring is twice that per unit load of the single spring, otherwise, if the single " Semi-Elliptic " deflects 1 in. per ton, the " Elliptic " will deflect 2 ins. per ton.

D. This illustration shows an ordinary " Semi-Elliptic " used as a " cantilever " (automobile trade definition). With the arrangement shown, all the deflection is concentrated at one end, thereby rendering possible great flexibility. This pattern is nearly unknown in railway practice, but it is largely employed in the highest type of passenger automobile.

E. Shows a true cantilever spring, in which one end is fixed, and the other is weight carrying. The security and solidity of the end fixing of this pattern is a matter of considerable importance, and generally speaking, except with certain makes of light road vehicles, springs of this design are not used.

F. Illustrates a " double-cantilever," the plates of the spring being continuous, but having two bearing points in the length. It is on rare occasions used in railway

Spring Arrangements.
Beam Diagrams.
Engine Frame
Engine Frame
Centre Pivot
Bogie Frame.
Tramcar Body
4-Wheel Truck
Auto. Chassis
Auto. Chassis
Greatest Bending Moment
Buffer Beam.
A
B
C
D
E
F
W
W/2

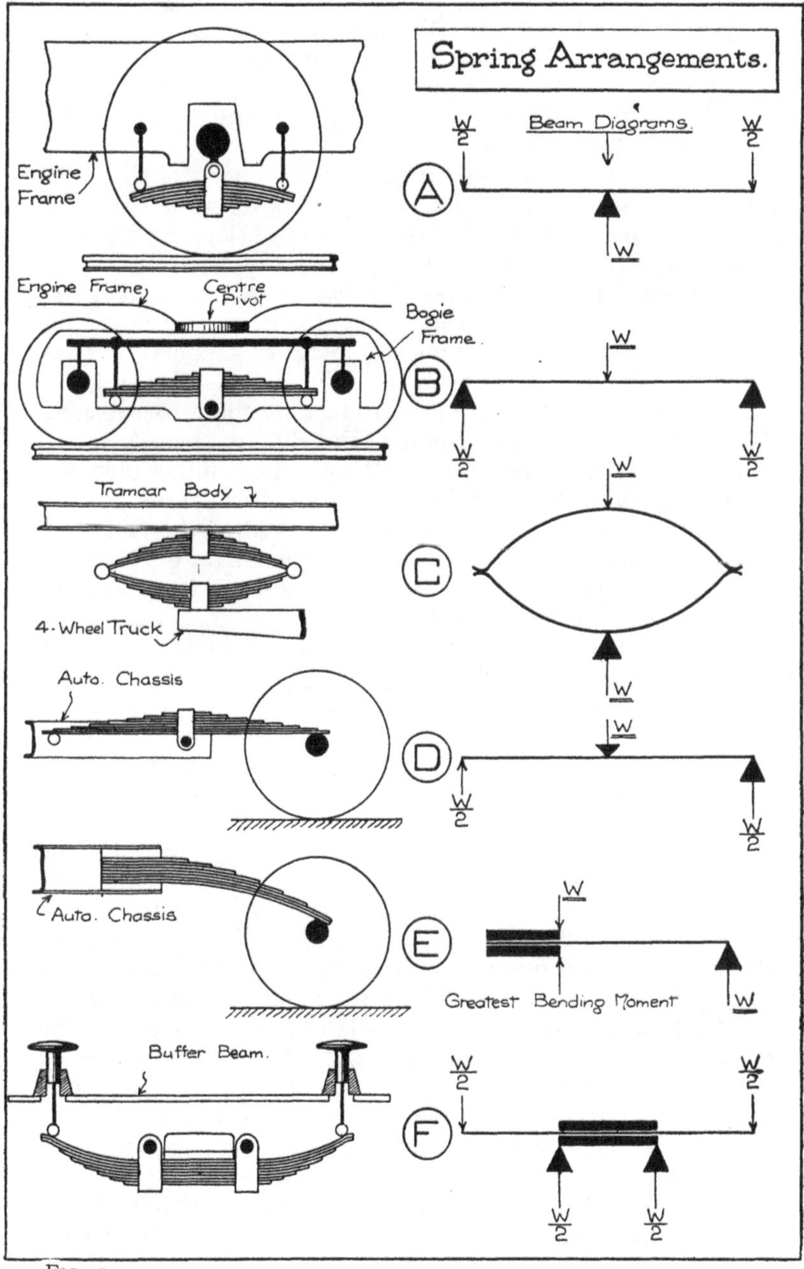

FIG. 1.

practice, but not, in the exaggerated form shown, in road practice. It is, however, a matter of considerable interest as to what extent a wide bearing surface at the middle transforms an ordinary "semi-elliptic" into a "double-cantilever."

Having thus shown practically and conventionally, the various spring arrangements in their relationships with the mechanical beam, it will be well to touch briefly on the few elementary calculations, and definitions pertinent to those calculations. It is clear that the first point turns upon the weight carried and the length of the beam between the point of weight application and the point of support. Specifying the length of the beam as indicated as (L) and the weight supported as (W), the product of these expressions ($W \times L = WL$) is known as the "bending moment" at the point of support. The product (WL) is the bending moment for the simplest form of beam, namely the cantilever. A beam supported in the centre and weighted at the ends can be regarded as a double cantilever, and the maximum bending moment is then ($WL \div 4$), which is acting at the centre. This can be taken as equivalent to the simple "semi-elliptic" spring, which, for calculation purposes, will be the only one at present regarded. This bending moment is the greatest at some position remote from the end supports and consequently, in a symmetrical semi-elliptic spring, it is greatest in the middle of the length. Fig. 2 illustrates these moments, and also includes in the diagrams the shearing force on a beam, which is due to applied weight. This is included to render the aspect theoretically complete, but in the consideration of spring beams (as distinct from structures) shearing force can be neglected, as the available total steel section at all points is always such that the stress due to shear is very low, due to the primary need of design having to be devoted to arranging the sectional strength so as to keep within practical working limits of skin stress due to the bending moment. Perfect beams, that is, beams designed in such a way that throughout their length the sectional area is proportionate to the bending moment, will deflect, under a centrally applied load, to a true circular arc. Such beams are shown on Fig. 2, of uniform thickness throughout and varying in width symmetrically and proportionately to the bending moment diagrams. Beams not of this perfect design (which alternatively may vary in thickness and remain

Simple Beams.
Cantilever
Fixed one end
Concentrated load other end → W = 5
Beam
Supported both ends
Concentrated load in centre of length
W = 10
Re-action each end = 5
L = 15
L = 30
Bending Moment at ※
WL = 5 × 15 = 75
Bending Moment at X
WL ÷ 4 = 10 × 30 / 4 = 75
75
Z1
Greatest Bending Moment 75
Z2
Z3
Bending Moment Diagrams.
Ordinates Z show magnitude of Bending Moment along beam lengths
Shearing Diagrams.
Shearing Force W = 5
W = 5
W = 5
The Deflection of Perfect Beams.
5
10
The plans below are of the triangle (half-rhombus) and double triangle (rhombus) forms respectively, the widths corresponding relatively at all parts with the magnitudes of the bending moments along the beam lengths All portions are therefore equally stressed, and the deflected forms shown are circular arcs.
Plan Views of Perfect Beams of Uniform Thickness.
W1
Max Width
W2
W3

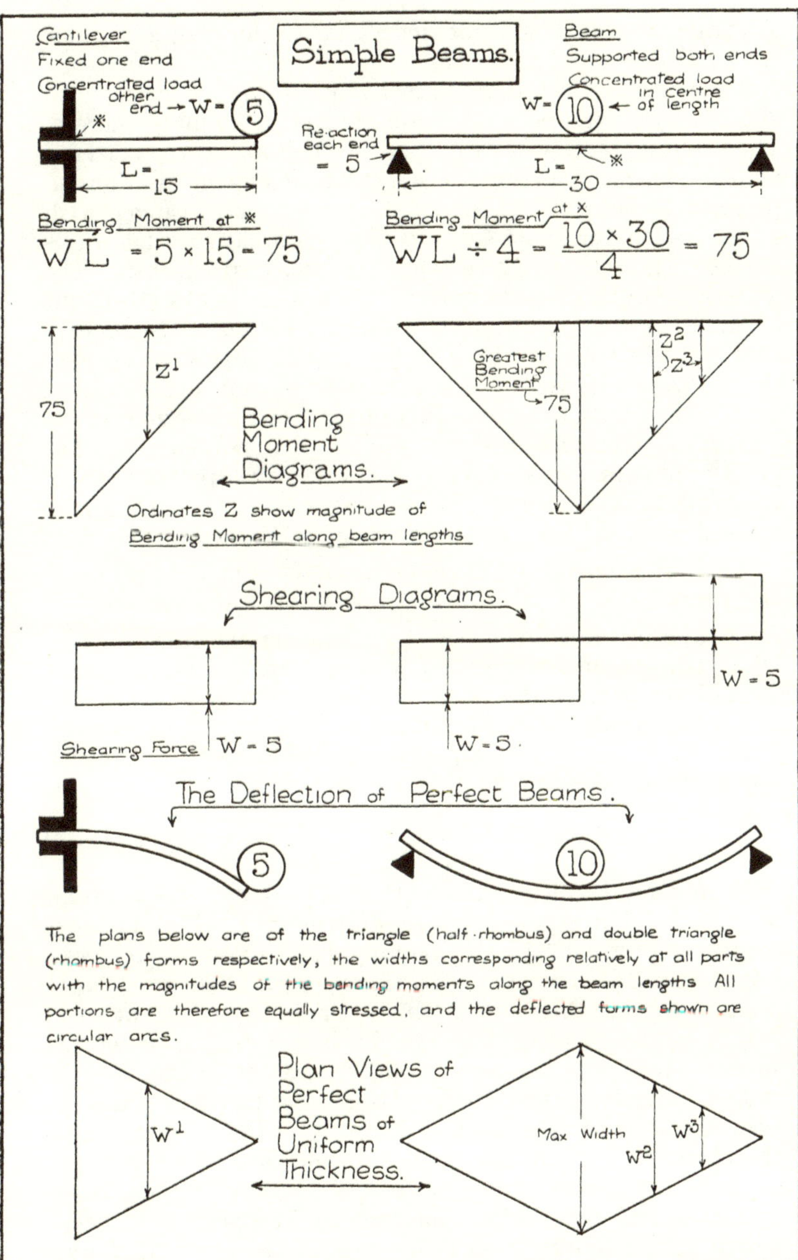

FIG 2.

of constant width) will not deflect in the form of a circular arc as will be explained.

Knowing the bending moment, it is then necessary to obtain some expression of beam strength to equate thereto. The beam strength will turn upon two factors, (1) the material employed and (2) the section and arrangement thereof. (A rectangular or flat section can obviously be arranged with either its long or short sides vertical, and clearly such alternatives respectively give very different strengths.) The material employed in laminated springs is always steel, but of different qualities and treatments, having consequently varying allowable working limits. Such limits or stresses will be called (f) tons per square inch (or any other desired unit). The second factor is a property of the section and is known as "the modulus of section," with symbol (Z). The product of these two factors ($f \times Z$) gives an expression which can now be equated to the bending moment, so that in a general way, a first equation has been evolved, namely,

$$\text{Bending Moment} = \text{Stress} \times \text{Modulus of Section},$$

or . . .

$$BM = f \times Z. \quad \text{(I.—1)}$$

As has been explained, the modifications of the bending moment are numerous, due to type of beam and arrangement of loading, but for the purpose of calculations dealing with laminated springs, it is not immediately necessary to consider forms other than the simple, centrally loaded beam, with the (greatest) bending moment as ($WL \div 4$), or

$$BM = \frac{WL}{4} \quad \text{(I.—2)}$$

To understand clearly the method of obtaining the modulus of section (Z) it is necessary to study Fig. 3, which includes diagrams pertinent to the rectangular section in universal use for laminated springs. In sketch 3—A is shown a beam bent to the arc of a circle, which results in an alteration of length to the surfaces, the outside, larger radius, or tension surface, becoming obviously longer, and the inside, smaller radius, or compression surface, becoming obviously shorter. Clearly, if one surface has lengthened, and the surface remote therefrom has shortened, some intermediate and imaginary internal layer must have retained its original length. If the material under consideration has the properties of steel, which, for all practical purposes, has similar physical characteristics in both tension and compression, otherwise, under equal stress, extension or compression proceed at the same rate, this imaginary internal layer will be midway between the remote surfaces, and will correspond therefore with the geometrical midway of the section. In the case of material having properties similar to cast iron, which is stronger in compression than in tension, and accordingly stretches more

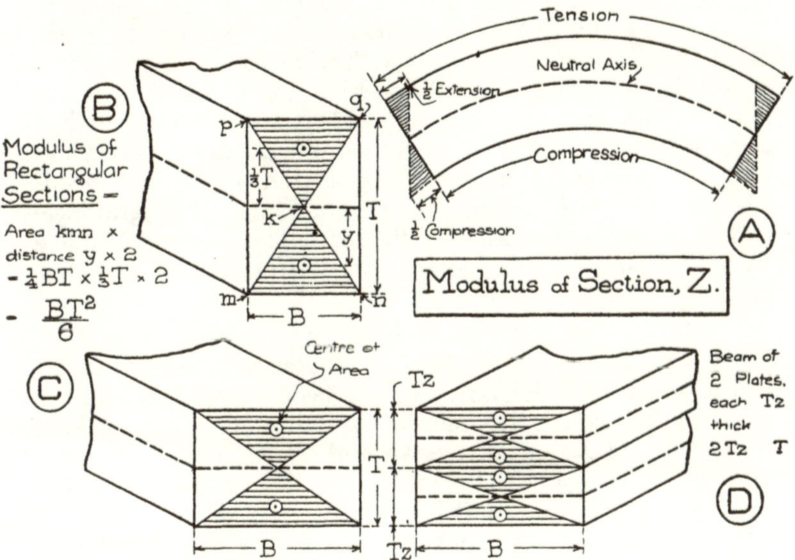

FIG. 3.

than it compresses under equivalent loading, this internal layer will be nearer the tension surface than the compression surface. With steel, as pointed out, the position will be midway of the thickness, and to this non-altered length in a bent beam is given the name " neutral axis," and this axis is used as a plane of reference in connection with the obtaining of (Z). Sketch 3—B shows the method of calculation, to which the following reasoning applies. Assuming the beam to be composed of a series of attached layers, or laminæ, then, under deflection, the surface layers deform to a maximum, and the neutral axis layer to the absolute minimum of zero. Between this zero of the mechanical and geometrical axis and the maximum deformations of the remote surfaces, each intervening imaginary layer is deformed in length according to its distance from the neutral axis, and such deformation in length corresponds to a proportionate stress.

The deformation is assumed to be of regular variation on layers of regular variation as regards distance from the axis, and the stress in such layers—resisting or applied—also is assumed to be of similar regular variation. Accordingly are obtained stress diagrams represented by the triangles (pqk)—say, tension—and (mnk)—say, compression. Each triangle in area is equal to one quarter of (BT)—breadth × thickness—or (BT ÷ 4). The centre of resistance of such triangular system corresponds with what would be the centre of gravity of a thin triangular section and is distant $\frac{1}{3}$ of the height from the base line (coincident in this case with a remote surface) or $\frac{2}{3}$ of the height from the apex (coincident with the geometrical centre of the section). This distance is therefore $\frac{2}{3} \times \frac{1}{2}T$ from the neutral axis. The graphical resistance of the upper half of the section is its area multiplied by the distance of this resisting centre from the neutral axis, or

$$BT \div 4 \times \tfrac{2}{3} \times \tfrac{1}{2}T = BT^2 \div 12.$$

There are, however, two areas of resistance, a tension and a compression, of which only one has been considered, and the effect of the two is to double the resistance of the one, making thus the total graphical resistance of the section ($BT^2 \div 12$) × 2 or ($BT^2 \div 6$). The actual practical resistance of the section depends additionally on the material used for the beam, and the limiting stresses thereof, which function is expressed by the symbol (f), so that the total expression of the resistance becomes ($BT^2 \div 6$) × f, or, as generally written,

$$\frac{fBT^2}{6} = Z = \text{the modulus of section of a rectangular beam.} \qquad \text{(I.—3)}$$

The moduli of the sections shown in 3—B, C, D, then become as follows :—

Sketch 3—B. $T = 3, B = 2, Z = \frac{18}{6}$ or 3
,, 3—C. $T = 2, B = 3, Z = \frac{12}{6}$ or 2
,, 3—D. $T_z = 1, B = 3, Z = \frac{3}{6}$ or $\frac{1}{2}$ for one section and the two sections have therefore a combined Z of $2 \times \frac{1}{2}$ or 1.

For the same weight of material, a beam of two laminæ such as 3—D, has therefore only one-half the resisting power of the corresponding solid beam 3—C, in other words, its limiting stress will be reached with one-half the load. The maximum practical resistance is proportional always to the square of the thickness.

A Elevation
X
Y
T
15
5
W 1
Beam Elevation is a
RECTANGLE
At Plane XX
BM = 15 · 1 = 15
At Plane YY
BM = 5 · 1 = 5
On Centre Line
BM = 25 × 1 = 25
Z = B · T² ÷ 6
or as B = 20
A Plan
B = 12
B = 20
B = 4
Z = B × T² ÷ 6
or as B = 12
Z = B × T² ÷ 6
or as B = 4
Beam Plan is a
RHOMBUS.
BM @ YY = 5 : 4 ∴ Z
BM @ XX = 15 : 12 ∴ Z
BM @ Centre = 25 : 20 ∴ Z
or
5 : 4 :: 15 : 12 :: 25 : 20
B Elevation
Beam Elevation is
PARABOLIC.
T = 20
T = 12
T = 4
T = 5
Two forms of Elevation Outline
16
9
25
25
At Plane XX
BM = 16 × 1 = 16
Z = B × T² ÷ 6
or as T² = 4² = 16
At Plane YY
BM = 9 × 1 = 9
Z = B × T² ÷ 6
or as T² = 12² = 144
B Plan.
On Centre Line
BM = 25 × 1 = 25
Z = B × T² ÷ 6
or as T² = 5² = 25
On Centre Line
BM = 25 × 1 = 25
Z = B × T² ÷ 6
or as T² = 20² = 400
Beam Plan is a
RECTANGLE.
Perfect Beams.
BM @ YY = 9 : 144 ∴ Z
BM @ Centre = 25 : 400 ∴ Z
or 9 : 144 :: 25 : 400

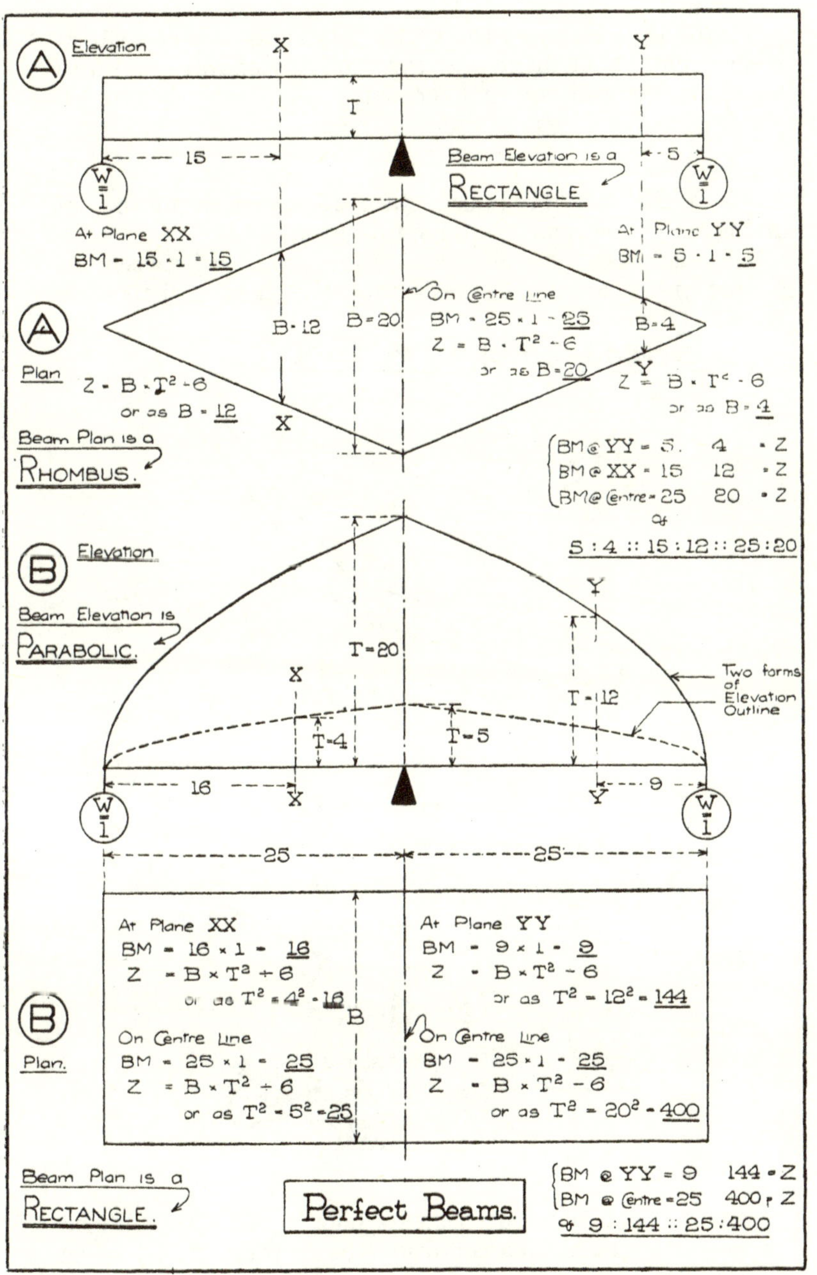

FIG. 4.

Having thus acquired the necessary expressions equivalent to the bending moment and the modulus of resistance respectively, the full equation becomes :—

$$\frac{WL}{4} = \frac{fBT^2}{6} \quad \ldots \quad \ldots \quad \ldots \quad (I.—4)$$

which is the well-recognised standard equation for beams analogous to the semi-elliptic spring form.

The relationship having been now established between the loading of a beam and its resistance, it will be well to show

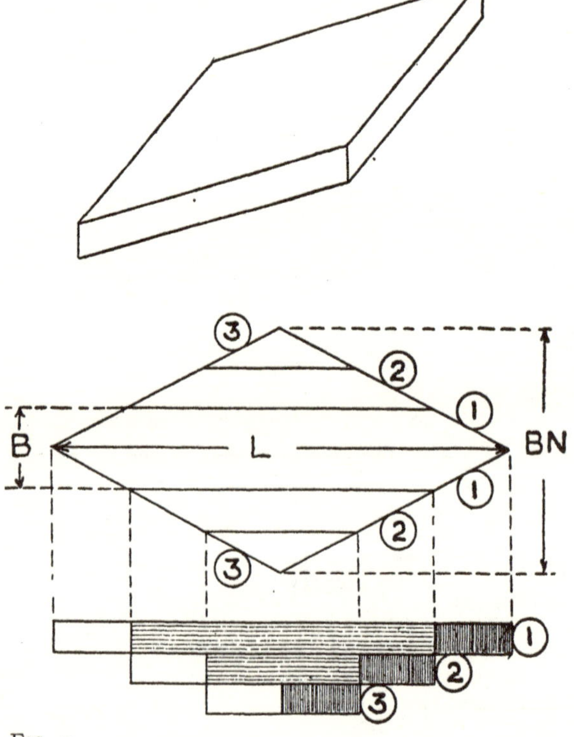

FIG. 5.

the two perfect types of beams which agree with uniform bending conditions, such conditions being present only when the unit stress from end to end of the beam length is constant, which fact indicates that throughout its length the beam section is exactly proportional to the bending moment at any given point. Fig. 4 show these types of beams, with sufficient matter thereon to make clear the general argument. The uniform thickness, varying width, beam, is shown at 4—A, and the uniform width, varying thickness, beam, at 4—B. This varying thickness has to be a curve of the parabola form, based on the variations of (T^2), which has been shown to occur in the formula for (Z). The design of the latter can clearly be practically applied to a beam of solid form, but is not practically applicable to a beam of superimposed laminæ such as a spring, as each separate plate would have to be of this form. The only manufacturing possibilities for normal springs become therefore limited to the beam of the " uniform thickness, varying width " type, and throughout this is taken as the ideal of perfect spring form. Fig. 5 shows the simple stage by which such ideal single plate beam is transformed into the multiple plate laminated spring, the expression (*bn*) which occurs frequently in spring formulæ being merely equivalent to the (B) of the single plate beam—(*bn*) meaning the width of the section of the spring steel multiplied by (*n*) which is the number of active plates in the spring.

CHAPTER II

THE DEFLECTION OF BEAMS

IN matter dealing with the deflection of beams, it is too often indicated that the curve of deflection is a circular arc. This is only true if the beam section and bending moment are at all points proportional, as has been shown in the foregoing chapter, and is quite incorrect if the designed beam varies in any way from this condition. In practice, as regards most beams, for bridges, cranes, etc., the deflection is relatively so small in proportion to the length, being not infrequently 1/600, that it is a matter of no importance as to what is the exact shape of the curve to which the elastic beam sets under load. Neither does the question of span versus true or " straight " length intrude itself, whereas this is a very serious matter as regards a highly flexible spring beam, with a maximum deflection probably equal to $\frac{1}{8}$ of its true length. Spring drawings have a tendency to give the " span " or chord between bearings, as the important dimension, and many formulæ accordingly take the span for calculation purposes, which is entirely incorrect, as a beam must be calculated always and continually on its " straight " or undeflected length between bearings; in other words, as deflected, the length must be taken around the arc or curve of the bent beam.

Inherent inaccuracies are bound to be present in all working formulæ dealing with the deflection of beams of high flexure, such as a spring, as the true ratios of arc to chord, and camber or versin to arc, are indeterminate. Many practical amendments can, however, be made, so as to obtain results sufficiently accurate for calculations necessary to design and manufacture, and such necessary points will be touched upon. One item worthy of note turns on the fact that whilst all spring deflections are measured as "ordinates" or right-angle lines to a working plane, the actual path of deflection is not an ordinate, and is greater than the measured deflection; and, with the extreme case of deflection through a quadrant, the difference is very considerable. Obviously as the ratio camber : span decreases, this variation between true and measured deflection also decreases, but the fact of its presence indicates one of the difficulties in obtaining academic accuracy in a matter of the nature of the high deflection of beams. Fig. 6 shows the path of a beam AB deflected to quadrant shape, AC, and gives the relative distance travelled actually, and as measured by the ordinates. It also shows the necessity of calculating on the true length of the beam as distinct from the span. The distance through which the beam end moves is four times the distance through which the half-length beam end moves, measured on the locus line BC, or, if distance A2 be called L, distance AB is $4 \times L$, and the locus BC is equal to locus (2 to 7) $\times 4^2$, or 16 times. Beam XY represents what would be equivalent to a single plate, or a beam of rectangular plan and uniform thickness, in which case, the sections from point to point are not proportional to the bending moment. Such beam has a deflection $\frac{2}{3}$ that of the beam AB for a given load, and contains much superfluous material, as the skin stresses at the point of fixture, X, are the maximum, but decrease continuously to the point Z, owing to the assumed beam being of uniform width. The resulting shape of deflection is shown as XZ, the curvature becoming less and less as the weight-supporting end Z is reached.

The formula which is generally taken as a standard for the deflection of beams of the type in Fig. 6, is as follows :—

$$\text{Deflection (D)} = \frac{WL^3}{3EI} \text{ or } \frac{2WL^3}{3EZT}. \quad \ldots \quad \ldots \text{(II.—1)}$$

This is the cantilever type, bending moment—with concentrated load at the end—being (WL). With the type of the semi-elliptic spring form, namely, a beam supported at each end and bearing a concentrated load in the middle, the bending moment is $(WL \div 4)$. As the deflection of a half beam or cantilever of relative length and relative loading will be the same, such equivalent terms are substituted, namely, ($\frac{1}{2}$L) and ($\frac{1}{2}$W). The above equation, for the end supported beam form, becomes therefore :—

$$\text{Deflection (D)} = \frac{2}{3EZT} \times (\tfrac{1}{2}L)^3 \times \tfrac{1}{2}W \text{ or } \frac{2WL^3}{48EZT}. \quad \text{(II.—2)}$$

On occasion, this formula is derived by taking into account the area of the bending moment diagram, but this would appear not sound, as this diagram is assumed a constant regardless of the proportionate sectional aspects of the beam, and such derivation therefore gives a constant deflection, whereas the deflection has been already shown to be very obviously different according to the design of the beam. The formula given for a beam which when deflected, forms a truly circular arc, is

$$\text{Deflection (D)} = \frac{WL^3}{32EI} \text{ or } \frac{2WL^3}{32EZT}. \quad \ldots \quad \ldots \quad \text{(II.—3)}$$

It will be noted that Formulæ II.—2 and II.—3 differ in respect only of the constant in the divisor, which in the one case is 48 and in the other case 32, indicating that the minimum deflection of a centrally loaded, end supported beam, is $\frac{32}{48}$ or $\frac{2}{3}$ of the maximum deflection, the extremes of the beam plans—the thicknesses remaining uniform—being the rectangle and rhombus respectively, corresponding to the deflection forms, imperfect and perfect (circular) shown in Fig. 6. Put in another way, the ratio is 100 : 150. A geometrical graph shows it as 100 : 150, and a similar ratio has been obtained from practical experiments. The hand book issued by the British Engineering Standards Committee in connection with Standard Steel Sections gives the Formula II.—2 which in practice is correct for rolled bars, the section of which remains unaltered from end to end of the span. Considerations of the variations between true deflections and ordinate deflections do not apply in the case of structural work owing to the great length of beams relative to their deflection, which is kept invariably as small as possible for obvious reasons. On the other hand, when dealing with beams of high deflection, as springs, where ratio camber : working length may be 1 : 6, the difference between true deflection and ordinate deflection is more apparent, but is never taken into account, ordinate deflection being always measured. This difference is appreciable to a serious degree, however, as this ratio increases in magnitude, until, as shown, when the quadrant is reached, where camber : working

Beam Deflection Shapes.
Concentrated Load here
A
B
1 2 3 4 5
6
8
9
D'
D''
10
C
AB of uniform thickness and width proportioned to Bending Moment.
True Deflection BC equals 1·20 times ordinate D''
True Deflection 4·9 equals 1·05 times ordinate D'
1 to 6 Locus of L/5 = 1/25 of BC, or 1/4 of 3 to 8
2 to 7. Locus of L/4 = 1/16 of BC, or 1/4 of 4 to 9
5 to 10. Locus of 4L/5 = 16 times 1 to 6 or = 4 times 3 to 8
True Deflection
Span
Length
Concentrated Load here
X
Y
D'''
XY of uniform thickness and uniform width.
Circular Curve of AB.
Z
Locus YZ is approximately 2/3 of BC
True Deflection YZ equals 1·11 times D'''
D'' : D''' :: 4 : 3
C

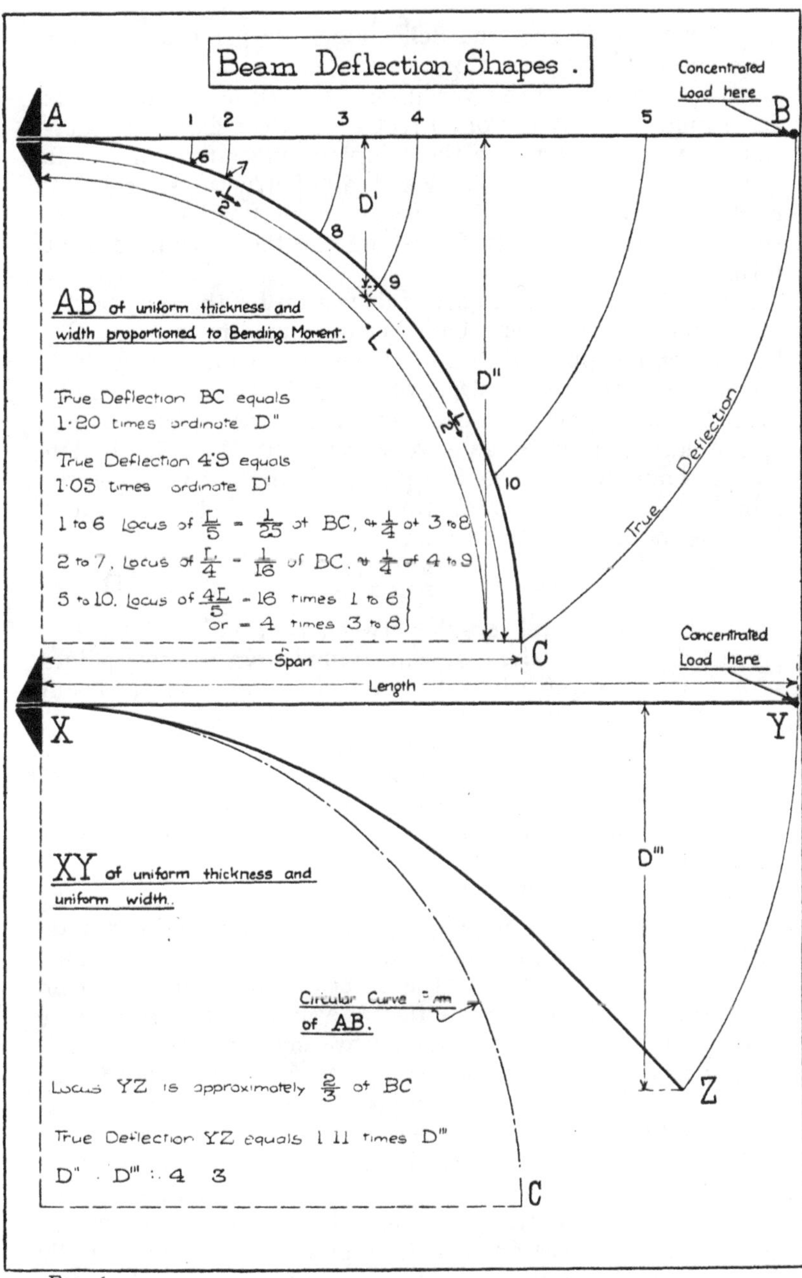

Fig. 6.

length :: 1 : 1·57, the true deflection is 20 per cent. greater than the ordinate deflection.

An appreciation of this variation of unit deflection on beams accordingly as they are of correct theoretical shape or otherwise, is very necessary to the complete understanding of the design of laminated springs, which of all structures, probably lend themselves the best to very exact correspondence between the practically necessary forms and the theoretically perfect forms.

Before leaving this subject of deflection for the time being, whilst remarks in general have been confined to the centrally loaded, end supported beam, two other arrangements call for a brief attention. One of these is the non-centrally loaded, end supported beam, equivalent to a semi-elliptic spring fastened to an axle at some point along its length which is not the middle, such pattern being not infrequently found in automobile and tramway work. The accepted bending moment for this form is

$$BM = \frac{W(Lx)(Ly)}{(Lz)}. \quad \ldots \quad \ldots \quad \text{(II.—4)}$$

where Lz is the total length between bearings, and Lx and Ly the distances of the non-central load from the two points of support respectively. The deflection of such a beam (of uniform section) under the load is then :—

$$\text{Deflection (D)} = \frac{2W(Lx)^2(Ly)^2}{3EZT(Lz)}. \quad \ldots \quad \ldots \quad \text{(II.—5)}$$

The other case is that of the true cantilever, Fig. 6, for which the deflection is as given in Formula II.—1, the bending moment being

$$BM = WL. \quad \ldots \quad \ldots \quad \ldots \quad \text{(II.—6)}$$

The whole of the foregoing formulæ are printed as standards throughout engineering text books, sometimes with and more frequently without, the intermediate steps of their ultimate derivation. In practice, as applied to constructional beams, etc., the results obtained are accurate within usual engineering limits ; but as applied to laminated springs, of normal design, no satisfaction is obtainable, chiefly owing to the fact that no account is in general taken of the differences in deflection, with the same bending moment, which occur between beams of constant central section modulus and material, due to varying stresses from centre to ends owing to the non-proportionality throughout the beam length between the bending moment and the section modulus.

CHAPTER III

STRESS AND STRAIN.—ELASTICITY

THE great principle underlying the spring is of necessity the elastic property of the steel of which it is built up. It has already been shown in Chapter I. and Fig. 3 that when a beam or spring plate is deflected under load, the external surfaces suffer extension or diminution in length, according to whether they are on the tension or compression sides; and, as it is the function of a spring plate to be continually altering these lengths due to varying deflections of the general spring, clearly the elasticity of the steel employed, otherwise, its possibility of continual travel within a limit, with the certainty of its returning to its original manufactured position, is a matter of the highest importance.

When a spring plate is deflected under ordinary working conditions, a certain "stress" is imposed on its surfaces, and throughout the sections, stress implying that a resisting force inherent in the plate has had to assert itself. This internal force may endeavour to resist tension, compression, shearing, or torsion, but whatever the nature or extent of the external force, the resistance thereto of the plate will be its equivalent. In beams of the spring form, generally speaking, only tension and compression forces need be considered, and these are invariably applied primarily by actual static or dead weight loading, and secondarily, by running shocks which are cumulative to the static loading. The stress thus imposed, of (X) tons per square inch or (X) kilogrammes per square millimetre is accompanied by a corresponding and proportionate "strain," or length alteration (which may be plus or minus, that is, extension due to tension, or diminution due to compression) this "strain" being always proportional to the stress within the limit of elasticity of the material. This important fact is generally known as "Hooke's Law," and was propounded in 1676. If the stress becomes excessive, the strain passes beyond the point at which the inherent elasticity of the material will return the surfaces (the most highly stressed and strained parts) to their original lengths, in other words, the limit of elasticity will be reached and passed; and the bar will remain permanently deformed, and in its "plastic" state, in which condition, stress and strain cease to be proportionate.

A general summing-up of the foregoing amounts to the following:—A certain load will produce on a spring plate a certain deflection, equivalent to a definite stress and corresponding strain. This strain is greatest on the outside surfaces, and the cohesive force of the material on these surfaces, resisting or re-acting against the effect of the actually applied load, is known as the "skin stress" or "extreme fibre stress." The whole of these items vary directly as the applied load, in other words, if the applied load is doubled, doubled also are the deflection of the spring (true deflection, if this is not practically coincident with the ordinate deflection), the strain of the surfaces, and consequently, the extreme fibre stress.

Properly to come to an understanding of all these features and co-relate them in such a way as eventually to produce the simple formulæ necessary for intelligent spring design, it is first compulsory to appreciate correctly the principles relating to the elasticity of steel. Research has reduced to practically accurate figures this property, which is known as "Young's Modulus" or "The Modulus of Elasticity" or (E), and really represents the fixed ratio between stress and strain within the elastic limit. Put simply, this modulus is a number, in the case of steel given variously as 29,000,000 (twenty-nine millions) to 36,000,000 (thirty-six millions) pounds, or, say, 13,000 to 15,000 tons. The metric equivalent is generally taken as 20,000 kilogrammes per square millimetre, which is 12,700 tons per square inch. This is for all practical purposes equal to the usual figure taken in English measure, of 13,000 tons. The modulus of elasticity is a constant for any steel, and is quite independent of quality or heat treatment, which is another way of indicating that the unit deflection of all qualities of steel subjected to all treatments, is a constant. A spring made from dead soft steel will have exactly the same unit deflection as one made from best quality spring steel, hardened and tempered, and it is well to keep this in mind, as it is not infrequently believed that a good quality of spring steel will be stiffer, or have a less unit deflection, than an inferior quality. Actually, such unit deflection will be the same, but the limit of deflection before permanent set is attained will be higher for the good quality than for the inferior quality, as, whilst the modulus (E) is the same, the " elastic limit " of the qualities varies.

The figure of 13,000 tons per square inch is, of course, purely theoretical, and expresses the load which would be required to pull in tension a bar, one square inch in area, to double its original length, assuming the bar to remain elastic throughout the pull. This cannot be performed in actual practice, but the quantity of 13,000 tons has been obtained from actual tests, in which tensile pieces have been weighted within their elastic limits, and the resulting stretches most accurately measured, and plotted against the loads. To elucidate further, if a piece 10 ins. long and one square inch area was found to stretch 1/100 in. (0·01) with a load of 13 tons, and it were possible to stretch it to 20 inches in the same condition, or to double its original length, it would require 1000 times (10 ÷ 1/100) the load of 13 tons, or 13,000 tons, which would then be taken as the modulus of elasticity for this material. The stress divided by the strain remains a constant, as follows :—

$$13 \text{ tons} \div 1/100 \text{ in.} = 1300$$
$$13{,}000 \text{ tons} \div 10 \text{ ins.} = 1300$$

Having now defined as clearly as possible, the modulus (E) it is necessary to touch upon the " elastic limit " of materials. At one time, this term generally implied all the phenomena present on the stress-strain curve at the point where it ceases to have a regular inclination, and breaks away from the main line. To-day three points are recognised, namely (1) Limit of Proportionality ; (2) Elastic Limit, and (3) Yield Point. These are indicated on the diagram Fig 7, and defined as follows :—

Limit of Proportionality.—This is the point on the first part of the curve where the stress ceases to be exactly proportional to the strain, and is the terminal point of the (assumed) straight line from the zero point. (A straight line can always be considered as part of a curve laying in the circumference of a circle whose radius is infinity.)

Elastic Limit.—This is a point slightly beyond the limit of proportionality, where commences a definite break in the upward line.

Yield Point.—This is the position generally observed in commercial testing, as it is the first obvious indication to the observer that the piece has definitely stretched. Whilst this is slightly above the points (2) and (3), for all practical purposes, it can be regarded as coincident, as the accuracy of ordinary commercial testing is rarely within 2 per cent. of absolute truth.

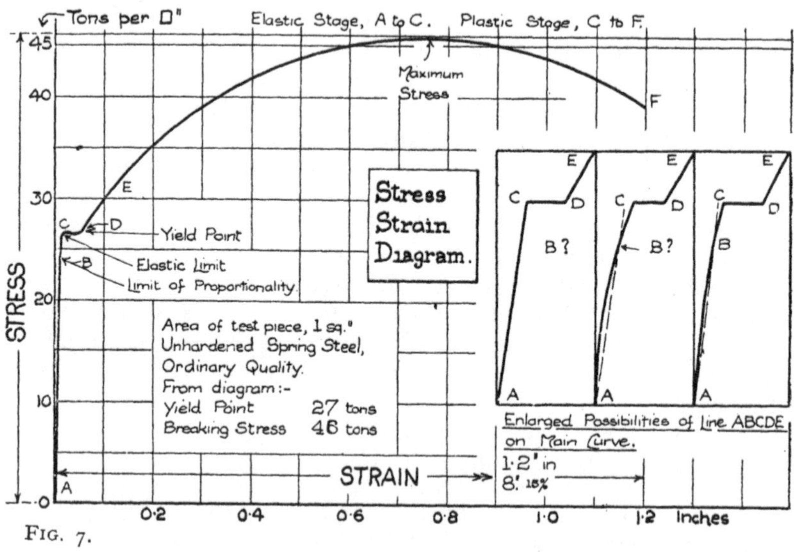

FIG. 7.

For the study of springs and beams in general, these three points will not be separated, and the term "yield point" will be used as indicating the break-away from the proportionate aspect of stress and strain. The steel at this point has just ceased to retain its elastic properties, and from the "elastic stage" passes into what is known as the "plastic stage" until actual rupture occurs.

The line joining the zero point to the "limit of proportionality" in Fig. 7 is shown straight and generally assumed to be so. There is, however, no special reason why "stress to strain" within the elastic limit should for all steels take the form of a mathematically straight line, or otherwise, the form of a ratio $x : y$. It conceivably might take some form based on a ratio $x : y^n$ (in which n is not 1) and accordingly the line would not be straight, but of some form other than straight, as indicated also in Fig. 7, which shows that the point " limit of proportionality " is really no well defined break away, but merely a point where the deflection of the elastic curve accelerates. The conception of an actual curve as an alternative to the usually accepted " straight-line " curve is somewhat at variance with Hooke's Law, but certain accurate observations of recent years have tended to show that such elastic curve line is obtained with particular steels and treatment.

Another point that might be here touched upon is that concerning the artificial raising of the elastic limit by repeated stressing. (The idea of repetition stresses must be kept very distinct in this connection from the method of " cold-working " such as wire-drawing, by which the elastic limit of materials can be raised). It has long been an observed fact that if steel test pieces are stressed to a point just in excess of the elastic limit, and then un-loaded, at the second application of stress, the limit will be higher than the original. Logically, therefore, this loading and unloading in accordance with the new limits obtained should be able to be continued until the elastic limit is coincident with the breaking point. This phenomenon would appear to be caused by the non-homogenity of metals (steel, for instance, is only theoretically homogeneous) which results in succeeding elastic limits being due to local deformations rather than uniform deformation, and accordingly the piece is in a condition for rapid and non-ductile fracture as stress repetitions raise the limit towards the ultimate breaking stress. Attention has been drawn to this known property in view of the suggestions sometimes made that the discrepancy between the calculated figures and known material qualities as regard spring loading and testing, is due to the test or working load closely approximating and sometimes exceeding the elastic limit, and by consequence raising this limit. The author considers, however, that this attempted explanation has only a partial bearing on the matter—which will be included in the proper place.

In connection with tensile testing, it might be noted that what is always called " tons per square inch " of breaking load, really is a conventional standard of " tons per square inch calculated on the original section." The fractured sectional area of a tensile test piece is always less than the original section, and this contraction of section becomes in a general way, more pronounced in the softer steels, for instance, on 25 tons per square inch material, the fractured area will probably be about one-half of the original area. The actual breaking stress on a highly contracted area such as this, is obviously therefore considerably higher than the unit stress on the original section. All testing is, however, standardized—advisably and correctly—on the basis of original section. This contraction of area commences with the entrance of the material from the "elastic stage" into the "plastic stage," in which stage most calculations of standard form cease to have accuracy.

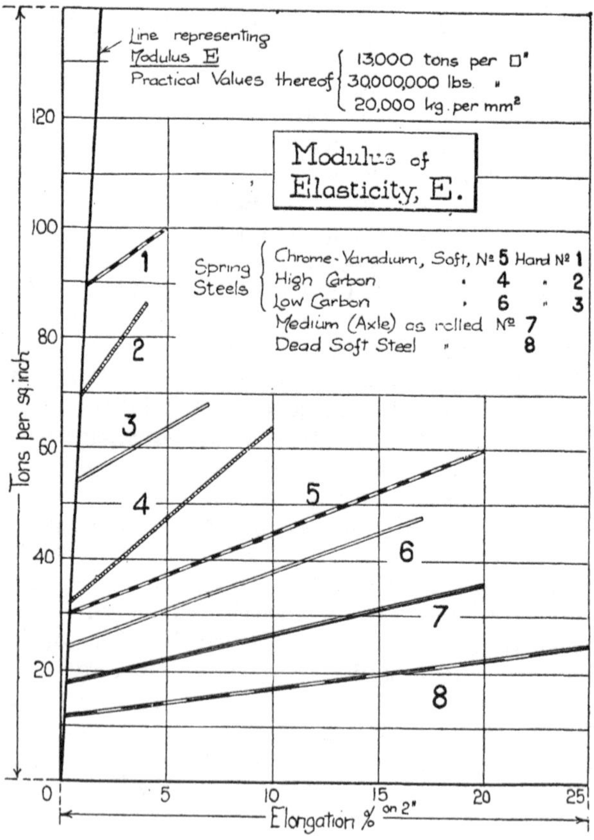

FIG. 8.

In further connection with the modulus of elasticity, it cannot be sufficiently emphasized that, as previously indicated, this modulus remains constant regardless of the steel qualities or treatments. If this were not so, every batch of springs would vary in unit deflection, as no two casts of steel are exactly alike, and it is also probable that springs from the same cast would vary according to the individual treatment they had received at the hands of the spring fitters. It sometimes happens that it is known that steel is rolled slightly thicker or thinner than nominal size, and under such circumstances, the deflection will not exactly accord with calculation. In order to obtain the variation due to this slightly imperfect rolling, it is not unusual to cut off a set of plates from the bars as received from the mill, and load test these, just as cut off. The result obtained is sufficient indication of the deviation of the steel thickness and consequent unit deflection, to render it possible to fix on the "as made" camber. The diagram, Fig. 8, will probably make clear this point regarding the constant value of "E" so far as steel is concerned. Up to the present no very obvious amendment to the recognised value of 13,000 tons has been shown to be needed for the "special" or "alloy" steels at present in considerable use for spring purposes, as those with the highest alloy content still retain over 95 per cent. of iron.

CHAPTER IV

DEVELOPMENT OF THE FORMULA FOR DEFLECTION

THE evolution of a theoretical deflection formula from first principles will now be proceeded with, step by step, as it is only by the careful tracing of such an important expression that the student will obtain and retain the confidence which is very necessary to efficiently handle such formulæ, and which is generally lacking with those who look at matters entirely from the " practical " aspect.

In Fig. 9 a beam deflected is shown, in which is known the original length (L) and also the deflection or camber (D) due to the load. (It is assumed that in the no-load position this beam was straight). The first necessary information is to obtain the " strain," plus and minus, of the tension and compression surfaces respectively, and the first step towards the obtaining of this, is to discover the radius (R) of the neutral axis (a true radius being also assumed, implying thereby that the beam is of perfect design, as, say Fig. 4—A). Unfortunately, no exact relationship exists between the arc of a circle, represented by (L), its version or camber (D), and its radius (R). Under such circumstance, it is necessary to take the nearest practical approximate relationship, and instead of the true arc to take the chord $\left(\frac{L}{2}\right)$ joining the middle of the arc to the end. This expression $\left(\frac{L}{2}\right)$ represents also approximately, the half-length of the beam, but it will be appreciated that for relatively small deflections the half-arc and the chord of the half-arc are practically equivalent lengths. The radius (R) can then be found as follows, the simultaneous equation being based on the relationship between the two sides and the hypotenuse of a right-angled triangle :—

$$\left.\begin{aligned} R^2 &= (R^2 - 2RD + D^2) + C^2 \\ \left(\frac{L}{2}\right)^2 &= \qquad\qquad\quad D^2 + C^2 \end{aligned}\right\} \text{subtract}$$

$$R^2 - \frac{L^2}{4} = R^2 - 2RD \quad \text{or} \quad \frac{L^2}{4} = 2RD.$$

$$\text{Therefore,} \qquad R = \frac{L^2}{8D}. \qquad \text{(IV.—1)}$$

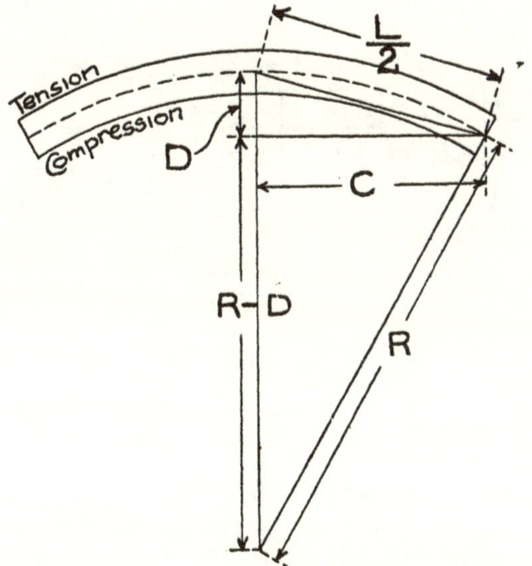

FIG. 9. RADIUS OF DEFLECTION.

The next step is to arrive at the strain (i) depending upon the deflection (D) the length (L) and the thickness (T) of the beam. Assume it is bent into the complete circle, as in Fig. 10. On the neutral axis accordingly, (L) will be equal to $2\pi R$. On the outside surface it becomes $2\pi\left(R + \frac{T}{2}\right)$ or $2\,\pi\,R + \pi T$. In other words the total extension (i) is equal to πT. Then by proportion :—

$$2\pi R : \pi T :: L : i$$

$$i = \frac{L\pi T}{2\pi R} = \frac{LT}{2R}.$$

Substituting (R) as in Formula (IV.—1) equal to $L^2 \div 8D$, then

$$i = \frac{8DLT}{2L^2} = \frac{4DT}{L}. \quad .. \quad .. \text{(IV.—2)}$$

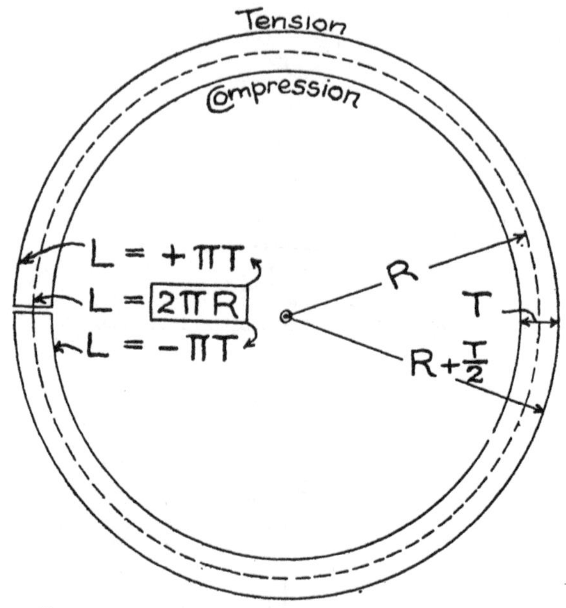

FIG. 10. CIRCLE OF DEFLECTION.

The elastic extension or strain (i) has thus been determined with regard to the deflection, length, and thickness. It is now necessary to find the unit stress, (f), the beam being assumed of unit section. Obviously with (i) known, and the modulus of elasticity (E) known, together with the neutral axis length of the beam (L), it becomes a matter of simple proportion only to find (f), as follows :

$$E : f :: L : i$$

$$\text{or} \quad f = \frac{Ei}{L} \quad .. \quad .. \quad .. \text{(IV.—3)}$$

Substituting for (i) as in Formula (IV.—2), equal to $4DT \div L$, then

$$f = \frac{E(4DT)}{L^2} \quad .. \quad .. \text{(IV.—4)}$$

Having attained this stage, a concrete illustration might be given with special regard to the British Standard Specification Test requirements, which, on a plate 30 ins. long by ½ in. thick demand a deflection of 2 ins. To find (*f*) :—

$$f = \frac{13000 \times 4 \times 2 \times \frac{1}{2}}{30 \times 30} = \frac{52000}{900} = 58 \text{ tons per sq. inch.}$$

This result, it will be noted, is based on fibre extension only. The use of standard formulæ based on the load carrying properties of such a plate give a very different result with substantial discrepancy, which will be gone into at a later stage. At the moment, it is sufficient to take cognisance of the fact that, by simply evolved formulæ, the result shows that the maximum stress on such a plate, when deflected through 2 ins., is 58 tons per square inch, which figure represents about the limiting elastic stress obtainable on ordinary qualities of spring steel, subjected to good heat treatment.

From the designer's point of view, the most important features of a spring resolve into two, namely, the deflection per unit of load (per ton, or per 1000 lbs. or per kilogramme) and the safe test load, on which is based the safe working load. In view of the previous matter which has been given in connection with standard beams, it is now possible to derive formulæ to give these, the spring being taken for the time, as of the most usual type, namely, end loaded and centre supported, Fig. 1—A, or end supported and centre loaded, Fig. 1—B. The equation for bending moment and moment of resistance in such case is that given by Formula I.—4 recapitulating :—

$$\frac{WL}{4} = f\frac{BT^2}{6}. \quad .. \quad .. \quad .. \quad \text{(I—4)}$$

Sufficient material has now been accumulated to enable the unit deflection to be arrived at, on the following lines, with (*d*) as the unit deflection, for the time being equivalent to " deflection per ton," and (D) as the total or test deflection under load (W), both (D) and (W) in the meantime being taken in accordance with British Standard Specification.

$$d \text{ (deflection per ton)} = \frac{\text{D (test)}}{\text{W (test)}}. \quad .. \quad .. \quad \text{(IV.—5)}$$

Substituting for (W) from Formula I.—4, equal to $4\,fBT^2 \div 6L$, then

$$d = \frac{6DL}{4fBT^2}.$$

Substituting for (f) from Formula IV.—4, equal to $4EDT \div L^2$, then

$$d = \frac{6DL \times L^2}{(4BT^2)(4EDT)} = \frac{6L^3}{16EBT^3} \quad \ldots \quad \ldots \quad \text{(IV.—6)}$$

For calculation in "English" units, this last formula becomes on the whole, easier to work if (T) is taken in sixteenths of an inch, $\frac{1}{4}$ in. = 4, $\frac{3}{8}$ in. = 6, and so on. The symbol for thickness in sixteenths will be taken as (t), and to convert Formula IV.—6 into this unit, it will be necessary to multiply it by 16^3, as the expression (T) in the denominator is (T^3). Formula IV.—6 then becomes, with expression (E) also translated :—

$$d = \frac{16^3 \times 6L^3}{16Bt^3 \times 13000} = \frac{16^2 \times 6L^3}{13000Bt^3} \quad \ldots \quad \text{(IV.—6a)}$$

and with this equation worked through, the following result appears :—

$$d = \frac{0{\cdot}118\,L^3}{Bt^3}. \quad \ldots \quad \ldots \quad \ldots \quad \text{(IV.—7)}$$

Before proceeding further, with calculations involving test load based on the above, it will be of interest to compare the deflection now obtained with the usual standards given, as in Formulæ II.—2 and II.—3. They all include the factors of (L^3) and (T^3) with (B), and when extracted out to appear in the same general form as Formula IV.—7, the following constants appear :—

Formula II.—2	Constant	0·079
Formula II.—3	,,	0·118
Formula IV.—7	,,	0·118

For the beam bent to the arc of a circle, such as has just been considered, the latter constant of 0·118 would seem correct, and represents maximum deflection with minimum material. In the other case, where the constant is 0·079, representing minimum deflection with maximum material, the beam would be equivalent to a solid round or rectangular shape of uniform section, between the bearing points.

To obtain a test load or equivalent deflection for the beam, and thereby arranging to stress the latter to a definite figure, which for example, will be taken as 58 tons per square inch, the following sequence arises :—

$$d = \frac{D}{W} = \frac{6L^3}{16EBT^3} \text{ (from Formula IV.—6).}$$

Multiplying and substituting for (W) from $W = 4fBT^2 \div 6L$

$$D = \frac{6L^3 \times 4fBT^2}{16EBT^3 \times 6L} = \frac{L^2 \times 4f}{16ET}.$$

Translating (E) and (f) into known terms, and completing,

$$D = \frac{L^2 \times 4 \times 58}{16 \times 13000 \times T} = \frac{L^2}{900T}. \quad .. \quad \text{(IV—8)}$$

In this simple form is expressed the test formula of the British Standard Specification for Laminated Springs, but it does not therein state specifically the skin stress obtained by this test deflection, which has been shown to be, working along usual lines, equivalent to 58 tons per square inch, with a beam of perfect form.

To obtain the maximum load (W) from the foregoing, the following process is necessary :—

$$W = \frac{D}{d} \text{ and } D = \frac{L^2}{900T} \quad \therefore \quad W = \frac{L^2}{900Td}.$$

Substituting for (d) from Formula 6, equal to $6L^3 \div 16EBT^3$

$$W = \frac{16EBT^3 \times L^2}{900T \times 6L^3} = \frac{16 \times 13000 \times BT^2}{900 \times 6L}.$$

These figures then cancel out into the following form :

$$W = \frac{38{\cdot}5\ BT^2}{L} \text{ or } \frac{0{\cdot}15\ Bt^2}{L}. \quad .. \quad .. \text{ (IV.—9)}$$

This load (W) is based on the uniform stress of 58 tons per square inch noted, and for lesser stresses, (W) becomes necessarily proportionately less.

An interesting diagram is shown in Fig. 11, indicating graphically the proof of the deflection of a beam under a unit load being proportionate to the cube of its length. Assume two small adjacent unit lengths of a cantilever, with a load (W) at the imaginary junction. Such load will cause a deflection (D) at this point which at the end of the second length assuming this to remain stiff, will be (2D). If the equivalent of (W × L), namely ($\frac{1}{2}$W × 2L) is placed at the extreme end, the deflection will again double, in other words, be (4D). This is, however, with one half of the unit load, so that if the unit load be placed at the extreme end, the deflection again doubles, and becomes (8D). Sketch 11—B gives part of a circular arc, showing how this conforms to deflection proportionating to (L^2), with a constant weight W, or L^3 if the bending moment remains the same.

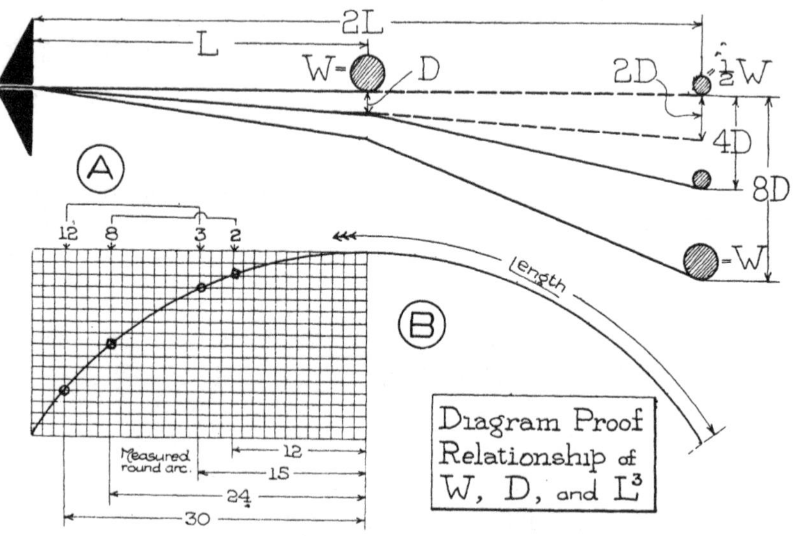

FIG. 11.

The dimensional factors employed relative to these formulæ have been (L), (B), and (T). No question at any time arises as to the unit deflection of a beam varying in direct proportion to the cube of its length (L^3), or varying in inverse proportion to the cube of its thickness (T^3). The factor (B) employed becomes of the form (*nb*) in a spring, (*n*) being the number of plates, and (*b*) the width of each plate. In this connection, however, it has been not infrequently stated that the deflection of a spring is not inversely proportional to the number of plates, in other words, if the number of plates, are doubled, the unit deflection is not halved. This apparent confusion of thought appears to have arisen owing to tests having been tried on one plate, then two, then three, and continuing, with the result that the unit deflection has not been found to be inversely proportional to the number of plates, or as $\frac{1}{2}$, $\frac{1}{3}$, etc. One authority (Landau, *Franklin Institute Journal*, April, 1918) gives the following particulars relative to such tests :—

1 Plate spring.	Strength =	1·00	
2 ,,	,,	1·60	instead of 2
3 ,,	,,	2·20	,, 3
4 ,,	,,	2·82	,, 4
10 ,,	,,	6·50	,, 10
20 ,,	,,	12·69	,, 20

The whole point of this discrepancy would appear to arise in the fact that the one plate which has been tried has been literally " one plate," in other words, it has been a beam of rectangle plan. The application of a second plate (of length L ÷ 2 to approximate spring form) has however converted the beam into an approximate rhombus plan, and the more plates that are added, the closer does this rhombus approximation become. With the constant for the rectangle plan of 0·079, and for the true rhombus as 0·118, it might be expected that the more plates introduced into the spring, the nearer would the ultimate results approach to those given by the 0·118 constant. When worked out on the basis of the particulars given, the following table appears :—

1 Plate spring,	strength	= 1·00.	Constant	0·079
2 ,,	,,	= 1·60	,,	0·099
3 ,,	,,	= 2·20	,,	0·107
4 ,,	,,	= 2·82	,,	0·111
10 ,,	,,	= 6·50	,,	0·121
20 ,,	,,	= 12·69	,,	0·124

The above figures generally confirm the foregoing expectation, the highest, at 0·124 being probably due to considerations of a practical nature, such as varying thicknesses, etc. (The constants are referred to on a basis of a full rectangle section of steel, not the standard concave section). It can be taken with full confidence, that if an 8-plate spring is stiffened with 2 additional plates, making it a 10-plate, its deflection per load unit will be 8/10 of its former deflection ; and likewise throughout normal spring design, that deflection is always inversely proportional to (n). This remark, of course, excepts cases of such additional plates being all full-length or " top " plates.

Summarising the determining factors of deflection from the formula, it can be said, in connection with the loading of beams such as springs, that :—

Length. For a given load, the deflection will be directly proportional to the cube of the length, (L^3), or if a spring is doubled in length, its deflection will be 2 × 2 × 2, or 8 times, its previous deflection, under the same load.

Width. For a given load, the deflection will be inversely proportional to the width, that is, if the width be doubled, the deflection is halved.

Thickness. For a given load, the deflection will be inversely proportional to the thickness (T^3) or (t^3). If a plate be doubled in thickness, for the same load the deflection will then be $\frac{1}{8}$, or $\frac{1}{2} \times \frac{1}{2} \times \frac{1}{2}$,

Number of Plates. In normal (multiple plate) springs, the increase in number of plates is equivalent to an increase in width. For a given load, therefore, the deflection will be inversely proportional to the number of plates, otherwise, if the number of plates be doubled, the deflection will be halved.

Load. The deflection of a spring or beam is always directly proportional to the load—if the load be doubled the deflection also doubles.

Skin Stress. The skin stress is directly proportional to the deflection, which is directly proportional to the load. If either of these be doubled, the skin stress is therefore also doubled.

CHAPTER V

THE CIRCULAR ARC

BEFORE proceeding further with the study of the deflection formula, it is advisable to give some little attention to certain aspects which are introduced by the lack of exact relationship between a circular arc, its chord, and its camber (versin or deflection). Springs should always be formed to a true arc, but owing to considerations of design, this is not always possible, as many designs are by no means perfect, and in accordance with irregularity of skin stress throughout the spring, owing to superfluous material being present, so is the deviation from the true arc shape. In Fig. 12 are shown three springs, A–B–C, of varying designs, but all carrying

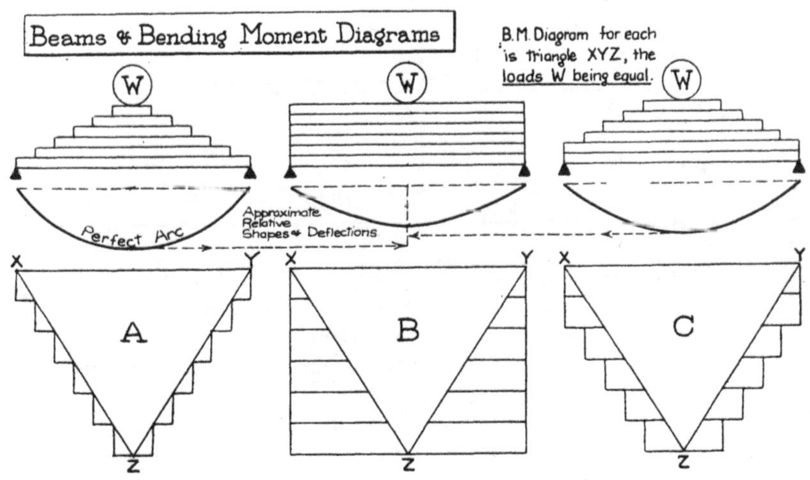

FIG. 12.

the same central load, and giving accordingly the same bending moment diagram. The plans of these springs superimposed on the bending moment diagrams clearly show the proportional or non-proportional aspects of the design as the case may be, and whilst (A) has at all parts of its length material proportioned to the bending moment, and will therefore uniformly deflect at all parts as a circular arc and to maximum deflection, (B) and (C) have superfluous section which will not deflect uniformly, and which will act to reduce the central deflection. Incidentally, these diagrams show the inaccuracy of working deflections off bending moment diagrams. However, the purpose of these is to induce accurate designing, which implies uniform stress, and the setting under load from the straight line to the true arc, or alternatively from the true arc to the straight line.

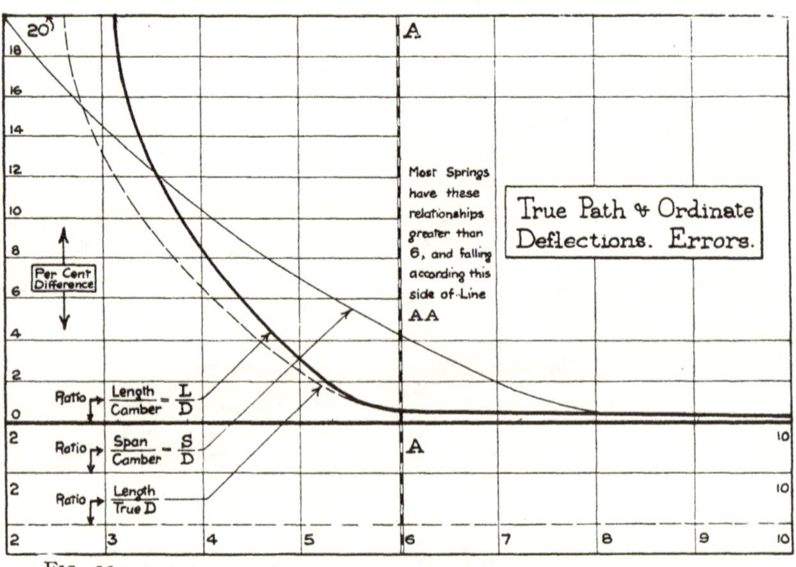

FIG. 13.

It has already been shown in Fig. 6 that the inaccuracy due to taking the ordinate as deflection, as against the true deflection path, can amount in an extreme case to 20 per cent. Practically, however, such an extreme case is seldom reached, and furthermore, the measurement of the true path is not too easy, which limits deflection considerations to the measurement of the ordinate. Fig. 13 presents curves showing the percentage errors between ordinate and true path up to the quadrant, with base lines as ratio-straight length : camber, and span length : camber respectively. The most important curve, relating to the really useful ratio, namely, straight length : camber, shows that at value 6 the discrepancy between the true deflection and vertically measured deflection, is under 1 per cent., and therefore negligible, as most springs have a higher ratio figure than this. The curve relating to span length : camber does not attain the low level of error until value 8 is reached.

It is frequently necessary to determine the radius for a spring or a bent plate, from given particulars, which may be either straight length and camber, or span length and camber. If the former, the expression given in Formula IV.—1 is sufficiently accurate for most practical purposes, but it will probably be of interest to show here also, the percentage errors from the extreme case of the semi-circle down to the minimum error. A curve figuring these particulars is included in Fig. 14.

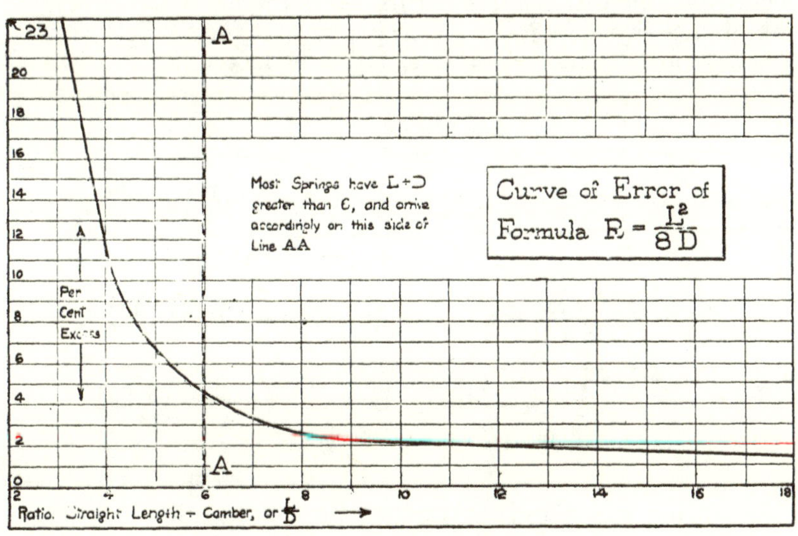

FIG. 14.

When span length and camber is to be worked to, a more accurate formula is available, as follows :—

$$R = \frac{D^2 + (\frac{1}{2}S)^2}{2D} \quad \ldots \quad \ldots \quad \ldots \quad (V.—1)$$

This is correct between extreme cases, as it is necessarily based purely upon the relationship of triangles.

The chief thing it is important to be able readily to obtain in practical spring making, is the additional length needed to specify the " straight length " of a back plate, when the span and camber figures are given on a design. Undoubtedly the best solution of this problem (subject to certain modifications as dealt with hereafter) is provided by the matter given in Molesworth's Pocket Book. Herein are given lengths of the arc on the basis of subtended angles, on the basis of ratio between camber (versin) and span (chord), and also on the basis of increment lengths, according to the ratio of versin : chord. A formula is also stated as below :—

$$L = \frac{8K - S}{3}. \quad \ldots \quad \ldots \quad \ldots \quad (V.—2)$$

In this (K) equals the chord of half the arc. The results obtained by this means have a maximum inaccuracy of $1\frac{1}{2}$ per cent., but the disadvantage of its use is the necessity of first obtaining—by a square on the hypotenuse of the triangle with one side as ($\frac{1}{2}$S) and the other as (D)—the factor (K).

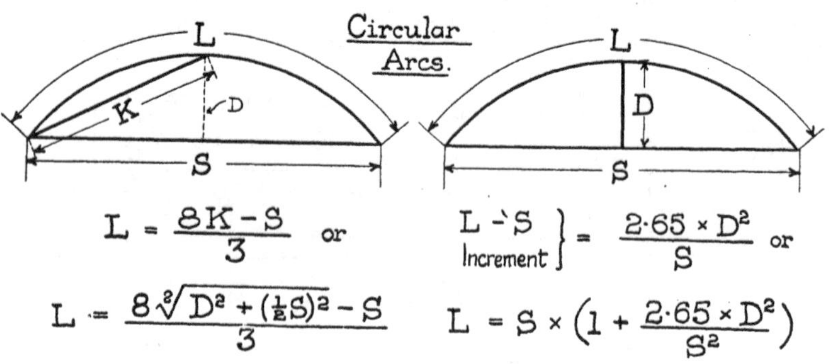

FIG. 15.

A further formula gives the increment to be added to the chord length (S) by which the arc length (L) can be obtained, and this is expressed in the following terms :—

$$\text{Increment} = \frac{2 \cdot 65 \times D^2}{S}. \quad \ldots \quad \ldots \quad (V.—3)$$

If this is put in the following form, it will be found to agree closely with the table given in Molesworth to the limit of usual spring ratios :—

$$L = S\left(1 + \frac{2{\cdot}65 \times D^2}{S^2}\right) \quad .. \quad .. \quad (V.—4)$$

The maximum inaccuracy obtained by the use of these expressions, with the extreme case of the semi-circle, is 5 per cent. Fig. 15 shows diagrams illustrating the foregoing formulæ.

For all practical purposes, probably V.—3 is the easiest to use, and it lends itself to a curve construction for rapid reference. It must, however, be admitted, that calculation methods of obtaining back plate lengths are not readily appreciated by spring makers, as certain other points obtrude themselves which militate against the accuracy of purely theoretical means. Consequently, many prefer to retain the shop method of bending heavy lead wire, to the leading dimensions of the design, and finding the more or less accurate length for the trial back plate by straightening the wire. With certain empirical and simple modifications, however, the calculation methods can be used with confidence.

The chiefest points naturally that will tend to cause discrepancy are the type and shape of the heads of rolled-eye or solid-end back springs. Fig. 16 illustrates this point, from which it will be seen that the true camber is not the drawing camber, which should be given in all cases as shown, namely, from the top of the back plate to the centre of the eyes. Furthermore, after discounting the effect of this additional camber over the true camber, the difference in the type of head has to be taken into account, as the rolled-eye back can be, and is generally, curved in fitting nearly up to the eye, whereas the solid-end back, particularly if it has a long " sweep " as shown in the drawing, cannot be curved at the end, and remains straight. The result is, therefore, that the rolled-eye back plate takes slightly more length of material for similar dimensions, than the solid-end back plate. According to the design of the spring, as regards plate lengths and other characteristics, the general " sweep " may vary, but this cannot be taken into account, as it is a very indeterminate feature, and depends on the judgment of the fitter. Back plates are required within relatively fin- limits of length, $\frac{1}{16}$ in either way being the usual for the higher class springs, and the maximum variation permitted is never above $\frac{3}{16}$ in. either way. Clearly, very slight differences in " sweep " on plates 40 ins. to 60 ins. long can affect these limits, and whether back lengths are calculated or obtained by shop methods, it is very necessary to have the first one curved to the correct " sweep " of the spring to ensure the accuracy desired.

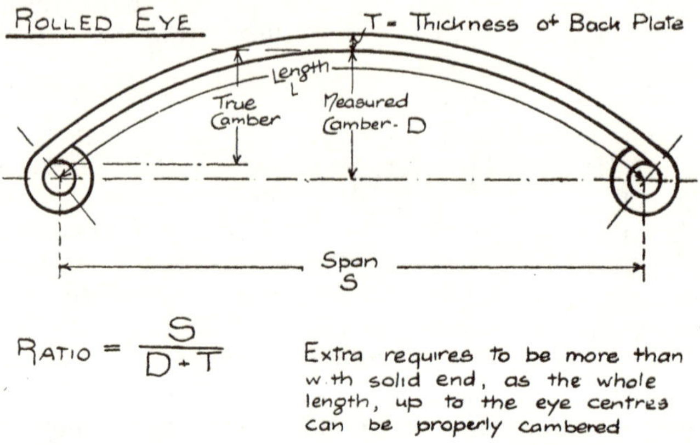

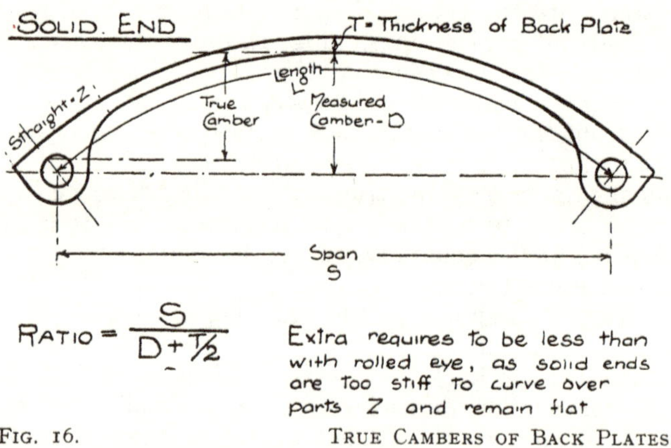

Fig. 16. True Cambers of Back Plates.

The empirical corrections for rolled-eye and solid-end backs can be made as follows, and worked in with Formula V.—3. For rolled-eye backs, the camber (D) must be augmented by the back plate thickness (T) and for solid-end backs the camber (D) must be augmented by one-half the thickness ($\frac{1}{2}$T). Examples for a back-plate, 40 ins. long, and

4 ins. camber, $\frac{1}{2}$ in. thick, with respectively plain ends, rolled-eye ends, and solid ends, are given herewith :—

Plain Ends.

$$\text{Increment} = \frac{2{\cdot}65 \times 4^2}{40} = 1{\cdot}06 \text{ in.}$$ Back length cc. say $41\frac{1}{16}$ in.

Rolled Eye Ends.

$$\text{Increment} = \frac{2{\cdot}65 \times 4\frac{1}{2}^2}{40} = 1{\cdot}34 \text{ in.}$$ Back length cc. say $41\frac{5}{16}$ in.

Solid Ends.

$$\text{Increment} = \frac{2{\cdot}65 \times 4\frac{1}{4}^2}{40} = 1.20 \text{ in.}$$ Back length cc. say $41\frac{3}{16}$in.

It will be noted that the above empiricisms do not take into account the diameter of the head, which has a bearing on the matter. As, however, all back plate ends are within certain well defined limits—for instance, normal design of solid-ends is between 2 in. and $3\frac{1}{2}$ in.—the effect of taking this point into consideration is purely mathematical, as it is not usual, and would not be practical to give shop lengths in any smaller units than $\frac{1}{16}$ in. or millimetres. Charts have been made which introduce this aspect, but it is open to question as to whether any serious value is thereby represented, particularly when it is fully appreciated that an extra hammer blow will alter the length $\frac{1}{32}$ in. or more.

CHAPTER VI

PRACTICAL CONSIDERATIONS OF THE THEORETICAL ASPECT

The theoretical considerations affecting the accuracy of the formulæ under review have now been generally considered, and it will have been noted that the chief item therein is due to the impossibility of extracting exact co-relationships between the arc, deflection, and chord, of the loaded plate or spring. This has been shown, however, to be usually rather of academic interest than of strictly practical interest, as with the necessities otherwise imposed on spring design, such becomes restricted within certain limits, with the result that the majority of laminated springs have a ratio of straight length : camber which reduces the inaccuracy due to errors between true deflections and ordinate deflections to a very small figure. As has been shown, however, the effect of the approximation of Formula IV.—1 and Fig. 9 if localized is far from negligible, as this expression worked out on the extreme, of the semi-circle, gives a radius figure which exceeds the true radius by 23 per cent. This has the effect when worked through the formulæ, of reducing (*i*) and consequently (*f*). Resultantly, owing to the position in which (*f*) appears in the formula leading to No. IV.—6, (*d*) tends to increase, to represent the true deflection path rather than the ordinate deflection path, and thereby negatives the inaccuracy due to the latter. Therefore, 23 per cent. extreme inaccuracy in the one case balances approximately 20 per cent. in the other case. In any event, the maximum of inaccuracy is only obtained when ratio of straight length to camber is abnormal, and as indicating that it can be forgotten from the strictly practical point of view, the following list shows ratios-straight length : camber—of normal design, with the theoretical extreme :—

Ratios, Straight Length to Camber.

Springs taken " as made " or unloaded.

Theoretical Extreme	(Semi-Circle).	L : D : : 3·14 : 1·00,	Ratio	3·14
Locomotive Springs	(Minimum)	L : D : : 6·00 : 1·00	,,	6·00
,,	(Maximum)	L : D : : 1·00 : 0·00	,,	∞
Railway Passenger Stock Springs	(Minimum)	L : D : : 10·00 : 1·00	,,	10·00
,,	(Maximum)	L : D : : 25·00 : 1·00	,,	25·00
Railway Freight Stock Springs	(Minimum)	L : D : : 6·00 : 1·00	,,	6·00
,,	(Maximum)	L : D : : 30·00 : 1·00	,,	30·00
Tramcar Springs	(Minimum)	L : D : : 6·00 : 1·00	,,	6·00
,,	(Maximum)	L : D : : 10·00 : 1·00	,,	10·00
Automobile Springs	(Minimum)	L : D : : 6·00 : 1·00	,,	6·00
,,	(Maximum)	L : D : : 20·00 : 1·00	,,	20·00

In the worst case, of ratio 6·00 to 1·00, the inaccuracy of Formula IV.—1 is 4 per cent., and of the ordinate deflection compared with the true deflection, 1·5 per cent. These would practically balance out if true deflections were measured, but as they are not, the maximum error remains as approximately 4 per cent.

A more important amendment presents itself in connection with the modulus of section (Z). Spring steel is invariably rolled as a concave section, the edges (which represent " nominal thickness ") being thicker than the middle, for the purpose of facilitating fitting the plates, as from the point of view of workmanship, it is necessary to have the plate edges touching throughout their length when the spring is assembled and clamped up, and if the section were perfectly flat, this would not be easy of accomplishment, as the cambering of the plates, particularly when small radii were needed, would in itself tend to buckle the section across the width, and interfere with the perfect edge fitting necessary. In addition to this concave on each of the width dimensions, it is also usual to roll spring steel with round edges. Certain practice (usually Continental) employs square edged sections, but the round edge is more generally favoured in universal practice. This consideration is linked up to a certain extent with the methods of rolling, and will be more fully touched upon when dealing with the manufacture of the spring.

The effect of the combined concave and round edge section is clearly to reduce the amount of material available for resistance to the bending moment, but no definite figure can be given for this amount of reduction as there is no general standard practice. Fig. 17 illustrates the actual dimensions of sections selected at random and carefully measured, and below appears the nominal modulus of section compared with the actual modulus as derived from the modified "as used" sections :—

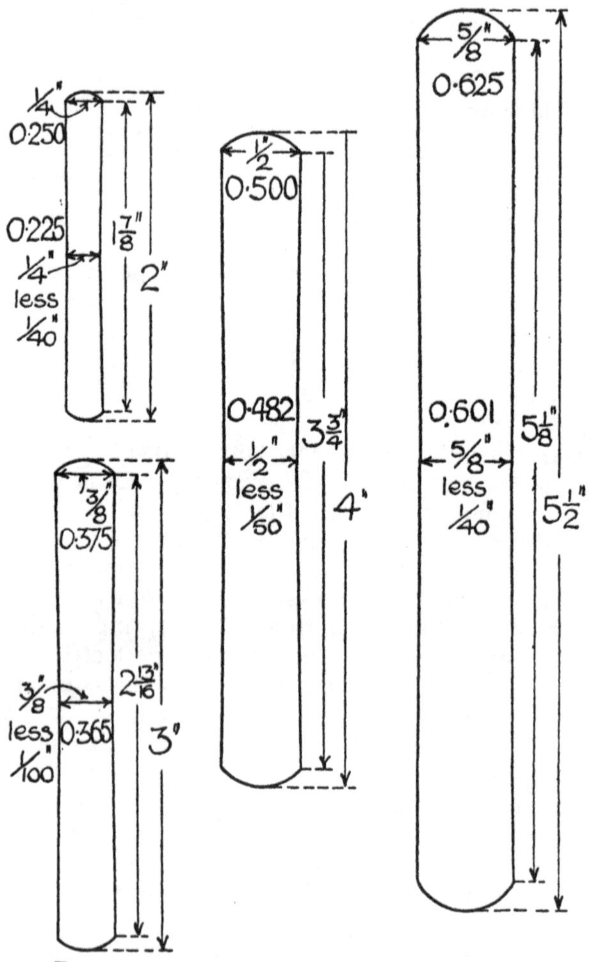

Fig. 17. Actual Spring Steel Sections.

Section 2 ins. × $\frac{1}{4}$ in.	Nominal Z, 0.0208.	Actual Z, 0·0189.	91 per cent.
,, 3 ins. × $\frac{3}{8}$ in.	,, 0·0703.	,, 0·0675.	96 per cent.
,, 4 ins. × $\frac{1}{2}$ in.	,, 0·1666.	,, 0·1590.	95 per cent.
,, $5\frac{1}{2}$ ins. × $\frac{5}{8}$ in.	,, 0·3580.	,, 0·3400.	95 per cent.

At times, when the rolls are worn, the concave of a section is reduced from the standard of the particular plant, and the section correspondingly increased, so that it is not possible to dogmatize on an exact relationship of the actual section to the nominal. Generally, however, a compromise of 95 per cent. is not far out, which increases to 97 per cent. if square edged plates are used. In the finished formulæ following, 95 per cent. will be taken as the necessary figure for modifying the constant, as, with this explanation, there is no longer further need to be concerned with the variation between nominal and actual moduli of sections, it being the most simple course to amend proportionately the formulæ constants.

Potential inaccuracies on a pure theoretical basis have now been dwelt upon, and also the only practical deviation which can be reduced to approximate figures. Many other remaining practical considerations occur, however, which all influence the deflection possibilities of springs, but which cannot be reduced to any dimensioned terms. Remarks are made on these as follows :—

Steel Sections.—Apart from the deviation from nominal section due to concave and round edges, deviations of necessity are present owing to the impossibility of hot rolling steel to micrometer sizes. Spring steel represents the highest type of bar rolling, but nevertheless, section tolerances have to be claimed and permitted, and consequently further modification of the modulus of section takes place. However, within practical engineering limits, these can generally be discounted, the chief size affecting deflections being the thickness, which operates in deflection calculations as "thickness cubed." A difference of 2 per cent. on the thickness of a plate $\frac{1}{2}$ in. thick, or 0·01 in. in actual thickness, makes a difference of nearly 6 per cent. in unit deflection ; and 4 per cent. on $\frac{1}{2}$ in., or actual thickness difference 0·02 in., amends the deflection by 13 per cent. A tolerance of $\frac{1}{100}$ in. on $\frac{1}{2}$ in. thickness is sufficiently limited, and if taken as "plus and minus" can make the unit deflection difference between extreme thinness and extreme thickness as 12 per cent.

Length Variations.—These are not of sufficient importance to make any serious effect on the unit deflection, as $\frac{1}{16}$ in. either way is about the maximum tolerance permitted, and on springs within the usual range, say 20 ins. to 75 ins., such variation is clearly negligible from all points of view. On springs with plain backs, however, certain effects appear, owing to the bearing surfaces altering with the deflection of the spring. Generally speaking, however, this can be neglected. A similar variation can occur with back plates having large " eye " ends, as shown in Fig. 18.

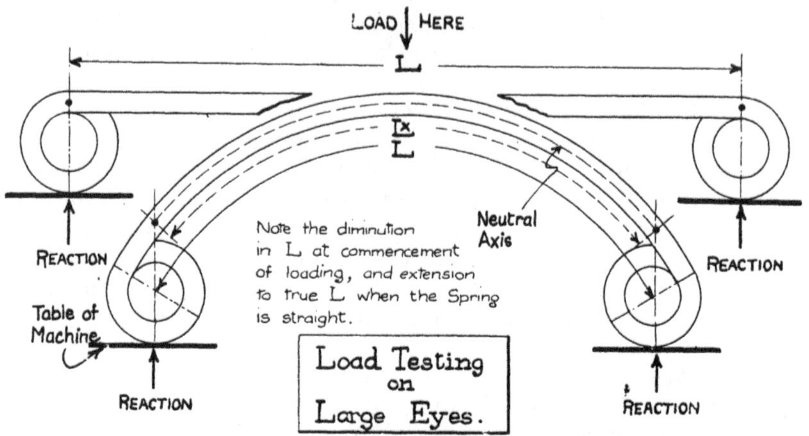

FIG. 18.

Tightness and Width of Hoop.—When a centre hoop is used, as is general with railway and tramway springs, and not unusual with automobile springs, the general tendency is to " stiffen " the spring, otherwise, to decrease its unit deflection. This has been recognised in many formulæ by their requiring the length (L) of the spring to be taken as the real length less the width of the hoop. The results obtained by the use of such formulæ are, however, not too accurate, as this deduction causes the apparent stiffness via calculation to be excessive. The assumption for the use of (L) minus hoop width is, of course, that the spring is working from the edges of the hoop, and acting really as a double cantilever. This assumption will in due course be examined, and its inaccuracy indicated but meanwhile, it will be admitted that a wide hoop, squeezed tightly on all sides under heavy pressure, will sensibly reduce unit deflection. The majority of hoops, put on under normal conditions, do not affect to a great degree the stiffness, as a maximum, probably 5 per cent., and as an average about 2 per cent., and their effect will be excluded therefore in deflection calculations.

Width of Axle-Pad.—Automobile springs are, in general, not fitted with hoops, but clamped to axle pads. The tightness of the fastening, shape of spring plate fitting in the middle (which is of four varieties), shape of pad, and the amount of camber in the spring when working, all affect unit deflections, and present sufficient variations when taken separately. Taken in combination, with the tightness of the fastening an unknown quantity, and the other features of numerous possibilities, variations obviously become legion. Considerations can be given in correcting formulæ to one only of these, namely, the pad width, the clamps being always assumed tight.

Off-sets of Plates, and Lengths of Short Plates.—According to the design of the spring in these respects, so will the unit deflection alter. Generally speaking, however, the effect of such can be neglected from the practical point of view, although as a matter of interest, the method of exact calculation is proposed in the following chapter.

Altering Leverage of Weight.—With all springs, the leverage effect of the weight is always varying according to its intensity, in other words, according to the deflection of the spring. The effect is a minimum when the spring has its maximum camber, and a maximum when the spring is at or approaching "straight." In cases of springs taking their weight from pinned "hangers" or "shackles" this automatically corrects —and in other springs, the effect can generally be neglected.

Friction Between Plates.—This item is credited with a higher importance that it deserves. It certainly is present to some extent, but if spring plates are well fitted and properly oiled before being put into service, its effect can be neglected. On new springs, such as are tested for deflection results, numerous trials show that on the most flexible patterns, the difference between the loading results with the plates dry, and then with the plates well oiled, is under 5 per cent.—in many instances no variation can be detected.

A summarisation of the foregoing points which tend to prevent the introduction of formulæ of absolute exactness gives the following :—

THEORETICAL

(A) Inaccuracy due to lack of exact ratio in circular arcs, chords, and versins. Error 4 per cent. on springs of ratio L ÷ D = 6.
Neglected in formulæ.

(B) Inaccuracy due to measurement of ordinate deflection instead of true deflection path. Error 1·5 per cent. on springs of ratio L ÷ D = 6.
Neglected in formulæ.

(C) Reduction of (Z) or modulus of section from the nominal (Z) owing to concave and round edged steel. Corrected in formulæ by amending constant, taking (Z) of the usual "as-rolled" section as 95 per cent. of the nominal (Z).

PRACTICAL

(D) Tolerance necessary for the rolling of spring steel. Neglected in formulæ. If bars are known to be substantially under or over thickness, calculations must be made on the basis of the (t^3) of the actual thickness.

(E) Variations in length, due to manufacture or design, from the straight length (L) taken for calculation.
Neglected in formulæ.

(F) Tightness of centre hoop or band. If very tight, tends to stiffen spring.
Neglected in formulæ.

(G) Width of centre hoop or band; or axle pad. When very wide, will stiffen spring under certain conditions of shape.
Neglected in general formulæ. Correcting formulæ are given and can be used.

(H) Offsets of plates and lengths of short plates.
Neglected in formulæ. If considered necessary correcting formulæ can be used.

(I) Friction between plates. Maximum on normal springs when new about 3 per cent.
Neglected in formulæ.

(J) Altering leverage of load. Generally corrected by suspension gear.
Neglected in formulæ in all cases.

In connection with (G) if the axle pad or inside bottom of the hoop remains flat, and the spring retains positive camber when under maximum load, the deflection will be in accordance with calculation from the standard formula. If, however, the axle pad or hoop bottom is shaped to the light spring camber, or alternatively, they remain flat whilst the spring may work with negative camber, the effect will be to shorten (L) to a certain extent, and stiffen the spring as the result.

As will be seen, any attempt to obtain a manufacturing formula capable of dealing with all the foregoing points with accuracy is useless, as the number of variables included would be such as to retain it entirely in text-book pages, whereas it is desired to produce such an expression as will be employed with confidence in everyday shop practice, and it is claimed that this has been accomplished with the main formula (which deals with unit deflection) and the subsidiary formulæ which will be set forth in the following chapters.

A further point which also tends to affect results as regards the comparison of actual loadings with calculated, turns upon the type and use of the weight testing machine. At the moment, this will not be gone into at great length, but, as a preliminary remark, it is clear that to calculate deflections per ton to three places of decimals, as for instance, 0·195 per ton, (for a loco. spring) and then load test it ton by ton to perhaps 15 tons, measuring results with callipers and a wooden rule to sixteenths " full " and " bare," is not conducive to the production of a very excellent curve if the load points are carefully plotted out. Various self-recording devices have been tried on loading machines, but the majority include some weak point which makes for inaccuracy, such as the stretch of wires, or the spring of the loading machine. Probably one of the simplest and most accurate methods is that sketched in Fig. 19, in which joggled clips of sizes suitable to fit all hoops, can be readily affixed, and a standard rule inserted. The camber of the spring is measured after scragging, and the clip and rule set to this, the deflection of increment loads then being read direct, in $\frac{1}{100}$ in. from the bevelled edge of the clip to the rule. For the testing of unhooped springs, a rule fixed in a stand, and a pointer on a block immediately under the loading head, that is, between this head and the spring, is a good rigid form. It is, however, clear from the foregoing that the testing machine and operating personnel can make considerable amendments to accurate deflections, both by machine operation, and final measurements. Fig. 19 includes one small aspect also of the lengths of back plates, by showing a badly formed head, which makes the weighting length more than the actual centre to centre length. Strictly speaking, all springs should be loaded in a similar fashion to their harnessing when finally in position, but this is not usual, as the simplicity of the arrangement generally adhered to, namely, placing the spring reversed directly on the machine table, with the heads as bearings, considerably outweighs the advantage of the more meticulous accuracy which would be obtained from swinging solid end and rolled eye backs from pins in hanging shackles.

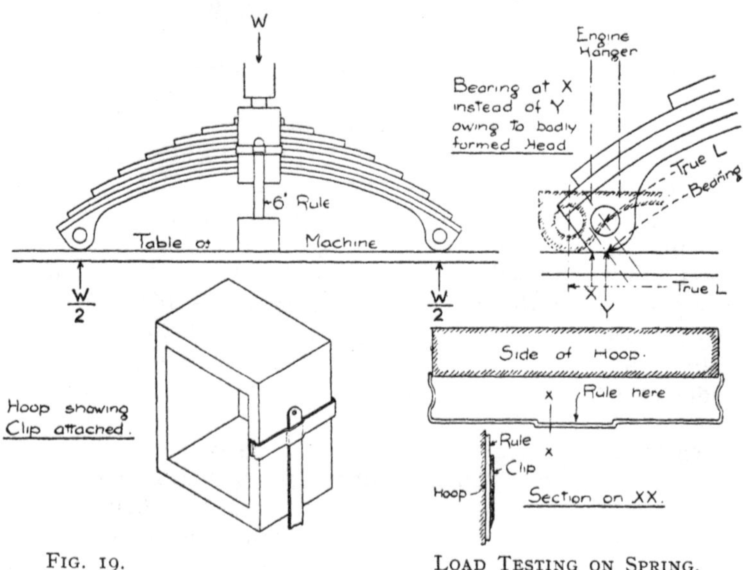

FIG. 19. LOAD TESTING ON SPRING.

Considerably discrepancies sometimes appear in springs made for the automobile trade, owing to the fact that the spring maker tests them as made in the shop, and has no information regarding the width of the axle pad on the vehicle. In position, therefore, on a wide axle pad, such springs have a less unit deflection, and also a smaller safe deflection. All loading of such springs should be carried out with clamps corresponding to the axle pad in position and well tightened up, that is, if it is of great necessity that very accurate results are obtained. Generally speaking, however, owing to the high ratio value of full load to empty load in automobile work, it would not appear of great value to insist on such precautions as this for ensuring extreme exactness.

One last point in connection with the loading of springs of the heavy and flexible type, such as long railway carriage springs, for non-bogie stock; railway (transverse) buffing springs; and heavy automobile rear springs. If cambers are measured on the scrag, and the spring be inverted in the usual way on the loading machine table (see Fig. 19), the amount of deflection for the first ton (say) will be found the smallest of all the unit deflections, owing to the fact that the weight of the spring acting at the centre, causes appreciable deflection on its own account; for instance, on a British type buffing spring, deflection 1·60 per ton, weight approximately 240 lbs., the centre deflection due to the spring weight is about $\frac{3}{16}$ in. These remarks only apply if dead measurements are taken from the table, as would probably be the case with this type of spring. If they are taken by an indicator adjusted along a rule when the spring is in loading position, such indicator being set to the camber as measured on the scrag, the discrepancy does not appear.

CHAPTER VII

THE PRESENT FORMULÆ

SEMI-ELLIPTIC SPRINGS

THE formula for deflection (No. IV—7) evolved in Chapter IV., and which governs the remaining working formulæ, does not entirely agree with the usual unit deflections obtained in practice, entirely owing to the fact that very few springs are of the correct theoretical design. Any formula of use must necessarily be relatively simple, and to attempt to include therein all corrections which are likely to be required according to the variations in designs which have to be worked to, would result in the evolution of a product of much academical interest and no practical value.

Two variations can be dealt with fairly easily by modification or translation. The first necessary one is required owing to the fact that the modulus of section of standard spring steel averages, as previously stated 95 per cent., of the modulus of the nominal section, owing to the concave rolling. The amendment necessary for this is made simply by altering the numerical factor in the formula No. IV.—7, as follows :—

No. IV.—7 stood as $d = \frac{0{\cdot}118\ L^3}{Bt^3}$.

As amended, factor 0·118 becomes higher in the proportion of 95 to 100 and the formula then stands as

$$d = \frac{0{\cdot}124\ L^3}{Bt^3} \quad .. \qquad .. \qquad .. \text{(VII—1)}$$

This represents the deflection of the beam or spring of uniform strength (made of standard spring steel), that is, with the section altering to suit the varying bending moment—in the case of springs, the developed plan showing a rhombus. No. VII—1 is the "maximum unit deflection" and No. VII.—2 is the "minimum unit deflection" with the factor 0·124 reduced by one-third on account of the higher resistance to bending of the "uniform section" beam or spring, with a rectangle plan as developed :—

$$d = \frac{0{\cdot}083\ \mathrm{L}^3}{\mathrm{B}t^3} \quad \ldots \quad \ldots \quad \ldots \text{(VII.—2)}$$

With the above Formulæ VII.—1 and VII.—2 for extreme possible cases, deflections per unit load of all spring designs must be somewhere within these limits, and every design should respond accurately by variation of the multiplying factor in the numerator.

As a matter of fact, this multiplier amends in strict relationship with the characteristics of the stress diagram, but as it would hardly be a practical matter to take out a stress diagram with the necessary accuracy for every design of spring, and bearing in mind also the manufacturing possibilities of error militating against intense exactness, some average figure must be arrived at which will include all general spring designs, and neglect all variables.

The study of the stress diagrams for extreme and intermediate cases is a matter of importance, as it gives the influence on unit deflections due to the position and amount of the material in the spring. Fig. 20 shows springs and diagrams representing the two extreme cases. An examination will show that the resistance of Diagram A—the theoretically perfect (and practically impossible) spring of rhombus plan—is 30 × 20 = 600 area, multiplied by 15 (centre of resistance) which equals 9000. (Empirical units are employed.) Diagram B, showing the rectangle plan, plates being all one length, gives a resistance of 30 × 20 ÷ 2 = 300 area, multiplied by 20 (centre of resistance) which equals 6000. It has already been shown (Chapter II.) that the central deflection under equal loads of such extreme spring beams as the above, is respectively as 150 : 100, from which it follows that central deflections are in proportion to the resistance of the stress diagram. Relatively uniform stressing throughout the spring beam—as the result of the efficient employment of the material used—leads to a large resistance diagram, and a

Stress Diagrams, A & B.
A. Elevation.
A Plan
T
Stress
X
Z
A Bending Moment Diagram
N
Plan Outline
Centre of Resistance
K
A Stress Diagram
XY = Greatest Bending Moment / Greatest Unit Stress
S
Y
B. Elevation
B Plan
KN
T X
Z
Plan Outline
B Bending Moment Diagram
Stress
Centre of Resistance
B Stress Diagram
Line SY = 30
Line XY = 20
S
Y

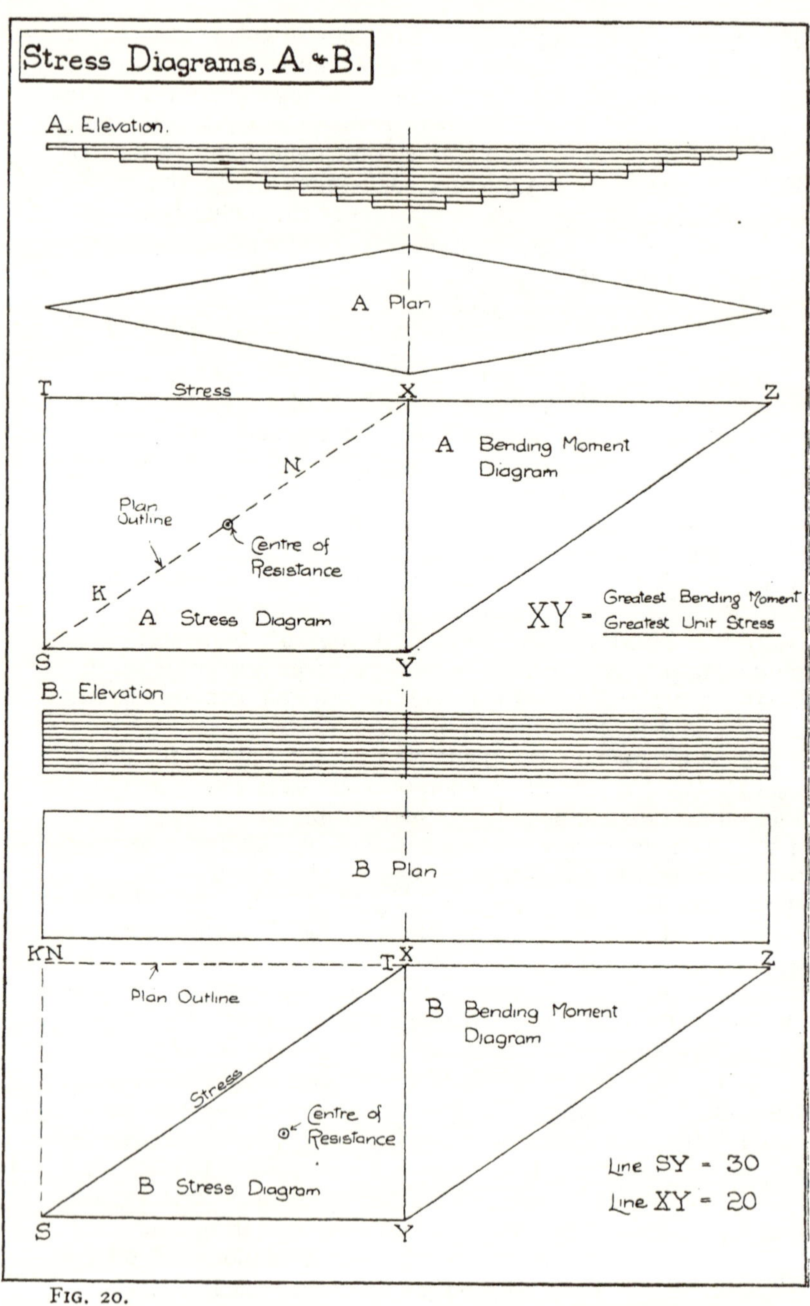

FIG. 20.

Stress Diagrams, C & D.
C. Elevation.
C. Plan
N
X
Z
T
C. Bending Moment Diagram
Stress
Plan Outline
Centre of Resistance
K
C. Stress Diagram.
XY = Greatest Bending Moment / Greatest Unit Stress
S
Y
D. Elevation.
D. Plan.
N
X
Z
Plan Outline
T
D. Bending Moment Diagram.
Stress
K
Centre of Resistance
D. Stress Diagram.
Line SY = 30
Line XY = 20
S
Y

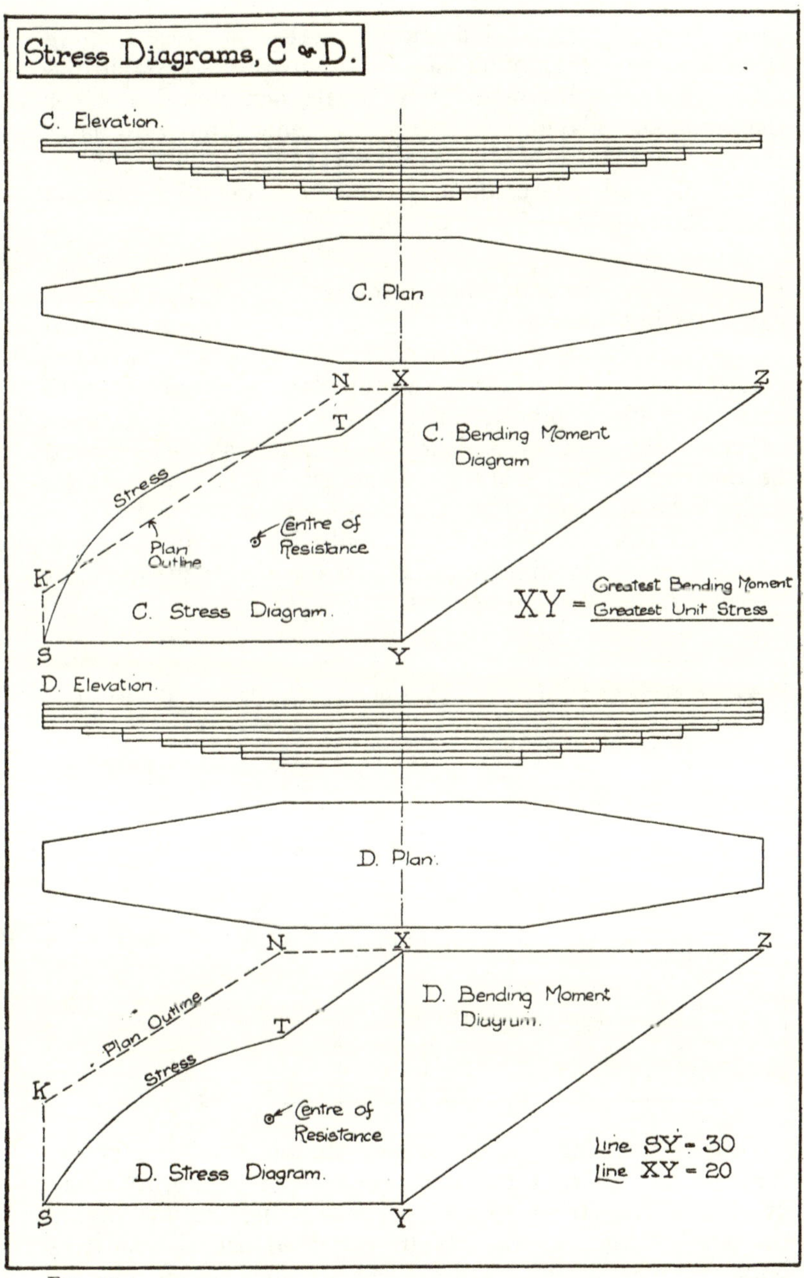

FIG. 21.

proportionately large deflection, whereas stresses ranging from zero to a maximum cause a comparatively small diagram, with a proportionately small deflection. Further diagrams of this nature appear in connection with later remarks, but it is advisable in the meantime to illustrate two additional and intermediate aspects to these two extreme cases, shown in 20—A and 20—B.

The diagrams Figs. 21—C and 21—D illustrate springs including the same number of plates (10) as the extreme examples but being each of what might be termed "normal design." Diagram C has a short plate of the minimum correct practical length, with two top plates of the same length and the resulting product of the area × centre of resistance is 7200. Diagram D has a short plate twice as long as it need be, and four top plates of the same length, the resistance of the diagram being 6440. The four results now obtained might be tabulated as under, based on Formulæ VII.—1 and VII.—2, which have the modified constant to agree with the modulus of usual spring steel sections :—

Type of Spring.	Area × Centre of Resistance.	Product.	Deflection Constant.
A. Extreme. Theoretical Perfection	600 × 15·0	9000	0·124
C. Normal Design. Light	405 × 17·8	7200	0·100
D. Normal Design. Heavy	344 × 18·7	6440	0·089
B. Extreme. Uniform Section	300 × 20·0	6000	0·083

In these drawings, the stress diagram and bending moment diagram are plotted together, each representing therefore one-half of the spring, which is quite as satisfactory as taking the whole spring, as the results obtained are comparative only. The nomenclature of the various parts is as follows :—

XYZ	is the bending moment diagram, constant always for the same length of spring and the same assumed central load.
STXY	is the unit stress diagram varying in shape according to the design of the plate lengths.
SKNX	is a proportionate plate plan outline.
XY	represents, on the one side of the diagram, the bending moment in definite units, inch-tons for example, whilst for the other side of the diagram, it represents unit stress at each point along the spring length, say in tons per square inch.

The shape of the diagrams C and D are of extreme interest, the stress dropping uniformly from X to T owing to the length of the short plate (which is here full width for its entire length) and then following a curve line to the zero at the bearing point.

The obtaining of the centre of resistance of irregular areas such as STXY by ordinary means, is a tedious process, and it is worth while noting that the combined areas of the diagrams XYZ (bending moment) and STXY (stress) are closely approximate to the product in the table "area × centre of resistance" ÷ 10. The bending moment area remains constant at 300, and if to this are added the STXY areas, the following results are obtained :—

Diagram	A.	600 & 300 = 900.	Product in table	= 9000
,,	C.	405 & 300 = 705.	,,	= 7200
,,	D.	344 & 300 = 644.	,,	= 6440
,,	B.	300 & 300 = 600.	,,	= 6000

Another procedure still simpler, and much easier to work, as it does not involve the taking out of a stress diagram, and gives results sufficiently close for all practical purposes, is, for abnormal springs, to calculate deflection according to weight—the basis calculation for unit deflection being the standard formula to be given. The method of obtaining the weight (w) of any spring will be given later. Meanwhile, it is sufficient to show the following process. The weight of a spring (approximate) for the use of the deflection constant 0·10, otherwise a "normal" spring, is :—

$$\text{Weight in pounds, } w = \frac{2LbnT}{11} \text{ or } \frac{Lb(\text{total depth})}{5\frac{1}{2}} \quad \text{(VII.—2a)}$$

Take the actual or calculated weight of the spring to be load tested and obtain the ratio between the weights as percentages, thus, spring weight to Formula VII.—2a : trial spring weight : : 100 : X. For example, assume a spring with (L) as 40 ins., (b) as 5 ins., (n) as 10, and (T) as $\frac{1}{2}$ in. From the above the weight should be 182 lbs. Assume the heavy spring, with many top plates, or long short plates to weigh 210 lbs. The ratio would be 182 : 210 : : 100 : X, which gives X as 116, showing the heavy spring to be 16 per cent. more in weight than the normal spring. If this figure 16 be divided by 3000 = say 0·005, and this be deducted from the constant 0·10, result 0·095, the latter is sufficiently near to give the required revision of deflection. Put briefly, it is the excess per cent. of spring weight over that derived from Formula VII.—2a, divided by 3, and deducted from 0·10. It is known that the real deduction figure is 0·005 and the idea of dividing by 3000 can be dropped for practical purposes.

The simplification of all possible stress diagrams, the number of which is infinity, to one general average, has, it is claimed, been accomplished by the employment of the constant 0·1 in the numerator of the now-to-be-stated formula. On the formula for unit deflection rest the remainder which deal with test loads and working loads, and consequently the adoption of any expression for this unit deflection requires careful consideration. The constant now given, however, is practically accurate for 95 per cent. of the springs which pass through the manufacturer's hands, and it has been in use in Sheffield works for some years—with the result that if the load test readings of a spring do not agree with its calculated unit deflection, according to this constant, an immediate study is made of the steel rolling or other possible variables.

STANDARD FORMULA FOR DEFLECTION PER TON

$$d = \frac{0{\cdot}1 \times L^3}{nbt^3} \quad \ldots \quad \ldots \quad \text{(VII.—3)}$$

The simplicity of the above needs no elaboration, as for slide-rule work, the constant is neglected. Very few springs will be found to vary to any practically measurable extent from the unit deflection results obtained from this formula, if the steel is rolled within correct limits. In a spring with a high percentage of full-length plates—say 4 out of 8,—the spring will be somewhat stiffer than the formula result.

On the other hand, a spring with only one full-length plate, and a short plate length of L $\div$ n, will be found to be weaker than the calculation. Such springs are, however, the exceptions rather than the rule, and with relatively stiff unit deflections, such as railway springs, combined with the usual method of taking load test results, it is probable that up to several tons, little variation will be observed. This key formula is illustrated by Fig. 22.

The supplementary formulæ for test load and working load based on the above—and at this moment, in English standard units—are as follows :—

TEST LOAD. $$W = \frac{0{\cdot}177 \times t^2bn}{L} \quad \text{(VII.—4)}$$

WORKING LOAD. $$W = \frac{Xt^2bn}{L} \quad \text{(VII.—5)}$$

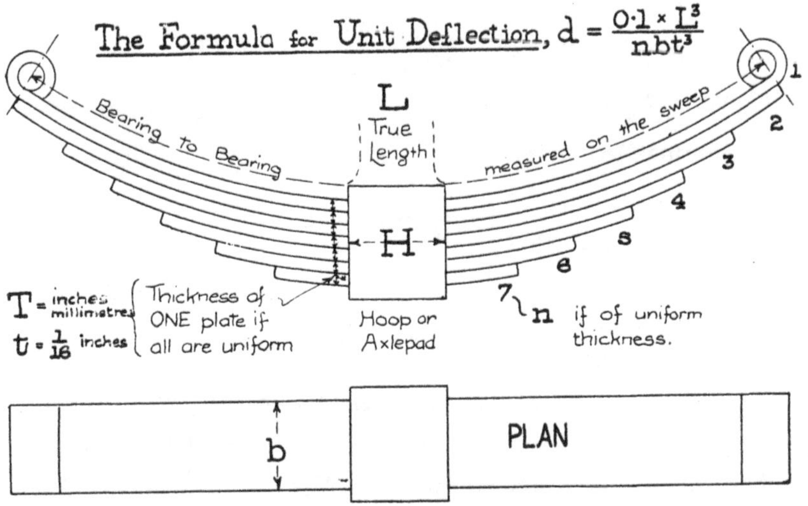

FIG. 22.

The above hold only if all plates are the same thickness; if varying thicknesses occur the test load will be less, as the British Standard Test is based on the thickness of the thickest plate.

The factor (X) in the expression for the working load represents the result of "test load $\div$ working load," and should generally be as follows :—

For all Railway and Tramway Springs.

Recommended practice. (X) should be 0·08, that is—

$$\frac{\text{Test Load}}{\text{Working Load}} = 2\tfrac{1}{4} \text{ (practically)}$$

Actual practice varies (X) between 0·06 and 0·12 giving

$$\frac{\text{Test Load}}{\text{Working Load}} = 3 \text{ and } 1\tfrac{1}{2} \text{ (approx.).}$$

Automobile Springs.—Front.

Recommended practice. X should be 0·06, that is—the

$$\frac{\text{Test Load}}{\text{Working Load}} = 3 \text{ (practically).}$$

Actual practice varies (X) between 0·05 to 0·10, giving

$$\frac{\text{Test Load}}{\text{Working Load}} = 3\tfrac{1}{2} \text{ and } 1\tfrac{3}{4} \text{ (approx.).}$$

Automobile Springs.—Rear.

Recommended practice. X should be 0·07, that is—the

$$\frac{\text{Test Load}}{\text{Working Load}} = 2\tfrac{1}{2} \text{ (practically).}$$

Actual practice varies (X) between 0·06 and 0·12, giving

$$\frac{\text{Test Load}}{\text{Working Load}} = 3 \text{ and } 1\tfrac{1}{2} \text{ (approx.).}$$

The "test load" referred to must be understood throughout as that due to the (D) of the British Standard Test, with deflection calculated on Formula No. VII—3, or $\frac{D}{d}$.

The (L) must be taken as straight length, with no deductions for hoops or axle-pads. Generally speaking, there is no need to consider hoops up to 3 in. wide, and not much value in taking account of them above this width, unless they are so fitted to the spring as to bed along to the short plate when the spring is free.

Springs having plates of varying thicknesses need to be translated into terms which include plates of one thickness only. It is usually the easier to put thick plates into terms of the thinnest plates, always of course, in accordance with the cubes of the thicknesses, one $\frac{1}{2}$ in. plate for instance, $t^3 = 512$, being equal to 8 plates at $\frac{1}{4}$ in., where $t^3 = 64$. In metric measurements, one 12 mm. plate $T^3 = 1728$, is equal to 1·728 plates at 10 mm. where $T^3 = 1000$. A conversion table for all plate thicknesses is given in another place.

Semi-elliptic springs divide into three main classes only for calculation variables, based on the standard formula, No. VII.—3. These are illustrated in Fig. 23 and are briefly described as follows :—

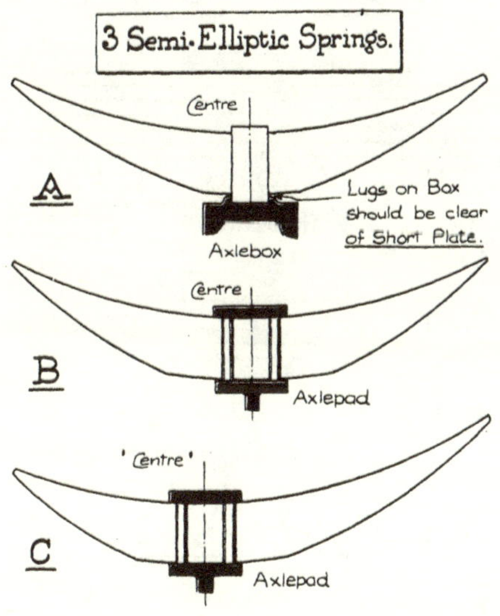

FIG. 23.

A. Ordinary semi-elliptic, hooped, as usual for all railway and tramway work, and to a certain extent for road work.

B. Semi-elliptic with wide axle-pad bearing, as chiefly employed for automobiles of all classes.

C. Asymmetrical semi-elliptic, limited to certain designs of automobiles, and chiefly to some types American pleasure cars ; used to a small extent in tramway work.

The formula for Type A can be taken as the standard without amendment. For Type B it is advisable to make correction, if it is desired to obtain the actual deflections of the spring as it will be clamped in position, as distinct from the deflection results which may be obtained on the load testing machine at the manufacturer's works. As a matter of fact, this type of fastening, namely, clamping to a wide pad, converts the spring into a double cantilever, but it would be incorrect to take each cantilever length as the distance (straight) from the shackle or hanger pin centre to the edge of the pad. The shape of the pad has also, obviously, an

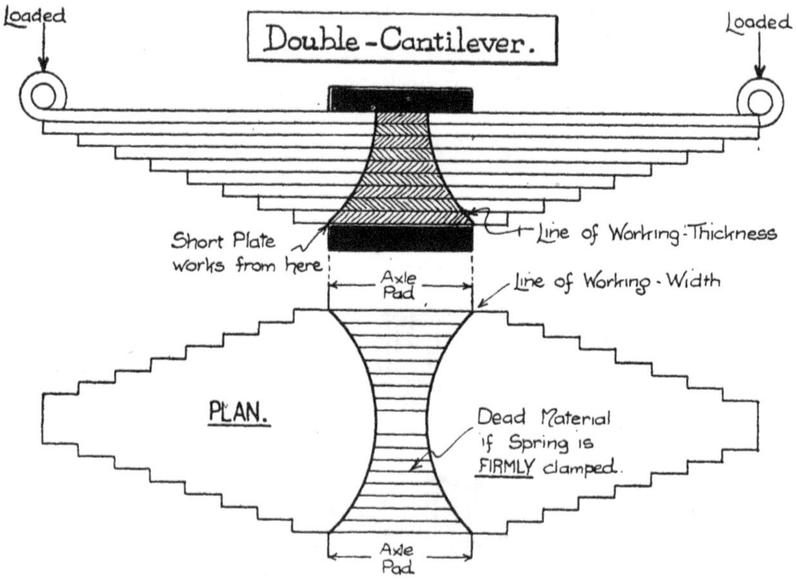

FIG. 24.

influence on the unit deflection. If it is shaped to the short plate, with the spring light, unit deflection will commence with the short plate bearing on the pad edges, whereas if it does not thus bear, the earlier deflections per increment will be higher than the deflections obtained as the spring short plate contacts with the bearing pad. This is particularly a point to be noted if the spring is likely to go " past straight " under bump loads, as if this has not been provided for in the original calculations, short plates will probably fracture or set from the bearing edges of the hoops or axle-pads. No cantilever, except it be of infinitely small thickness, = o, works throughout from its bearing edge, and the line of greatest bending moment travels from this edge towards the centre of the spring, as shown in Fig. 24. The exact positioning of this line is somewhat indeterminate, its governing features being the thicknesses and numbers of plates of which the spring is composed, with a further modifying element turning upon the deflection. Clearly, therefore, the subtraction of the whole width of the axle-pad from the length (L) before performing the calculation for deflection, will result in obtaining a figure which shows the spring to be substantially stiffer than it is. For instance, assume a 42 in. spring, with a 6 in. bearing, no unusual combination of dimensions. If the 6 in. be deducted, the (L) becomes 36 ins., and the cube thereof is 46656, as against 74088, which is 42^3. Actual loading results on a 6 in. pad, with the spring firmly clamped thereto, will give a figure intermediately, and a practical approximation for the necessary correction to meet such cases, is to take (L) as the full straight length (L) less the axle-pad width, and add 25 per cent. to the result obtained. The comparative figures would then stand as follows :—

L as 42 in., full straight length, cube =74088—result too high.

L as 36 in., full straight length less width of axle-pad, cube =46656—result too low.

L as 36 in., cubed, and plus 25 per cent. =58320—practically correct.

This last result is equivalent to taking (L) as a straight length of $38\frac{3}{4}$ ins. The formula stands as follows :—

Unit Deflection for Double Cantilever semi-elliptic type ;

$$d = \frac{0{\cdot}125(L - H)^3}{nbt^3} \quad .. \quad .. \quad .. \text{(VII.—6)}$$

In cases of this nature, it must be borne in mind that the allowable test deflection should be taken as $1\frac{1}{4}$ times the test deflection permissible worked out with $(L - H)^2$ instead of (L), otherwise :

Test Deflection for Double Cantilever, semi-elliptic type :

$$D = 1{\cdot}25 \times \frac{(L - H)^2}{900T} \quad .. \quad .. \text{(VII.—7)}$$

Test Load for Double Cantilever, semi-elliptic type :

$$W = \frac{0{\cdot}177t^2bn}{L - H} \quad \ldots \quad \ldots \quad \ldots \text{(VII.—8)}$$

This test load only holds good if all plates are of the same thickness.

The reasons for the designs of the asymmetrical spring, Type C, are indicated in the diagram, Fig. 25. In certain

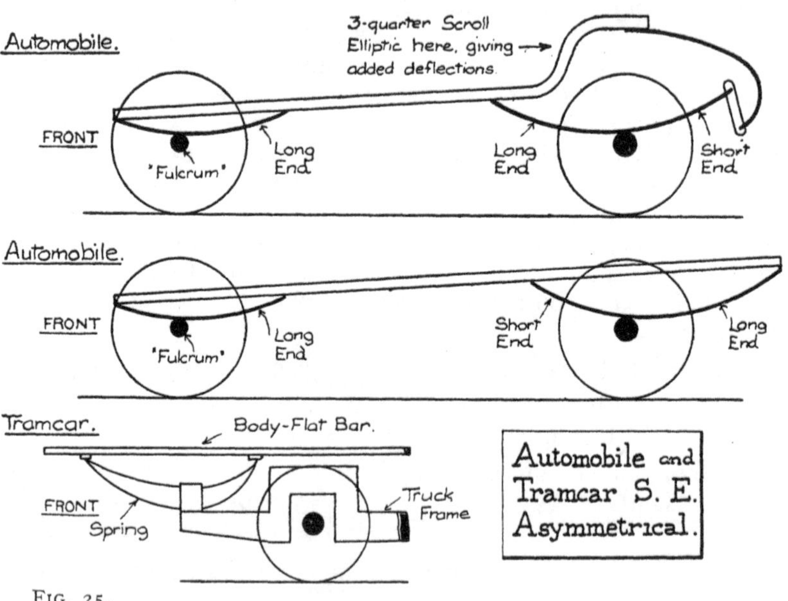

Fig. 25.

automobiles, it is assumed that the chassis moves on the front axle as a fulcrum, and the rear spring is accordingly designed with each half of different lengths. On other occasions, the fulcrum is assumed to be the leading dumb-iron, and in such case, both front and rear springs are asymmetrical. It has also been recommended that front springs should be mounted at a point apart from their true longitudinal centre, so that there is less spring movement to influence the steering, and that where there are no radius rods, rear springs should be mounted in the reverse way, that is, with the short " half " at the end of the car. Some types of tramway (4-wheel) trucks used the pattern owing to exigencies of design, where a short balanced spring would be too stiff, and a long balanced spring—to extend the spring base of the body to its maximum length—could not be fitted on the inside of the spring base. Generally speaking, however, such unbalanced springs are not to be recommended.

Springs of this type do not generally have their practical "centre" far removed from their true centre, a distance about $\frac{1}{10}$ of their length eccentric representing the maximum usual

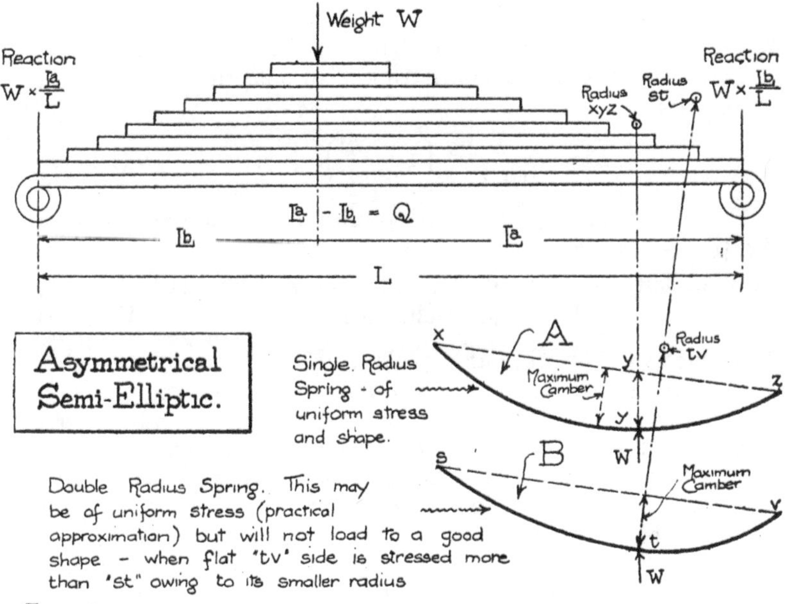

FIG. 26.

in practice; a 36 in. spring for instance, having one side 15 ins. long and the other 21 ins., = 3 ins. out of centre, or $\frac{1}{12}$ of the length. They should be fitted as one-radius springs, Fig. 26—A, as they will then be a good shape straight; and they should not be made as at 26—B, with sides of different radii. The calculations can easiest be made on the basis of the ordinary balanced semi-elliptic, and then proportionated to the point of loading. For instance, if the unit deflection of the ordinary balanced spring—taken, of course, at the true and working centre—is 3·60 ins., and the unbalanced spring is $\frac{1}{12}$ of its length out of centre, for all practical purposes the deflection under the point of fixation will be $\frac{1}{6}$ less than the maximum, or $\frac{1}{6}$ of 3·60 ins. = 0·60 ins. This figure ratio, $\frac{1}{6}$, equals the difference in length of the two arms, namely, $\frac{7}{12}$ less $\frac{5}{12}$. It can be said, therefore, that the formula for unit deflection of such spring is as below :—

Unit deflection for asymmetrical semi-elliptic :

$$d = \frac{0{\cdot}1 \times L^3}{nbt^3} \times \left(1 - \frac{Q}{L}\right). \qquad \text{.. (VII.—9)}$$

where (Q) is the difference in length of the two arms.

Test deflection for asymmetrical semi-elliptics, under point of fixation :

$$D = \frac{L^2}{900T} \times \left(1 - \frac{Q}{L}\right). \qquad \text{.. (VII.—10)}$$

Test load for asymmetrical semi-elliptics, under point of fixation—

W = same as Formula No. VII.—4 for ordinary semi-elliptics. .. .. (VII.—4)

Variables such as wide axle pads, full-length top plates, etc., will be dealt with as for the ordinary balanced spring. Such asymmetrical springs can be treated as double cantilevers, but it is not worth while, as up to the variation in arm-lengths generally employed, the difference between such treatment, and calculation on the basis of the true semi-elliptic, is not great. This can be seen by examination of the following tables :—

Lengths of Arms. Total L = 20.	Reaction on each "Centre" W = 20	Relative Deflection each end.	Total Deflection of ends.	Ratio $\frac{Q}{L}$
10 & 10	10 & 10	10000 & 10000	20000	0
9 & 11	11 & 9	8019 & 11979	19998	2/20
8 & 12	12 & 8	6144 & 13824	19968	4/20
7 & 13	13 & 7	4459 & 15379	19838	6/20
6 & 14	14 & 6	3024 & 16464	19488	8/20
5 & 15	15 & 5	1875 & 16875	18750	10/20
4 & 16	16 & 4	1024 & 16834	17858	12/20
3 & 17	17 & 3	459 & 14739	15198	14/20
2 & 18	18 & 2	144 & 11664	11808	16/20
1 & 19	19 & 1	19 & 6859	6878	18/20
0 & 20	20 & 0	0 & 0	0	1

A curve showing the table is included in Fig. 27, and from this, and the above figures, it is abundantly clear that the unit deflection need not generally be modified from the balanced spring formula, with the exception of the correction necessitated by measuring the spring at the eccentric point of attachment, where the manufacturing camber is given.

A point in connection with unbalanced springs that must not be overlooked, is that the reaction at each end is in inverse proportion to the length of the arms; if the arms are (L*a*) and (L*b*) long respectively, and the weight at the intermediate fixture point is W, the weight on cantilever end (L*a*) will be W × (L*b* ÷ L) and at the end of cantilever (L*b*), = W × (L*a* ÷ L). See Fig. 26.

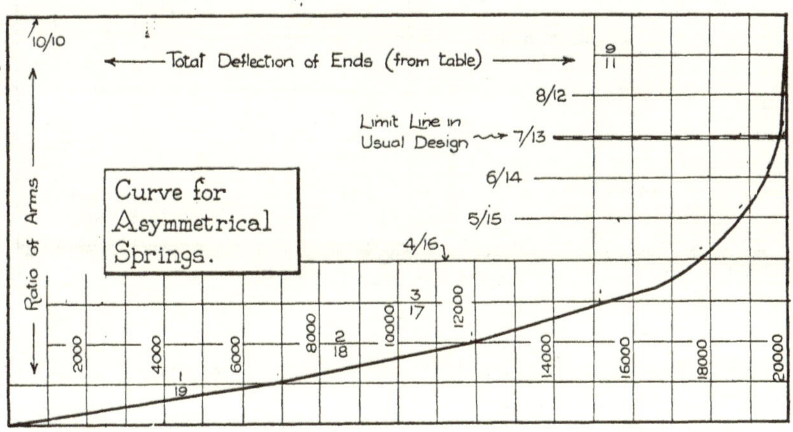

FIG. 27.

It might be of interest to consider an additional variable, common to the three main classes, namely, numbers of full length plates, and consequent amended offsets; or alternatively, excessive lengths of short plates, resulting in amended offsets; both as compared with the nearest practical approach to the theoretical rhombus. Strictly speaking, springs of these characters should not be designed, and it is hoped that, as the result of study of the various aspects included in the present chapters, such designs may soon cease to be, but meanwhile, numerous examples are extant for which calculations may be demanded. In the latter case, of abnormal short plates, the best thing is to make out a stress diagram on the lines of Figs. 20 and 21. In the former case, however, of full length plates over a normal or usual number, which is usually two or three on springs including from 8 to 16 plates, a tolerably accurate result may be obtained by modification of the constant. The following revised constants can be used. which have been drawn up from stress diagrams, and an illustrative graph is given in Fig. 28. Up to 30 per cent. of the total, constant 0·1, can be taken as correct.

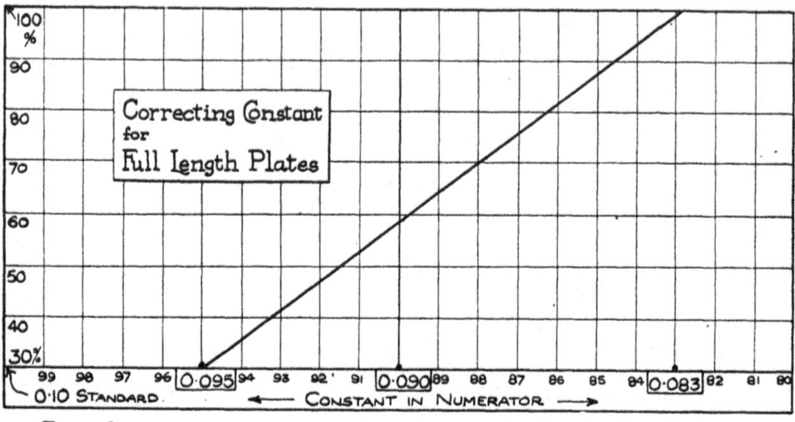

Fig. 28.

Full length plates	30 per cent. of total	..	constant	0·095
,,	40 per cent. ,,	..	,,	0·093
,,	50 per cent. ,,	..	,,	0·091
,,	60 per cent. ,,	..	,,	0·090
,,	70 per cent. ,,	..	,,	0·088
,,	80 per cent. ,,	..	,,	0·086
,,	90 per cent. ,,	..	,,	0·085

In dealing with springs of non-uniform thickness plates, equivalent thicknesses must obviously be taken into consideration in order to correctly employ the modified constants. For instance, if a spring had two full length top plates of $\frac{1}{2}$ in. thick, two full length following plates of $\frac{3}{8}$ in. thick, and eight succeeding plates with ordinary offsets, of $\frac{5}{16}$ in. thick, the percentage to be taken would have to be based on the equivalent number of the thinnest plates in the spring, thus :—

2 plates, $\frac{1}{2}$ in. thick	=	8 @ $\frac{5}{16}$ in. (approx.)
2 ,, $\frac{3}{8}$ in. ,,	=	4 @ $\frac{5}{16}$ in. ,,
8 ,, $\frac{5}{16}$ in. ,,	=	8 @ $\frac{5}{16}$ in.
Total	=	20 @ $\frac{5}{16}$ in. (approx.)

With a spring of this design, therefore, 12 out of 20 plates, or 60 per cent. are virtually full length, not 4 out of 12, which are the actual numbers, and only give 33 per cent. The calculation constant to be employed for such a badly designed spring would therefore be 0·90 instead of 0·10.

CHAPTER VIII

THE PRESENT FORMULÆ

SPRINGS OTHER THAN SEMI-ELLIPTIC

THE further types of laminated springs to be considered, with the formulæ relating thereto, are as follows :—

1. Quarter-elliptics, or true cantilevers. These are employed on certain types of tramcar trucks (4-wheel), and to a moderate extent for light automobiles. (Fig. 1—E.)
2. Semi-elliptics used as cantilevers, almost entirely limited to high class passenger automobiles. (Fig. 1—D.)
3. Three-quarter elliptics—being a combination of a true cantilever with the semi-elliptic. This type is used, to a decreasing extent, in tramcar trucks (4-wheel) and passenger automobiles.
4. Full-elliptics or " bolster " springs, which are in almost universal employment in railway work for bogies on coaching stock, not uncommon in tramway trucks (4-wheel), and used in a very limited degree for automobiles. (Fig. 1—C.)
5. Full-elliptics, or " bolster " springs, of special character, known as " endless back " springs (illustrated hereafter) having some of the characteristics of the ordinary " bolster," (4). These are very rarely used in railway work (coaching stock), and for no other vehicle purpose.

As regards the quarter-elliptic, or cantilever, the unit deflection formula resolves itself into an amendment of the semi-elliptic formula for double-cantilevers, No. VII.—6, and becomes :—

Unit deflection at free end, true cantilever:

$$d = \frac{2L^3}{nbt^3}. \quad .. \quad .. \qquad \text{(VIII.—1)}$$

The factor (2) in the numerator is the product of 0·125 and 16, the former being the practical constant admitting of the cantilever working as in Fig. 24. The latter is the corrector for (L) which as compared with the semi-elliptic is now one-half, which gives $(\frac{1}{2}L)^3$ as $L^3 \div 8$. The factor (8) must occur in the numerator, multiplied by $2 = 16$, as the full unit weight comes on one free end only instead of two, as with the semi-elliptic.

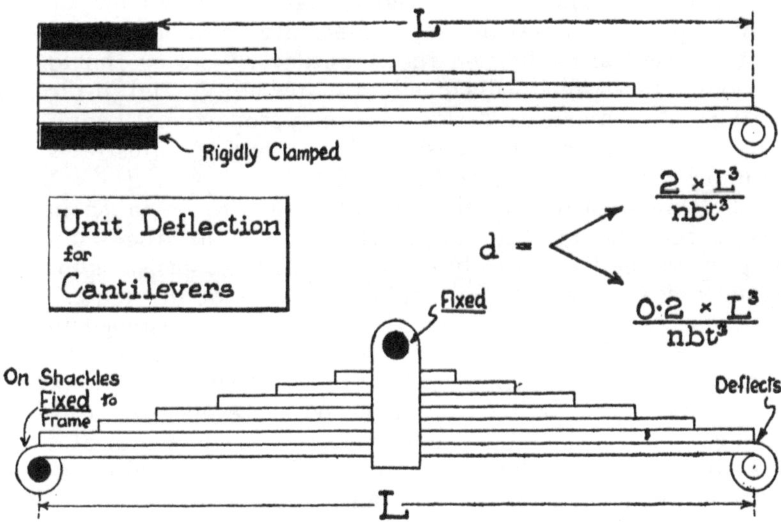

FIG. 29.

The remaining formula relating to this type, is as follows:—
Test reflection for true cantilever:

$$D = \frac{5L^2}{900T}. \quad .. \quad .. \qquad \text{(VIII.—2)}$$

The test load formula will be as No. VII.—4, with the constant as 0·044 instead of the semi-elliptic figure, and then:

With X as $2\frac{1}{4}$, constant becomes 0·020.
,, X as $2\frac{1}{2}$, ,, ,, 0·018.
,, X as 3, ,, ,, 0·015.

The next type of spring to be treated is the semi-elliptic used as a cantilever, and frequently known as a cantilever. The general scheme of calculation for such springs can clearly be along the lines of the more usual arrangement, but the designing deflection for such pattern is, of course, at the free end of the spring, and the formula for this is the following simple modification :—

Unit deflection at free end, semi-elliptic as " cantilever " :

$$d = \frac{0{\cdot}2L^3}{nbt^3}. \quad .. \quad .. \quad \text{(VIII.—3)}$$

For test purposes, springs as these should be regarded as semi-elliptics. The working load can also be based on the semi-elliptic characteristics, by taking into account the fact that the central loading on the " cantilever " spring should be double that which will come on the free or axle end (if the spring is of symmetrical design; if other, proportional figures have to be taken into consideration.) Fig. 29 illustrates formulæ for the two foregoing patterns.

The three-quarter elliptic calculation has to be taken in two parts, one for the cantilever, and one for the semi-elliptic, on the lines already given. The two suspensions for which these are sometimes employed have wide diversities; as in the case of automobiles, the free ends are fastened to the chassis, whereas in tramcar work, the free end of the semi-elliptic has a bearing only. In the former case as a rule, only a part of the cantilever seriously operates as a spring. Fig. 25 showed a usual arrangement, which indicates that by the use of the three-quarter elliptic, an additional static deflection under the load can be obtained.

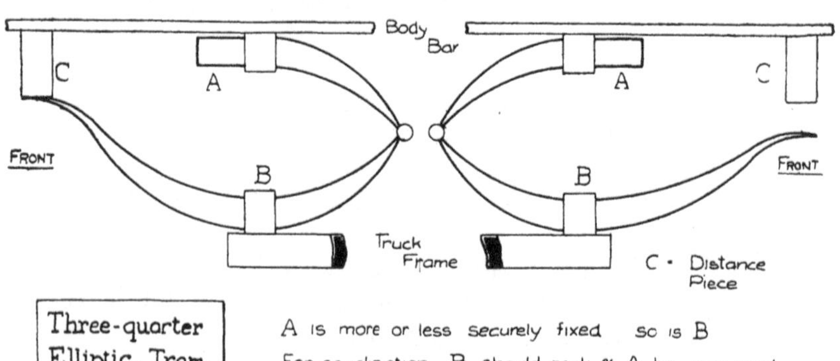

FIG. 30.

In Fig. 30 is a sketch of a common tramcar type, which cannot be regarded as very satisfactory, as the spring is not of the equalizing variety owing to the two fixed points and one floating point. Unless cambers are carefully designed, the front end will be taking either too much or too little of the load, as shown in the possibility illustrated by the right hand sketch, where, owing to a low camber on the semi-elliptic, the outermost end is not on the body bearing point.

Quarter-elliptics by themselves, or included in the three-quarter type, deserve very special attention as regards the fastening arrangement at the fixed end (these being equivalent to a beam " encastre " or " built in ") as the force tending to pull apart the fastening means is high, and directly slackness introduces itself, the whole aspect of unit deflections is revised, as a certain amount of " dead " deflection arises, due to the leverage of the weight on the fastenings, as indicated in Fig. 31.

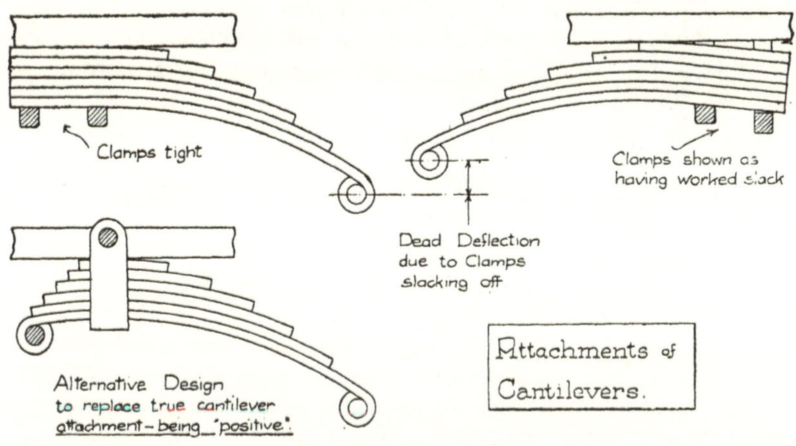

FIG. 31.

A method of obtaining an equivalent spring to the true cantilever, which includes a more positive holding arrangement of the " fixed " end, is also shown in Fig. 31,—as an inverted semi-elliptic of unbalanced pattern, with one very short arm. This is naturally, a somewhat more expensive device, owing to a hoop being required, but is very much more satisfactory. The resulting deflection is practically equal to that of the true cantilever or quarter elliptic, which it would replace, as whilst the long arm as such would deflect slightly less than the true cantilever arm, owing to the fact that the (L) of the latter " works backwards," an additional deflection is obtained as the result of the slight movement of the " eye " end of the short arm.

Full-elliptic springs (bolster springs) of the ordinary type are simple of calculation, the main formula for unit deflection being as for the semi-elliptic, and multiplied by 2, which gives the unit deflection for the full-elliptic—each end of each of the half-springs deflecting at the same rate. The deflection of a " group," such as is usual in railway practice, of perhaps 3 or 4 full-elliptics, is inversely proportional to the number of full springs. For instance, suppose a group of 3 full springs, the deflection of the half-spring being 0.60 ins. per ton. Then :—

Deflection of full spring—one unit—is 0·60 ins. $\times$ 2 = 1·20 ins.

Deflection of group, three units—is 1·20 ins. $\div$ 3 = 0·40 ins. Clearly also, the test deflection for a full spring should be twice that for the semi-elliptic. Formulating these points, the following appear :—

Unit deflection for full-elliptics,

$$d = \frac{0{\cdot}2L^3}{nbt^3G}. \qquad \ldots \qquad \ldots \qquad \text{(VIII.—4)}$$

Test deflection, full elliptic,

$$D = \frac{L^2}{450T}. \qquad \ldots \qquad \ldots \qquad \text{(VIII.—5)}$$

In the above, (G) is the number of units in the group. All spring units constituting the group are herein obviously assumed to be of the same design—in accordance with universal practice.

The " endless back " spring—(5)—is, as previously pointed out, in a very special class, and the calculation therefore represents features of interest. A working description of this spring will be given in the proper place, and meanwhile a diagram illustrating it is shown in Fig. 32. It can be regarded as a combination of two semi-elliptics, with two double superimposed cantilevers. The assumed lines of demarcation of these springs are indicated on the diagram, and whilst it may have debatable features, the variation of an inch or two one way or the other for the true bearing points is a matter of no moment, provided always that the overall length presents the same addition. (See the characteristics of added cube lengths, in connection with asymmetrical semi-elliptics, in the previous chapter.) On the basis of one elliptic only of this type, the total deflection will clearly be twice that of each separate semi-elliptic, plus that of the cantilever ends—which carry half the weight on the hoop.

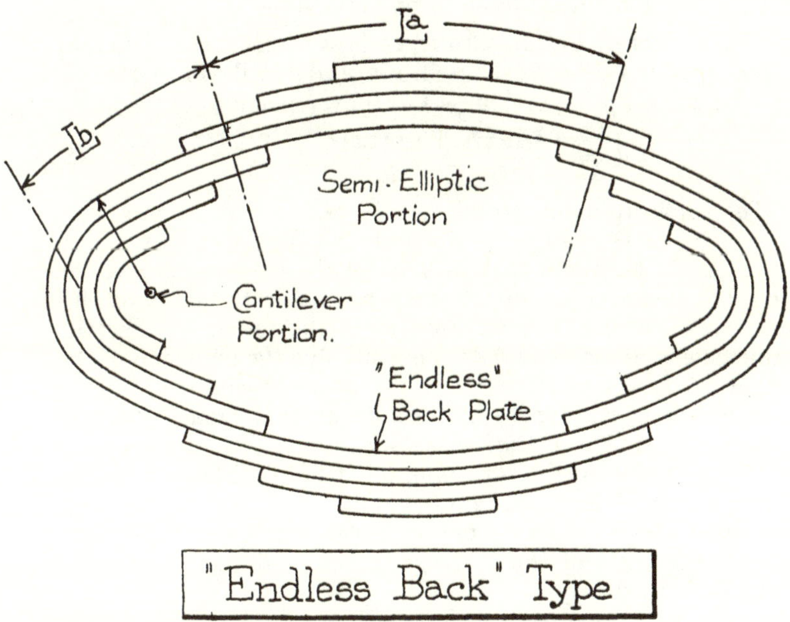

FIG. 32.

Generally the cantilever plates are the same thickness and quantity as the semi-elliptic plates, and if the design be on these lines, the following formula gives the deflection for any group of such springs. (L*a*) represents the semi-elliptic portion, and (L*b*) the cantilever portion. The letter (*n*) is the number of plates in either one cantilever or one semi-elliptic unit. Then :—

Unit deflection for full elliptic group, endless back type,

$$d = \frac{0{\cdot}2(La)^3 + 1{\cdot}6(Lb)^3}{nbt^3G}. \quad .. \qquad \text{(VIII.—6)}$$

Test deflection for endless back elliptic,

$$D = \frac{(La)^2 + 4(Lb)^2}{450T}. \quad .. \quad \text{(VIII.—7)}$$

The above formulæ only hold good if the plate section combination for the semi-elliptics and for the double cantilevers involved are identical. There is no reason why they should not be designed differently, with either stronger or weaker plates in the cantilever portions, so as to provide for greater or lesser unit deflections. In such unbalanced cases, the following formulæ have to be applied :—

Unit deflection for full elliptic, endless back type. Different plate combinations in semi-elliptic and cantilever parts:

Semi-elliptic portion + Cantilever portion,

$$d = \frac{0{\cdot}2(La)^3}{nbt^3G} + \frac{1{\cdot}6(Lb)^3}{nbt^3G}, \quad .. \quad \text{(VIII.—8)}$$

Test deflection for group, as above,

D = same as Formula No. VIII.—7 (VIII. 7)

It should be noted that in view of the employment of the double cantilever action of small radius at the ends of this pattern of spring, the working length is taken as the true dimensional average, and does not involve the inclusion of the empirical factor referred to in the previous chapter, due to cantilevers operating backwards from their true point of support, as shown in Fig. 24.

In conclusion of the foregoing let it be reiterated that any "test-weight" formulæ quoted only hold good on uniform thickness plates, and with non-uniform plates, test deflections are on the basis of the thickest plates, and the test weight reduces accordingly.

CHAPTER IX

FORMULÆ OF OTHER AUTHORITIES

NOTE.—Standard symbols not used in this chapter.

As a matter of interest, a number of other much used formulæ will be examined, and it will assist the comparison if a standard spring is calculated out by means of each. For this purpose, an ordinary bearing spring 40 ins. straight length, centre to centre, with 10 plates, 4 ins. × ½ in. will be assumed. On the standard unit deflection formula given in Chapter VII., this presents a deflection of 0·31 in. per ton, with test deflection of 3·55 ins. equal to a load of 11·4 tons. Buckle of spring 3 ins. wide.

MOLESWORTH

This is the usual authority accepted in this country and in the well-known pocket book, the following formulæ are included :—

Relative to Laminated Springs in general.

$$\text{Safe working load in tons} = \frac{13 \cdot 3bt^2n}{L}. \qquad \text{(IX.—1)}$$

In the above L = length of spring from edge of hoop to centre of bearing.

t = thickness of each plate in inches.

b & n have the usual meanings.

From this formula, the safe working load of the spring sketched out is 7·2 tons, whereas, with a test load of 11·4, if used for railway service, 5 tons would be a reasonable figure. The implication is therefore, that the result obtained by the use of this formula gives rather a test load—for the steel of the day—than a safe working load, which is partly confirmed by the statement that S,· the ultimate stress, is 40 tons—this factor being used in the derivation of the expression quoted.

The unit deflection formula is as follows :—

$$\text{Deflection in inches per ton} = \frac{L^3}{4000bt^3n}. \quad \ldots \quad \text{(IX—2}$$

The use of this expression (symbols as before) results in a deflection for the standard spring of 0·20, which is a long way from the truth. Two things assist this result, one being the length factor, taken as minus the buckle, and the other being the high modulus of elasticity employed in the formation of the expression, namely, 16000. The universal figure to-day is 13000, which amends 0·202 to 0·249. If the full length is taken of 40 ins. instead of 37 ins., the cube difference (64000 and 50653) will amend again the 0·249 to 0·317, which is practically correct.

Relative to Locomotive Springs.

The deflection formula in this connection is as follows :—

$$\text{Deflection in inches per ton} = \frac{0{\cdot}14S^3}{T^3BN}. \quad \ldots \quad \text{(IX.—3)}$$

In the above, S = span of spring.
T = thickness of each plate in $\frac{1}{16}$ in.
B = breadth of spring.
N = number of plates.

It will be noted that in the case of this formula, there are two variations only from the standard suggested, No. VII.—3. Both occur in the numerator, the constant being 0·14 instead of 0·10, and the span (S) being employed instead of the straight length (L). In certain highly cambered springs, it is clear that the larger constant multiplied by the smaller (S^3) will about balance 0·1 (L^3), but with lightly cambered or straight springs, the unit deflection result will be about 40 per cent. higher than will be found to occur with the actual spring.

The "safe load" formula "for locomotive springs" is given as :—

$$\text{Safe load on spring in tons} = \frac{BT^2N}{11{\cdot}3\,S}. \quad \ldots \quad \text{(IX.—4)}$$

A note states that recent practice increases 11·3 to 14 or 15. A calculation on this latter basis for the standard spring assumed, gives a "safe load" for the same (taking S as equivalent to L) of 4·25 tons, which is practically $\frac{3}{8}$ of the test load, and a reasonable figure. The "safety factor" (X) is 2·7 which is somewhat high, whereas if the 11·3 in the original formula is taken, the "safe load" becomes 5·6, and (X) is then 2·04, which is nearer recommended practice. These "Rules for the Strength of Locomotive Springs" quoted in Molesworth are those of D. K. Clark.

The author must here acknowledge his indebtedness to this formula for unit deflection of D. K. Clark, which has been the basis of the standard formula suggested for this purpose (VII.—3). The extremely practical feature of introducing the factor for thickness in sixteenths of an inch, substantially simplifies the calculation, and the old formula has probably been in past times, the most used of any of those which purported to deal with the subject.

A. A. Remington

In a paper read before the Institution of Automobile Engineers (London) in 1922 on the subject of laminated springs, this authority included numerous formulæ which are more or less co-related, and intended to be specially applicable to semi-elliptic springs for automobiles.

The following is the deflection formula :—

$$\text{Total deflection} = \frac{WL^3}{4bEnt^3}. \qquad .. \qquad \text{(IX.—5)}$$

In the above, W = the load in pounds.
L = length in inches (presumably "straight")
t = thickness of one plate in inches.
b & n have the usual meanings.
E = Modulus of Elasticity (modified value) and according to the Paper this "modified value" is 22,500,000.

The formula stated above is admitted to be the standard for the deflection of beams, and as such, is another expression of Formula No. II.—2. A "modification" of (E) has been made to bring the result in line with known results, as an alternative method to the employment of a "constant." A mathematical suggestion which more or less legitimises such method will be shown later, but certainly without a suggestion of this description, it cannot be regarded as sound to alter a standard value such as (E) in order to avoid (K). Anyhow, with (E) taken as 22,500,000 as suggested, the unit deflection of the assumed spring becomes 0·318 in., which is not greatly removed from the 0·31 in. obtained by the standard formula.

Goodman

In the standard work of this author *Mechanics Applied to Engineering*, the following formula appears :—

$$\text{Total deflection} = \frac{3WL^3}{8Enbt^3}. \quad \ldots \quad \ldots \quad \text{(IX.—6)}$$

In the above, W = the load in pounds.

L = length in inches (presumably "straight")

t = thickness of one plate in inches.

b & n have the usual meanings.

E = Modulus of Elasticity, to be taken as 26,000,000.

This latter (E) is again a modification to endeavour to bring the formula into line with known results. The deflection calculation from this expression for the standard spring, gives 0·415 in. per ton, which is a very long way from the truth. The discrepancy is due to the "true circular" deflection being taken as the basis, otherwise the "perfect spring," as Formula No. II.—3. This gives a result approximately 25 per cent. over that obtained with normal design, and accordingly 0·415 in. would reduce to 0·335 in. The reason for the amendment of (E) is not clear, and if this be taken as standard value, namely, 13000 tons instead of 11600 tons (26,000,000 lbs.) the final result is bound to return approximately to that obtained by the standard formula, and actually arrives at 0·30 in. In order to account for the discrepancy which would be obvious in testing between the deflection given by the formula under review, and the result actually obtained, the author suggests that the disagreement is due to friction between the plates, but that this element would operate to the extent of reducing the deflection 25 per cent. is fallacious.

"Machinery's Handbook"

This includes some of the most comprehensive matter on plate springs—apart from the specialized data books which are used in America. Generally speaking, there is a tendency to wrap the subject in rather an air of mystery owing to the quantity of tables and multiplying constants usually given in U.S.A. books which touch laminated springs—as the problem is by no means so complicated as to demand all this matter. Taking the simplest expressions dealing with semi-elliptics from this well-known handbook, it will be found that factors (P) and (F) are the unknowns, as follows :—

$$\text{P (maximum carrying capacity)} = \frac{2Snbh^2}{3L} \quad \text{(IX.—7)}$$

and

$$\text{F (deflection under P)} = \frac{SL^2}{4Eh}. \quad \text{(IX.—8)}$$

In the above, S = fibre stress, to be taken as 80,000 lbs.
L = span length, less the buckle width.
h = thickness of one plate in inches.
E = Modulus of Elasticity, to be taken as 25,400,000

It will be seen that a considerable modification is introduced into (E), in line with the practice of Remington and Goodman, as elsewhere in the handbook (E) is stated to be 30,000,000.

If the foregoing formulæ be combined to obtain one which will give a unit deflection, the following is the result :—

$$\text{Deflection per ton} = \frac{3L^3}{8Enbh^3}. \quad \text{.. .. (IX.—9)}$$

The (E) to be taken is 11,400 tons (25,400,000) and the result for the standard spring (with no camber included) is then 0·335 in. which can be regarded as fairly correct.

It will be noted that the allowable fibre stress for the "maximum carrying capacity" is 80,000 lbs. per sq. inch. It is, however, not clear whether this "maximum carrying capacity" means the equivalent of a test load, or actual overloading likely to be arrived at in service. Standard U.S.A. practice tends to design for maximum static loads with fibre stresses of 40,000 to 60,000 lbs. for automobiles, and 60,000 to 80,000 for railway and tramway springs.

The above formula, No. IX.—9, is the same as Goodman's as regards its symbols, and the discrepancy between the results obtained from the two is due to the fact that the (L) of Goodman's is the full length, whereas the (L) of *Machinery's Handbook* is the span, and also minus the buckle width. The relative values of the cubes of these varying lengths account for the deflection differences. The (E) is only slightly varied between these two formulæ, as can be noted.

A modified formula is given—unique in its way—purporting to deal with springs having a number of full-length top plates. This is as follows, the unit deflection being taken out from the two related formulæ :—

$$\text{Deflection per ton} = \frac{3L^3}{4\,(2 + r) \times Enbh^3}. \quad .. \quad \text{(IX.—10)}$$

Factors herein have the same meanings as those in the preceding formulæ from this source, (r) being the ratio of full-length plates to total number of plates ; for instance, with 3 full-lengths in 12 plates, (r) would be 0·25, and the bracketed expression would then equal 9, the corresponding multiplier in the Formula No. IX.—9 being 8. With all full-length plates, (r) would be 1, and the multiplier 12, which would give a deflection lower than would really be obtained—as (r) should approximate 10 under such conditions. The assumed spring would probably have two full length plates in 10, which would make the multiplier 4(2 + 0·2) = 8·8· The unit deflection would then be 0·304 ins., which is very close indeed to the 0·31 in. obtained by the suggested standard formula.

Meyer

In this author's work on *Modern Locomotive Construction* (U.S.A.) a formula is given as follows :—

Deflection per " short " ton of 2000 lbs. in $\frac{1}{16}$ in.

$$= \frac{1{\cdot}5L^3}{bnt^3}. \quad .. \quad .. \quad \text{(IX.—11)}$$

All these symbols have the standard significance.

It will be observed that the expression of this formula is practically the same as D. K. Clark's (via Molesworth) except that it replaces the decimal fraction as constant, by 1·5, and accordingly obtains " short ton " deflections in $\frac{1}{16}$ in. As applied to the assumed spring, the result, for the " long ton " of 2240 lbs., is 5·25 sixteenths, or 0·327 in., which for all practical purposes would be found accurate, and is considerably nearer the truth than the basis formula from which it was evidently taken.

Leitzman and Von Borries

The highly technical work by these authors, *Theoretisches Lehrbuch des Locomotivbaues* (German) includes the following:

$$\text{Deflection per unit load} = \frac{6PL^3}{2nbEh^3}. \quad \ldots \quad \text{(IX.—12)}$$

In the above, P = the load (any unit).
L = one-half the span.
h = thickness of one plate.
E = Modulus of Elasticity—standard.
b & n have the usual meanings.

This formula is practically identical with those of Goodman and *Machinery's Handbook* except that it takes (L) as one half the span. If this were taken as the span (or more correctly the straight length) the numerical factors in numerator and denominator would become $\frac{6}{16}$ or $\frac{3}{8}$, as in the two other formulæ referred to. Owing to the (E) being taken as standard, and the (L) not neglecting the hoop width, the unit deflection of the assumed spring—by this formula—is 0·37 in., which is less than that of Goodman, and more than that of *Machinery's Handbook*. The reduction in the figure for (L), however, due to this being based on span instead of straight length, will, with normal cambered springs, give results which are not far removed from the truth, although this is not a sound correction due to the possibilities of say 10 in. camber in 60 in. length, or 0 in. camber in a similar length, and all variations between.

Sufficient existing formulæ have been touched upon to indicate the trend of such, as nearly all follow the same general lines. Some, it will have been noted, take (L) as span, and others as "less the buckle," both of which equivalents, it is submitted, are incorrect, the only true (L) being the "straight length" (L), which is then independent of any fabricated cambers. Others, in order to obtain a constant, to cause the general expression to agree with known results modify (E), which cannot be regarded as the happiest method of this attainment. The original D. K. Clark formula stands as the best basis, and doubtless gave reasonably accurate results in the days when it was evolved, as the flat spring in this country was comparatively unknown, and the combination of 0·14 with (L) as the span corrected each other.

It must be emphasised that none of the foregoing formulæ will give accurate results with the calculation of springs which are fixed by wide axlepad attachments, and accordingly checked in service as thus loaded. An approximate corrector has to be arrived at, which was suggested in Chapter VII. (Formula No. VII.—7). Experience has shown many automobile spring designers that the results they have obtained off various formulæ (including the standard VII.—3) have been considerably removed from the truth, as obtained from the spring attached to the vehicle, and accordingly, they have included correcting factors for their special designs of springs which enable them to ensure satisfactory calculations. Such factors, although taken along different lines, parallel, in the ultimate result, Formula VII.—7.

CHAPTER X

REMARKS ON STEEL

UNTIL about 1910, the steel used for springs was of ordinary carbon quality, sub-divided only into "oil-hardening" and "water-hardening," the former being of relatively high carbon content, say 0·80 per cent. to 1·00 per cent., and the latter of relatively low carbon quality, say 0·40 per cent. to 0·70 per cent. Resulting springs from these two qualities were not widely different, as the "oil-hardening" was not as intensely hard when quenched as the "water-hardening," and being frequently tempered to a lower heat, the results balanced as regards final limiting stresses. The chief object in the use of the "oil-hardening" quality turns upon its being safer under heat treatment in the hands of comparatively unskilled labour, as it cannot, as a rule, be cracked during hardening, owing to the low cooling rate due to the oil immersion; whereas it is not a difficult matter with certain "water-hardening" qualities to produce water cracks, which at times are not readily detectable, but lead to the early failure of the spring in service. During later years, various "alloy" steels, or, as they are sometimes unkindly called "doped" steels, have come into fashion, chiefly for automobile spring work, most of these being of "oil-hardening" character. Spring steels, their manufacture and analyses, are treated more fully in the next section of this volume, but it is necessary to touch briefly here on certain features in connection with them to provide coherent understanding.

The stresses to which designers arrange their governing features vary considerably in different practice, and the following table gives an average idea :—

MAXIMUM FIBRE STRESS UNDER STATIC WORKING LOADS

School :—	British.	Continental.	American.
Tons per square inch			
Railway Springs ...	25 to 35	26 to 42	30 to 40
Tramway Springs...	25 to 35	26 to 42	25 to 40
Automobile Springs	20 to 40	26 to 35	18 to 35
Kilogrammes per square millimetre.			
Railway Springs ...	40 to 55	40 to 66	47 to 63
Tramway Springs	40 to 55	40 to 66	40 to 63
Automobile Springs	31 to 63	40 to 55	28 to 55
Pounds per square inch.			
Railway Springs ...	56000 to 80000	58000 to 94000	67000 to 90000
Tramway Springs	56000 to 80000	58000 to 94000	56000 to 90000
Automobile Springs	45000 to 90000	58000 to 80000	40000 to 80000

[It must be understood that the figures in this table are such as would be obtained, or worked to, with the Standard Bending Moment Formula, I.—4, which is usually employed by designers interested in skin stress—either in its direct form as I.—4, or in some modification thereof, the stresses arrived at being therefore based on weight carried, and the orthodox values of the section modulus Z.]

The above gives extreme variations over all classes of steel employed, and takes no account of manufacturing considerations such as " nip," and the varying stresses caused thereby. The only circumstances in which these figures are a real indication of the actual maximum loaded fibre stress are when the spring is made up of uniform thickness plates, which are fitted together without any " nip " or daylight between each plate. The extreme classes of steel employed, at the one end, water-hardening carbon quality, and at the other end, some expensive " alloy " steel will have, when heat-treated into spring condition, elastic limits of 50 to 80 tons per square inch. Generally speaking, however, the designing figures are not specially studied in connection with possibilities of the steel qualities, the higher elastic limit steels being employed not so much from the point of view of taking care of the maximum static working load, as from the aspect of safely allowing much greater running shocks, and also giving a longer life before failing from fatigue. With the present manufacturing range of spring steels, the following designing figures—based on the understanding that standard calculation formulæ are employed—will be found to give satisfactory working results :—

WATER-HARDENING STEELS, with

Elastic Limit	50 to 60 tons per square inch
or	80 to 95 kilogrammes per sq. mm.
or	110,000 to 135,000 lbs. per square inch.

Railway Springs Tramway Springs	30 tons per sq. inch. 50 kg. per sq. mm. 70,000 lbs. per sq. inch.
Automobile Springs (Front) ..	20 tons per sq. inch. 30 kg. per sq. mm. 45,000 lbs per sq. inch.
Automobile Springs (Rear) ..	25 tons per sq. inch. 40 kg. per sq. mm. 56,000 lbs. per sq. inch.

OIL-HARDENING, AND "ALLOY" STEELS, with

Elastic Limit	60 to 80 tons per square inch.
or	95 to 125 kg. per sq. mm.
or	135,000 to 180,000 lbs. per square inch.

Railway Springs Tramway Springs	40 tons per sq. inch. 65 kg. per sq. mm. 90,000 lbs. per sq. inch.
Automobile Springs (Front) ..	25 tons per sq. inch. 40 kg. per sq. mm. 56,000 lbs. per sq. inch.
Automobile Springs (Rear) ..	30 tons per sq. inch. 40 kg. per sq. mm. 70,000 lbs.per sq. inch.

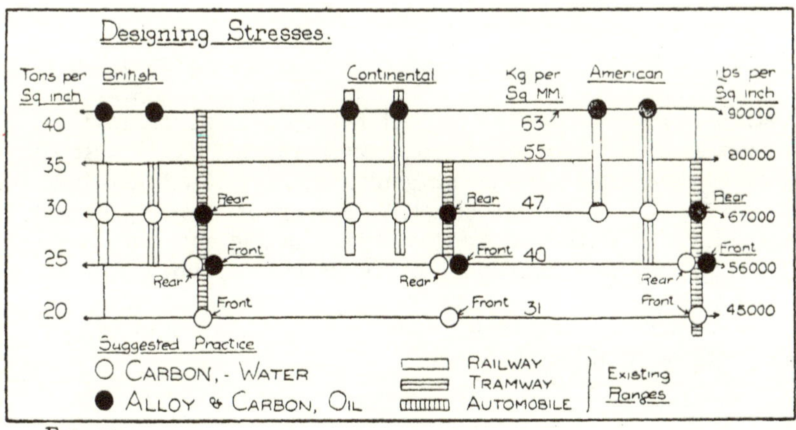

FIG. 33.

The diagram, Fig. 33, shows the position taken up by these suggested figures in relation to those given in the foregoing table as indicating general practice.

As mentioned, these limiting stress figures are only correct on the spring made of uniform thickness plates and fitted without " nip." The designer is not usually very cognisant with matters relating to this latter manufacturing property, and even if he were, no formulæ could be evolved to efficiently cover it, hand fitted springs being a personal reflection of the fitter. The " factor of ignorance " needed for the " nip " can, however, be considerably reduced by employing uniform thickness plates, instead of varying thickness plates, or " graded " springs. The suggested limiting stress figures are given for use, if desired, in certain existing formulæ in which they are necessitated.

The simplest method of designing is to take the British Standard test deflection as the basis figure (this and other test deflections are dealt with at length in the following chapter) rather than taking " allowable stress per sq. inch " or " per sq. mm." as the case may be. The Standard test will be assumed to be so many times the static maximum working deflection, according to the class of spring. The following figures are recommended for general design :—

Railway and Tramway Springs.

B.S.S. test ÷ maximum static deflection .. 2 to 2½

Automobile Front Springs.

B.S.S. test ÷ maximum static deflection .. 3 to 3¼

Automobile Rear Springs.

B.S.S. test ÷ maximum static deflection .. 2½ to 2¾

The figure of 80,000 lbs. or 35 tons per sq. inch (55 kg. per sq. mm.) so frequently worked to in one unit or another, represents a factor of comparison with the B.S.S. test of 1·92, when carbon water-hardening steels are used, and from 2·20 to 2·46 when carbon oil-hardening or alloy steels are employed. With heavy locomotives, having uncompensated springs, it is well to keep factor (X) as 2¼ to 2½, but if equalizing rigging is employed, or the springs are of the equalized types (such as an inverted bogie spring); factor (X) as 2 to 2¼ will be found satisfactory.

CHAPTER XI

FORMULÆ FOR SPRING TESTING

THE shop testing of the manufactured spring is an important factor governing its life, as if under-tested it is likely that inherent defects will not be detected, whereas if over-tested, a crippling effect is introduced which is the equivalent of several year's service existence. In this country before the days of the (revised) British Standard Specification of 1911, spring testing was run upon very crude lines, the most sensible being that requirement which desired every spring to be subjected to a test deflection equal to 50 per cent. in excess of the deflection caused by the working load. This implied that the designer was reasonably sound, but at anyrate, a definite DEFLECTION was stipulated. Not infrequently, however, specifications stated that "each spring is to be pressed until the plates are straight." The implication in this requirement was that all springs were cambered, which is by no means the case, Continental practice having always shown a leaning to springs fabricated straight, and working therefore with a reverse camber; even at times making springs with reverse cambers. With springs of these natures, an awkward problem in carrying out this specification requirement would at once present itself. With springs in general, however, the camber is not made a factor of length or plate thicknesses, but due to general combinations of the vehicle characteristics, available standard hanger lengths, ideas of designers as regards appearance, etc. It would therefore be found that springs of identical design as regards plate thickness and length, would be intended to have different test deflections because one would have perhaps 2 ins. less camber than the other owing to a larger wheel diameter, or a deeper axlebox, or a different design of shoe, or some other varying vehicle characteristic. As a rule, however, this crude empiricism of testing was not seriously pressed home to responsible engineers, owing to the general mist which appeared to envelop the matter, and accordingly some springs were under-tested and others over-tested, and a large number tested on the basis that the plates were " practically straight," an expression which not infrequently meant that they retained from 1 in. to 3 in. camber with the maximum test which the works actually applied. The introduction of the present British Standard test requirement is chiefly due to Mr. H. E. Wimperis, then on the staff of the Crown Agent for the Colonies, who by actual experiment and final calculation, determined upon the simple expression given in Formula IV.—8, which is now universally employed in the British railway spring shops, and included in specifications indenting material to British practice. Based on fibre extension, the skin stress due to this test is 58 tons per square inch (91 kg. per sq. mm. or 130,000 lbs. per sq. inch). Based on actual load carrying, however, it works out to 68·5 tons per square inch (108 kg. per sq. mm. or 153,000 lbs. per sq. inch) indicating an interesting, and by no means negligible, discrepancy, which can occur by the use of standard formulæ working along two different lines, the study of which aspect is reserved for later remarks. Anyway, the B.S. test has shown itself in practice satisfactory to all concerned, a fact sufficiently proved by its general adoption by British engineers and manufacturers. The basis of the experimental tests on spring plates was to obtain a deflection at which no measurable set could be observed when the plate was released, the implication then being that the test plate had not been stressed beyond its elastic limit. As, however, springs are invariably made on the high side as regards camber, it being easier to reduce the camber than to increase it—the former being readily done by the " scrag," whereas the spring will have to be more or less re-fitted to accomplish the latter—it was deemed necessary to introduce some safeguard which should provide against springs being made too much on the high side, requiring consequently a very heavy " scrag " test to reduce them to the correct camber. This safeguard is included in the provision which stipulates that at no time during manufacture must the Standard test deflection be exceeded by more than 15 per cent.

A spring fitted with "nip" and "pull" in the ordinary way always loses so much under the first test stroke. (The explanations of the trade terms "nip" and "pull" are illustrated in Fig. 34.) This loss will vary according to the amount of "nip" in the spring (the more the "nip" the more the loss), and the degree of hardness, or elastic limit, of the plates. No inspector will accept a spring below the specified camber, so that the necessary insurance against this test loss is made by fitting to a camber higher than that specified, which in the case of hand-made springs, will vary according to the individual fitter. A spring to finish at 3 in. camber, standard test 3 ins., might be brought to the scrag at $3\frac{1}{2}$ ins.

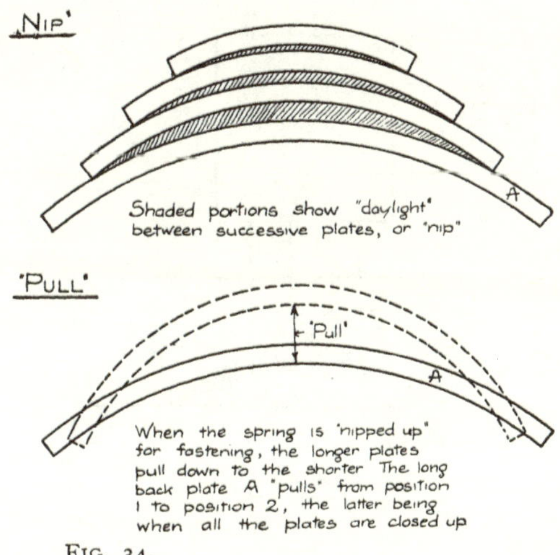

FIG. 34.

The 3 in. test plus the $\frac{1}{2}$ in. excess camber, $3\frac{1}{2}$ ins. in all, would be applied and the result would be perhaps a spring of $3\frac{1}{4}$ ins. camber. If 3 ins. was insisted on, that spring would be subjected to probably another $3\frac{1}{2}$ ins. or even 4 ins. test. to "scrag out" the unwanted $\frac{1}{4}$ in. The total "set" of the spring would thus be $\frac{1}{2}$ in., half of which might be called "normal," as due to the first test stroke, and the other half "abnormal," due to the after correcting strokes. An explanation of the "normal set" will be attempted later—the "abnormal set" (necessity therefor) should be eliminated from spring manufacture, as to obtain such set there is little doubt but that the extreme fibres of the spring plates are taken outside their elastic stage, and pass over into the plastic stage, thus being a continual drag on the elastic material yet remaining, which is already reduced in quantity owing to the defection of the overstressed external fibres. Such a spring will lose its camber and require re-setting or renewing earlier than a spring which has been made and tested in strict accordance with the present reasonable B.S.S. requirements.

In passing, it might be well to make clear the usual method of testing laminated springs in quantity, as they pass from the spring fitters' hands to the final finishing stages. Fig. 35 is a diagram plan of the steam scrag usual in this country. To the horizontal table on which the spring is placed, is

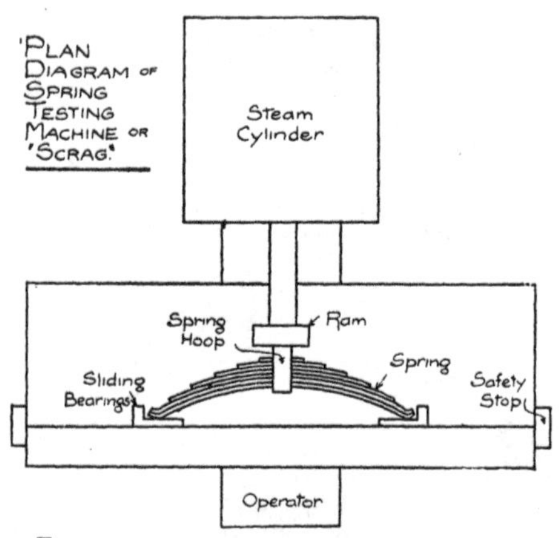

Fig. 35.

attached a large steam cylinder, capable of exerting up to 30 tons when employed for heavy spring work. Very rapid strokes can be given and the machine is very effective, in fact, it plays a very important part in the manufacture of hand-fitted springs. Similar machines are in use on the Continent, but they also employ belt-driven heads, with shaping-machine pattern drives. American practice invariably indulges in scragging machines of the " bull-dozer " type, with drives of the heavy forging machine variety.

It must be pointed out that it is a matter of no moment to the spring whether the whole of the test deflection (D) comes on what might be called the " positive " side of the spring, or partly on the " positive " and partly on the " negative " sides, or wholly on the " negative " side, these three possibilities being illustrated in the diagram, Fig. 36. It is a somewhat more awkward matter to test over on the negative side for large deflections—unless special tackle is to hand—but that is all, and to run over to the negative side does not reverse the stresses on the plates, as is sometimes thought. A spring has its tension side always on the top, and its compression side on the bottom, of the plates, regardless of the positive or negative signs of the camber.

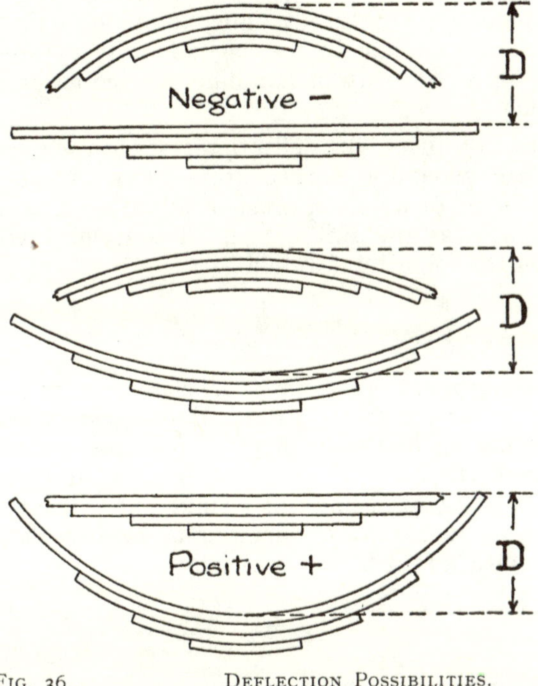

FIG. 36. DEFLECTION POSSIBILITIES.

The test given by the British Standard Specification is intended to apply to ordinary water-hardening carbon steels. To place therefore, " alloy " steels of much higher elastic limit on their proper test footing, the test deflection should be greater. Taking 60 tons per square inch as the maximum elastic limit which can be safely obtained from water-hardening carbon steels, and 80 tons per square inch as representing the elastic limit of a good alloy steel, properly oil-hardened and heat-treated, the divisor of the test formula should be reduced proportionately. For the ordinary steels it is 900, and it is suggested that for alloy steels as noted it should be 700, the formula thus reading

$$D = L^2 \div 700\ T. \qquad .. \qquad (XI.—1)$$

It is of no use reckoning on alloy steels satisfactorily dealing with heavier shock loads than ordinary carbon steels unless they are tested to the higher figure. Springs made from alloy steels, owing to the possible high elastic limit of such steels, require most careful judgment in fitting, as if the camber is too high, it is naturally much more difficult to reduce under the scrag than in the case of ordinary steel springs, and the only safe way of handling the reduction is to re-fit the spring.

On the Continent, the deflection tests of the manufactured spring are generally worked from more or less elaborate formula, none of which, it must be admitted, are as practical or as simple as the B.S.S. test. The usual French test is based on the following formula :—

$$D = \frac{iL^2}{6T}. \qquad .. \qquad .. \qquad (XI.—2)$$

the "allongement elastique" or elastic movement of the extreme fibres per metre length (i) being specified as 6, $6\frac{1}{2}$, or 7, according to the quality of steel used. In the above expression, (L) and (D) are in metres, (i) and (T) being in millimetres. Translating this formula into a bastard English form, so as to facilitate calculations for test deflections to this specification, it becomes as follows :—

$$D = \frac{iL^2}{6000T}. \qquad .. \qquad .. \qquad (XI.—3)$$

in which (D)—(L)—(T) are in inches, and (i) remains as the 6, $6\frac{1}{2}$, or 7 of the French specification. More attention will be given to the original formula in dealing with the testing of spring steel.

The Belgian State Railways' specification includes the following requirements for deflection of finished springs, the factor to be obtained being load instead of deflection :—

$$W = \frac{200b}{3L}\left[(n - 1)T_t^2 + T_a^2\right] \qquad (XI.—4)$$

in which the units can be worked in metric figures only. In this expression, (T_t) is the thickness of the back plate, and (T_a) the thickness of the following plates, Belgian practice employing either uniform thickness plates, or else one thicker back plate only. The length (L) is total length minus the width of the hoop. This is by no means a simple expression and appears to have an empirical basis rather than a calculated basis.

A French authority on automobile springs gives the following formulæ for maximum test stresses :—

On carbon steels. $W = \frac{60nbT^2}{L}$ (XI.—5)

On alloy steels. $W = \frac{83nbT^2}{L}$ (XI.—6)

both in metric units.

The alloy steels specially mentioned in connection with the latter are silico-manganese and chrome-vanadium, the former of which is largely used as "water-hardening" steel in France. The chrome-vanadium quality has almost of necessity to be oil-hardened, owing to the chromium content.

The German State Railways also demand a definite test load, based upon specified skin stresses according to the type of steel used. The standard section of plate employed is 90 × 13 mm. and a table is given in the official specification showing test loads on springs of varying lengths and with varying numbers of plates. These loads have been worked out from the following formula :—

$$W = \frac{2nbT^2k}{3L}. \quad .. \quad .. \quad \text{(XI.—7)}$$

This expression can be calculated in English or metric measurements as desired, the error being only 2 per cent. The factor (k) represents the specified test stress, of 100 kg. per square mm. or other variants due to the quality of the steel employed.

American practice generally bases itself upon the application to the spring of 1½ times the static working load, and assumes thereby, in a general sort of way, that the designer has worked to about 35 tons service stress (80,000 lbs.). The Standard Specification for Railway Springs of the American Society for Testing Materials, includes exactly the same formula as that given above for the German State Railways, with (k) as 127,500 lbs. per square inch, equal to 57 English tons, or 90 kg. per sq. mm. A special note is made as follows :—" For alloy-steel springs practice has not been sufficiently well established to enable definite fibre stress to be given. Unless otherwise agreed upon by the manufacturer and the purchaser, alloy-steel springs shall be tested under the conditions as to fibre stress specified for carbon-steel springs." It is provided that if spring drawings show no working load or working dimensions, they shall be subjected to $1\frac{1}{2}$ times an assumed load, based on a stress under this condition of 75,000 lbs., which causes (k) to become $1\frac{1}{2} \times 75{,}000 = 112{,}000$ lbs. or 56 tons. A special clause insists that the figure of 127,500 lbs. is the maximum allowable, and must not be exceeded in any test.

The scheme of the British Standard Test is paralleled in *Machinery's Handbook*, where an expression (for semi-elliptics) is given on the following lines :—

Deflection with maximum safe carrying capacity of spring

$$= 0{\cdot}00079\ L^2 \div \text{thickness} \qquad \text{(XI.—8)}$$

The (L) in this case is span less buckle width. Put in British Standard form, the above would be :—

$$D = \frac{(L - H)^2}{1260T}. \qquad \text{(XI.—9)}$$

Whether this is intended to represent a " maximum test load " or a " maximum working load " is not clear, but in view of the general figures taken for designing purposes in the U.S.A., this would appear to be a test load.

In connection with these test loadings for railway springs, certain fundamental diversities of spring practice between the British, Continental and American schools might here be referred to. Suspension gear on locomotives is sometimes arranged so that shocks received by any one axlebox (or spring) are transmitted via the rigging, to adjacent springs. In British practice, such " compensation " or " equalizing " is at a minimum ; Continental practice indulges in such arrangements to the degree which might be termed " simple equalizing " ; but American practice, with its very large engines, heavy axleloads, and staggered rail-joints, uses equalizing gears to the maximum extent, as many as six and eight springs being connected up through their hangers and intermediate levers (longitudinal and transverse) into one group. The result is that instead of one spring having to stand alone in receiving an abnormal shock or roll, the group of springs takes it through the medium of the " equalizers " or " compensating beams." (American and British terms respectively). Accordingly, " equalized " springs can be permitted to work with a smaller margin of " test load over working load " than similar springs which are uncompensated.

Tramway springs generally follow railway practice as regards test formulæ, but so far tests on the as-made automobile spring have not been as well standardized, except in the U.S.A., where the 1½ times working load is adhered to, with a designed working stress of 80,000 lbs., causing the test stress to be 120,000 lbs. (54 tons, or 85 kg. per sq. mm.)

In this country the following figures have been suggested for automobile springs :—

Front Springs. Carbon Steel.
80,000 lbs. = 35 tons per sq. inch = 55 kg. per sq. mm.

Rear Springs. Carbon Steel.
130,000 lbs. = 58 tons per sq. inch = 90 kg. per sq. mm.

Front Springs. Alloy Steel.
100,000 lbs. = 45 tons per sq. inch = 70 kg. per sq. mm.

Rear Springs. Alloy Steel.
160,000 lbs. = 71 tons per sq. inch = 112 kg. per sq. mm.

These test stresses are specified to be in all cases twice the maximum static working stresses.

For purposes of comparison, it will be of interest to assume two normal springs, and obtain their test deflections in accordance with the requirements quoted. One spring will be the standard wagon bearing spring of the Indian Railways, well known in this country, which consists of 13 plates, 4 ins. × ½ in. section, and 42 ins. straight length between bearing pins. The other spring is a usual heavy passenger automobile front, with 4 plates, 2¼ ins. × ⅜ in., and 9 plates, 2¼ ins. × $\frac{5}{16}$ in. Length between centres, straight, 46 ins. Fig. 37 shows the varying test deflections attained under the different formula conditions quoted, and also gives a percentage comparison of each formula with the B.S. test.

In practice, the best method of handling these formulæ is by comparison with the B. S. Test, to which they all (except one) bear definite relationship. The equivalent formulæ are then tabulated herewith.

COMPARISONS OF OTHER TESTS FOR SPRINGS WITH THE BRITISH STANDARD TEST

Authority.	Steel Quality or Remarks.	Divisor for L^2	Formula No.	From
British Standard ...	Railway Springs	900 T	XI.—10	IV.—8
French Railways ...	(*i*) as 6·0 mm.	1000 T	XI.—11	XI.—2
,, ...	(*i*) as 6·5 mm.	920 T	XI.—12	XI.—2
,, ...	(*i*) as 7·0 mm.	850 T	XI.—13	XI.—2
Belgian Railways ...	Impossible of exact translation. Practically equivalent to the B.S.S. on uniform thickness plates, but higher than the B.S.S. test when thicker top plates are included in the design.			XI.—4
French—Automobile	Carbon Steel. 90 kg. per mm.² 57·2 tons. per sq. inch.	1080 T	XI.—14	XI.—5
,, ...	Alloy Steel. 125 kg. per mm.² 79·4 tons per sq. inch.	780 T	XI.—15	XI.—6
German Railways. ...	Ordinary Steel. 100 kg. per mm.² 63·5 tons per sq. inch.	970 T	XI.—16	XI.—7
,, ...	Crucible Steel. 110 kg. per mm.² 69·8 tons per sq. inch.	880 T	XI.—17	XI.—7
,, ...	Special Steel. "Soft" quality. 120 kg. per mm.² 76·2 tons per sq. inch.	810 T	XI.—18	XI.—7
,, ...	Special Steel. "Middle-hard" quality 130 kg. per mm.² 82·6 tons per sq. inch.	750 T	XI.—19	XI.—7
American Railway ...	All steels.	1080 T	XI.—20	XI.—7
American General ...	All steels.	1140 T	XI.—21	XI.—7
British Automobile ... (suggested)	Carbon Steel. Front Springs.	1750 T	XI.—22	XI.—7
,, ...	Carbon Steel. Rear Springs.	1060 T	XI.—23	XI.—7
,, ...	Alloy Steel. Front Springs.	1360 T	XI.—24	XI.—7
,, ...	Alloy Steel. Rear Springs.	860 T	XI.—25	XI.—7

Comparison of Test Requirements Various Specns.
Belgian Rlys.
Carbon ○
Alloy ●
Both ◐
Suggested B.S. Test for Alloy Steels, $\frac{l^2}{t} \div 700T$
French Automobile Alloy Steel.
Railway Spring
Automobile Spring
British Auto. Alloy
8" 7" 6" 5 4" 3" 2"
B.S.
6·25"
B.S. Test Line 1
Line of B.S. Test, showing % comparative of other tests.
B.S. Test 100%
130 106 98 90 94 83 116 93 102 111 120 83 79 51 85 66 104 128
Rear
Front
3·92"
B.S. Test Line 2
B.S
Belgian Rlys.
1^{mm} = 6 6·5 7
French Railways
French Automobile Carbon-Water
$k^{kg.}$ = 100 110 120 130
German Railways
American Rly.
American Rly. + Auto.
British Auto. Carbon

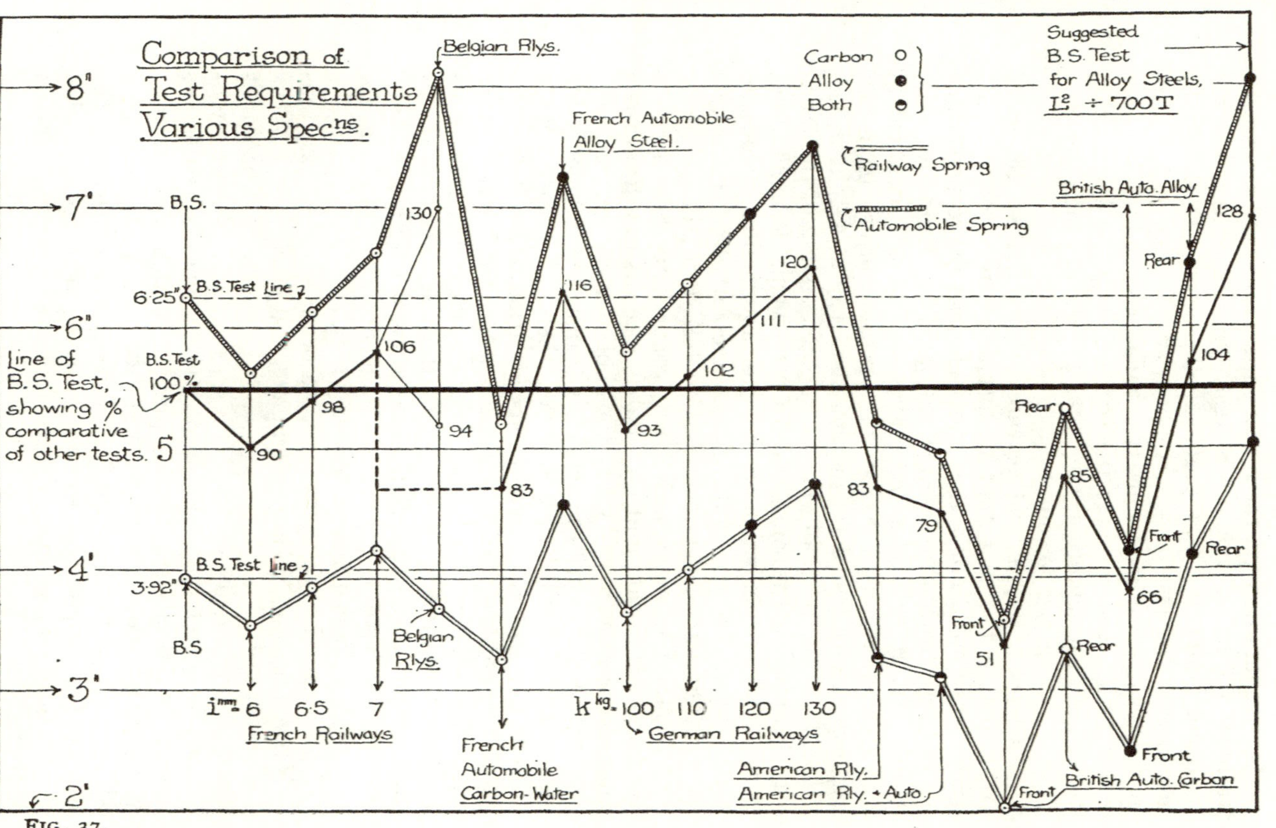

FIG. 37.

An examination of these comparisons clearly shows the rational nature of the B.S.S. test, which has attained the acme of simplicity and practical character. It should be noted that the test load formulæ (No. XI.—4, 5, 6, 7) are based on the bending moment calculations, and not on fibre extensions. The comparative stress due to the B.S.S. Test is, under such condition, 68·5 tons per square inch (153,000 lbs. or 108 kg. per sq. mm.)

One specially practical feature of the British Standard Test, lies in the fact that it specifies a test DEFLECTION, rather than a test LOAD, the latter being the expression which is derived from the Continental and American formulæ (these are, it will doubtless have been noted, identical in main construction). Whilst all types of springs have odd representatives tested for increment and total weight deflections, from the point of view of the bulk testing of the fabricated springs at the scrag, it is far more valuable to hand out a required test deflection to the scrag foreman than to hand out a required test weight. Spring men are used to working in cambers and deflections, and to tell them that a spring has to stand so many inches test conveys to their minds a fact which is readily understandable, whereas to inform them that a spring has to carry a test load of so many tons, conveys nothing to them until the necessary translation into deflection has been accomplished and given them. It must not be disregarded, however, that the specification of a test DEFLECTION does not standardize a test STRESS, as the latter depends upon the design of the spring. For instance, the test stresses on the extremities of design, with the test deflection of the B. S. formula, and the calculations along usual lines, namely, equation of bending moment to moment of resistance, as for a centrally loaded beam, are as follows :—

All plates full length, rectangle plan beam ..	82·5 tons
" Normal design," (K) between 0·095 and 0·105 (average)	68·5 tons
Perfect rhombus plan	55·0 tons

Logically, therefore, to standardize test STRESSES, test LOADS should be stipulated, but as previously indicated, from a practical point of view the test DEFLECTION is much the better, although it lacks theoretical consistency. Furthermore, the aspect of maximum fibre stresses on spring beams is a question by no means as simple as it appears, and is open to much argument.

CHAPTER XII

TESTS ON SPRING STEEL

THE testing of spring steel is introduced into this portion of the work, owing to the fact that certain requirements involve numerous calculations, and are considerably removed from "practical" shop tests. These latter are usually carried out along empirical lines in order to judge the suitability of varying steel casts for spring manufacture, and are generally based on cambering to a high figure a piece of spring plate, hardening and tempering, and then "scragging" flat, the amount of loss of camber, which is comparatively large, being taken as the measure of the value of the steel for the manufacture of different types or designs of springs. An example of a test of this description, to judge whether steel supplied (3 ins. × ½ in. section in this case) is suitable for a spring which is very highly stressed under working conditions, is as follows:—Pieces are cut from the "as-rolled" bars of 42 ins. long, and these are cambered to 9 ins., with a more or less true curvature. They are then subjected to the average hardening and tempering treatment which will be used in the spring, and twice tested with a 9 in. stroke—in the middle. (This indicates that the piece, if of perfect curvature, will not be "flat" under the test. Present B.S.S. test pieces are pressed between beams, as will be referred to hereafter.) After this test, the measured camber varies from 5½ ins. to 6½ ins. Steel giving final cambers between 5¾ ins. and 6 ins. is found most suitable for the springs in question, and used accordingly—results outside this range indicating that the steel is best employed in some other design, or else refused for spring manufacture.

The usual standard of steel testing is generally taken as the tensile test, but this is not really of surpassing value for spring steels, particularly in the case of the harder qualities. Such test can be regarded as satisfactory for axle, tyre, or structural steels, as the test is the measure of the material as it is in use. It could, of course, be carried out on hardened and tempered spring steel, as employed in the spring, but as the spring itself can be effectively tested, such supplementary tensile is superfluous. If such tensile is carried out on the "as-rolled" steel, it gives an indication of suitable quality, but such indication can be as readily found by the simple method generically known as the "camber test," in which pieces of the "as-rolled" plate are sheared off to definite length, curved to specified cambers, and then hardened and tempered, and tested through a specified deflection. The best known camber test in this country is the "80 times thickness" which was in existence before the production of the British Standard Specification, and is now included therein as follows:—"Pieces of steel 30 ins. long shall be cambered to a radius of 80 times their thickness, hardened and tempered, and, after being pressed straight once and the camber carefully noted, must stand being pressed straight again six times in quick succession without showing any permanent set." The above requirements can be expressed mathematically thus:—

$$D = 80\,T\left\{1 - \cos\left(\frac{10 \cdot 7}{T}\right)^{\circ}\right\}. \qquad \text{(XII.—1)}$$

A "set" always occurs in the plate after the first stroke has been received, and this "set" is now limited in the B.S.S. requirements. On water-hardening carbon spring steels the "set" will amount to 10 per cent to 15 per cent. of the original camber (equivalent also to the test deflection) and on oil-hardened steels of spring quality to 5 per cent. to 10 per cent., depending upon the degree of tempering and also of hammering which the plate has received to bring it to the correct camber or to remove twist.

In Fig. 38 are shown curves relative to these standard camber tests, as follows:—

Curve D is the main curve, lettered with the actual cambers, to the nearest practical figures, to which plates of the thicknesses shown on the left-hand ordinate, are bent.

Curve Y is the actual radius used, and

Curve X shows the close relationship between this actual radius; and the radius, for cambers figured, calculated from Formula IV.—1.

Curve K gives the ratio L/D, and

Curve Z shows the percentage error of the true radius against the calculated radius, the minimum, at ratio L/D = 16·08 ($\frac{3}{4}$ in. plate) being 0·5 per cent.,

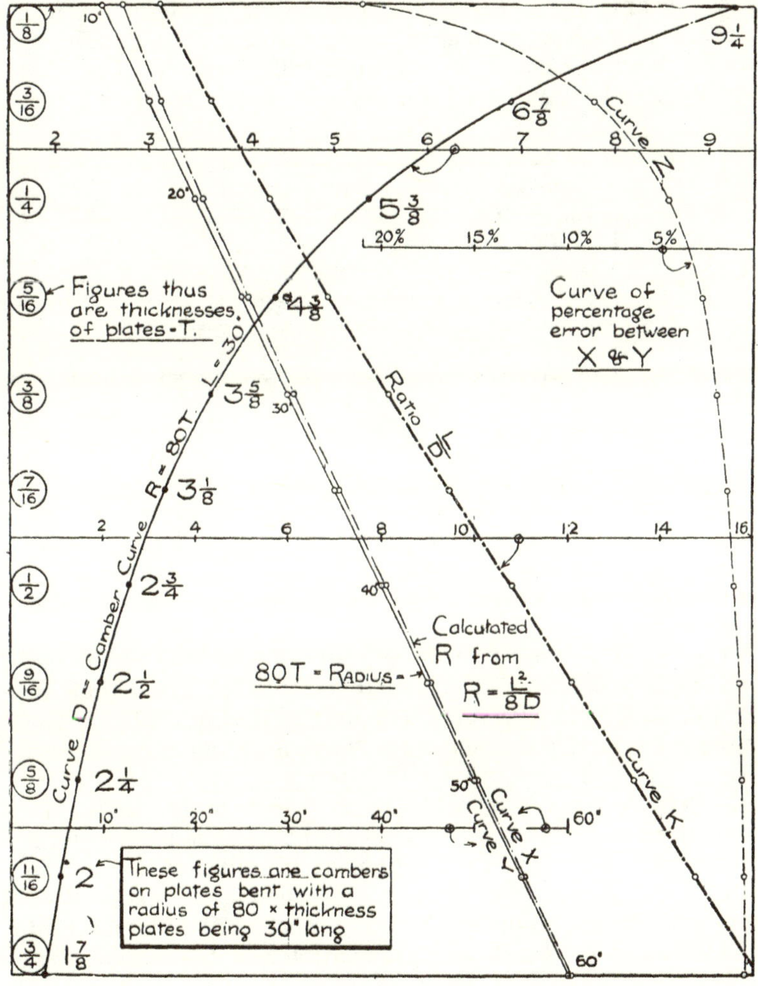

FIG. 38 B.S.S. CAMBER TESTS.

and the maximum at ratio L/D = 3·23 ($\frac{1}{8}$ in. plate) being 21 per cent. It shows, however, that for usual springs, with L/D as 6 to 20, the maximum is 5 per cent., and this has been generally neglected, having regard to other aspects which have been previously reviewed.

Camber test pieces to this specification are always pressed flat between two beams—one on each side of the test piece—or similar arrangement, as shown in Fig. 39. The effect of these beams is to bring the piece down practically flat, instead of with a tendency to over-stress in the middle, as would be the case if it were centrally loaded. The skin stress, based on Formula No. IV.—4, is practically 73 tons

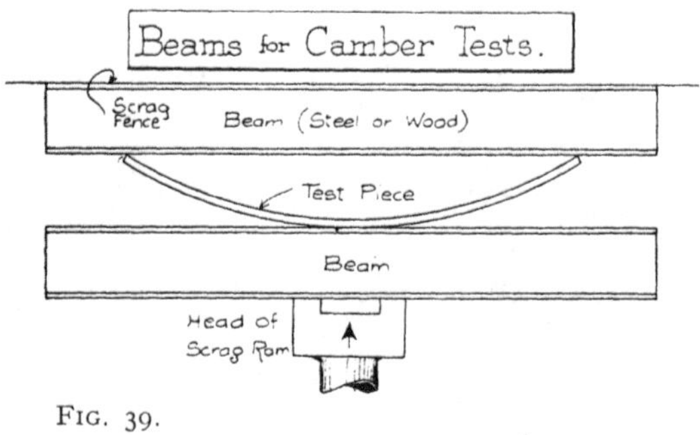

FIG. 39.

over the latter part of the test, and on the first stroke ranges from 79 to 81 tons. Actually it is rather doubtful whether such stresses are present, but the above represents normal calculation.

This B.S.S. test is of an extremely practical character, and readily made with the expenditure of little time or trouble, and generally, has everything to recommend it. The standard test of the French Railways is rather the converse—it is carefully drawn up and bears signs of study, but considerable time is involved in carrying out the necessary deflections and accurate readings, and when the results of all these have been obtained, no more information as to the suitability of the steel is available than could have been obtained by the simpler British method. Measurement of camber tests to the latter specification can be considerably facilitated if quantities are in hand—3 per cent. of the rolled bars are specified to be cut for test pieces—by the use of the gauge shown in Fig. 40, which can be instantly read, and simply adjusted over a range of cambers—which feature is necessary in order to avoid too great a movement of the return springs. In working through $\frac{1}{2}$ in. thick plates, made camber $2\frac{3}{4}$ ins., the gauge would be set to 3 ins. at rest, and so on for other thicknesses, the setting being always about $\frac{1}{4}$ in. above the " as made " camber.

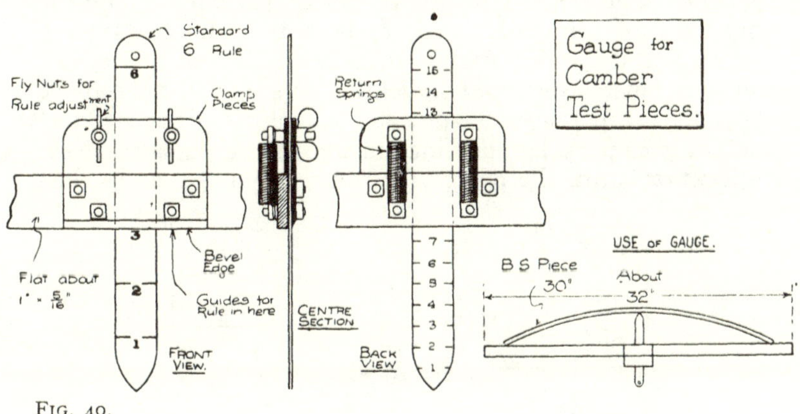

FIG. 40.

The French Railways specification referred to, desires the following tests to be carried out :—

Pieces have to be cut from the bars of a length of 1030 mm. ($40\frac{1}{2}$ ins.) and these must be cambered to the formula :—

$$D = \frac{iL^2}{6T} \text{ plus 40 mm.} \qquad \text{(XII.—2)}$$

In this formula, (i) is the " allongement elastique " or " elastic stretch " desired, of 6, 7, or 8, mm. per metre length, on the external fibres. The camber (D) and the length (L) are taken in metres, the latter being 1·030; the thickness in mm. The piece must then be hardened and tempered, after which the camber must be at least 10 mm. in excess of the (D) obtained by the formula, less the 40 mm. The piece must then be placed on special bearing carriages—see Fig. 41, on the load test machine, and must be deflected through a distance (D) less the 40 mm. and the load noted. This load must be maintained for 5 minutes.

The piece is then allowed to recover, following which it is submitted to a new test range based on formula :—

$$zD = \frac{i\,zL^2}{6T} \quad \ldots \quad \ldots \quad \ldots \quad \ldots \quad \text{(XII.—3)}$$

in which (zD) is the test deflection obtained with (zL) as the length, which is on this occasion, the true bearing length of one metre, the chariots having the end stops 15 mm. from the bearing points. The test piece is to be loaded through this new range at the slow rate of 500 kg. per minute, and the final load for the deflection is to be noted, and maintained for one minute. The piece is then again allowed to recover, which it must do without indicating loss, and is then to be subjected to increment loads commencing from the last test load of 50 or 25 kg. per increment, recovering after each application, until the point of permanent set is found, when

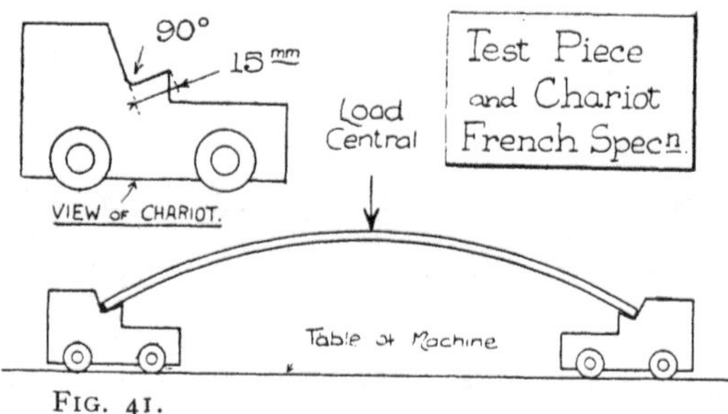

FIG. 41.

the immediately preceding load must be taken as indicating this. The set is to be deemed to take place when the piece ceases to recover its original camber by $\frac{1}{2}$ mm. Numerous results are now in hand, and specified formulæ as follows are then to be applied :—

To find the actual " allongement elastique " of the piece,

$$i = \frac{6DT}{L^2} \quad \ldots \quad \ldots \quad \ldots \quad \text{(XII.—4)}$$

This is merely a transposition of the previous formula, and the (D) to be taken herein is to be the (D) corresponding to the weight (W) which preceded the permanent set of $\frac{1}{2}$ mm.

To check the ratio D/W the following is to be used, load and deflection based on the test to Formula No. XII.—3 :—

$$\frac{D}{W} = \frac{L^3}{4EBT^3} \text{ (all dimensions in mm).} \quad \ldots \quad \text{(XII.—5)}$$

The two sides of this expression must agree within $\pm$ 8 per cent. The modulus (E) is to be taken as 20000 kg. per mm^2.

It will probably make clearer the whole process of this testing, if the actual particulars are given relating to a test piece of 100 mm. $\times$ 15 mm. section. The first camber to which this was made, on its length of 1030 mm., was 88 mm. (calculated) plus the 40 mm. or 128 mm. total. The first deflection applied was through 88 mm. and the load taken for this was 2030 kg. The piece recovered to a camber of 125 mm. showing a loss of 3 mm. The second test deflection to be applied was 83 mm., and the load required for this was 1980 kg. After the removal of this load, the "set" was zero, and the increment loading was therefore proceeded with, the increments being 50 kg., starting from 1980 kg. The indicating "set" of $\frac{1}{2}$ mm. was shown at 250 kg., so that the preceding figure of 200 kg. has to be taken as (W), the total load being 1980 + 200 = 2180 kg. The total deflection under this load was 94 mm. From the foregoing particulars therefore, the other requirements can be calculated, as follows:—

Formula No. XII.—4.

$$i = \frac{6DT}{L^2} = \frac{6 \times 0.094 \times 15}{1} = 8{\cdot}46 \text{ mm.}$$

The specification for this particular steel demanded 7·5 mm. as the "allongement elastique," and this result shows that the material was in accordance with these requirements. The next formula checks the ratio D/W as follows:—

Formula No. XII.—5. $\frac{D}{W} = \frac{L^3}{4EBT^3}$ all dimensions in mm.

The right-hand side of this

$$= \frac{1000 \times 1000 \times 1000}{4 \times 20000 \times 100 \times 3375} = \frac{1000}{27000}.$$

The left-hand side $= \frac{83}{1980} = \frac{1000}{23850}$.

Putting these in a percentage ratio form, they stand:—

27000 : 23850 : : 100 : 88·4

This result is outside the 8 per cent. tolerance allowed, owing to the piece being on the weak side. A further point to be noted is that the formula takes nominal dimensions of the section, 100 $\times$ 15 for instance, whereas the "as-rolled" section will be, as regards modulus, only equivalent to 95 per cent. of the nominal. With this amendment, the ratio would stand :—

25650 : 23850 : : 100 : 93·0

To obtain absolutely accurate results, however, it is clear that thicknesses should be taken with a micrometer, and deflections to small decimals, and this necessity, combined with the time features specified, entail a minimum of half-an-hour for each test piece.

Steel qualities for this specification used to be based on 6, 7, and 8 mm. "allongement elastique" per metre, but the present standard includes two qualities only, as below :—

"R" quality, to be 7·5 mm. per metre.
"S" " " 8·0 "

These requirements, it must be understood, are for the rolled bar only, and springs made from such rolled qualities of rolled bars, which are tested to the same formula (No. XI.—2) have the (i) for the fabricated test reduced 1 mm., "R" quality = 6·5 mm., and "S" quality = 7.0 mm.

The French Railway test is not yet, however, finished with. After the "essai de flexion" follows, on the same piece, a test of "contre-flexion," which stipulates that the plate must be deflected through a range equal to 1½ times the original camber, without fracture. (Original camber taken as first camber less the 40 mm. for workmanship etc.) Following this, on another piece, a shock test has to be made. The test piece for this is to be in exactly the same condition as for the finished spring plate, and is to be 200 mm. long. The bearings of the machine are to be 100 mm. apart, and the tup used is to be of a weight of 50 kg. (112 lbs.) with a radius of 25 mm. at the striking edge. Then the height of fall must be :—

$$\text{Height of fall in cm.} = \frac{\text{Section in mm}^2}{4}. \quad \text{(XII.—6)}$$

A different weight of tup may be permitted in which case the fall is to be based on the expression.

Weight of tup = height of fall = a constant. (XII.—7)

The height of fall is to be calculated from No. XII.—6, and multiplied by 50 (kg.) which gives the constant for any modification of tup weight. Under this test, " R " steel must stand 2 blows, and " S " steel 6 blows, without fracture. The blows are then to be continued until fracture results, and the number noted.

All the foregoing has been quoted to show what can be done in the way of specifications for spring steel, and whilst the whole thing is well and carefully drawn up, after being well and carefully thought out, one would not include it as a model for " recommended practice," as, it is not easy to see how the quality of material is better checked by these comparatively lengthy tests, than it is by the very simple and easy test of the B.S.S. This French railway test has frequently to be made here and in America and in view of this, the following shows a general model of the load and deflection testing in tabular form (figures are quoted for the test piece which has been given already) :—

Test No...	1	
Quantity of steel specified	R	
Width of section	100	mm.
Thickness of section	15	mm.
" Allongement elastique " specified	7·5	mm.
Deflection No. 1	88	mm.
Made camber	128	mm.
Load No. 1 for Deflection No. 1	2030	kg.
Camber after Deflection No. 1	125	mm.
Loss after Deflection No. 1	3	mm.
Deflection No. 2	83	mm.
Load No. 2 for Deflection No. 2	1980	kg.
Camber after Deflection No. 2	125	mm.
Loss after Deflection No. 2	0	mm.
Increment Loading, 1980 kg. + 50 kg.	0	mm. Set.
,, ,, 100 kg.	0	mm. Set.
,, ,, 150 kg.	0	mm. Set.
,, ,, 200 kg.	0	mm. Set.
,, ,, 250 kg.	0·5	mm. Set.
Deflection of load before " set "	94	mm.
" Allongement elastique " calculated	8·5	mm.
Error in D/W, calculated	11·6	per cent. minus.

The Belgian State Railway specify a tensile test on spring steel, as follows :—

Tenacity per sq. mm. .. 70 kg. (44·5 tons per sq. in.)
Elongation in 200 mm. .. 12 per cent.

To be taken on pieces well annealed.

A shock test is also desired, piece being 400 mm. long, and bearings of machine 250 mm. apart. The tup is to be 50 kg. weight, falling from a height as given by the following formula :—

$$\text{Height of fall} = \frac{BT^2}{43}\ (\text{mm.}) \qquad .. \quad (\text{XII.—8})$$

The piece for this test must be hardened and tempered, and must not fracture under one blow.

Tensile tests are sometimes still specified for British made indents, and are generally along the following lines :—

Tenacity per square inch .. 45 to 50 tons.
Elongation in a length of 4 ins. 15 per cent.

Pieces for this test are to be in the "as-rolled" condition.

The German Railway specifications include four varieties of spring steel (as indicated in the previous chapter on "Spring Testing"), and for each of these qualities, a tensile test is specified as follows :—

Quality of Steel.	Tenacity.		Elongation in 200 mm. per cent.	Sum of tenacity (kg.) and twice the elongation.
	Kg. per mm.²	Tons/sq. inch.		
Ordinary Quality	65 to 75	41·2 to 47·6	15 to 10	95
Crucible Steel	70	44·5	—	—
Special Steel "Soft" ...	75	47·6	12	105
Special Steel "Middle-hard"	85	54·0	12	110

Above tests are to be taken on the steel bars "as rolled." For the "ordinary quality" a test plate, hardened and tempered, resting on supports 600 mm. apart, must carry the following load without showing any permanent set :—

$$\text{Load to be carried, kg.} = 0{\cdot}12\ BT^2\ (\text{in mm.}) \qquad (\text{XII.—9})$$

This corresponds to a skin stress (calculated on the basis of bending moment $= fZ$) of 108 kg. per mm.^{2}, or 66 tons per square inch.

American practice does not, as a rule, take physical tests on spring steel bars, but is contented with analysis results, which, when every aspect is taken into consideration, are a reliable guide. They have the disadvantage, however, in the case of independent analyses, of requiring considerable time to carry out as compared with camber tests. With material from reputable firms, however, the Works' analyses can generally be relied upon to be reasonably accurate.

Impact, vibratory, and transverse "elastic limit," tests, have not yet entered into the general commercial testing of spring steels, which is probably a matter for congratulation.

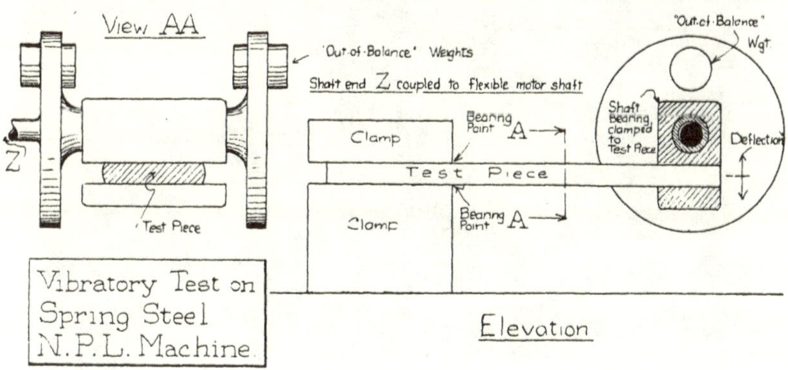

FIG. 42.

In the case of impact tests; pipe, slag, and sulphide streaks will cause an imperfect steel to show up better than a much sounder quality—which fact parallels what happens in service. Impact testing purely, would, if purchasing requirements were based on its results, probably lead to the gradual elimination of the "high-class" steels for laminated spring work, and create a demand for the more imperfect variety.

Vibratory tests are carried out in laboratory research, and give comparative results—which should be useful, but are more generally regarded as somewhat doubtful if attempted to be applied in practice—the main reason being that test pieces for such research are very carefully heat-treated, to very fine limits, under the supervision of high experts, whereas the commercial spring made from such qualities, will be, at anyrate in this country and the Continent, handled in more or less uncontrolled furnaces and manipulated by the skill of the spring fitter in ways undreamed of by the research people. An ingenious apparatus as used by the National Physical Laboratory for vibration tests is shown by diagram in Fig. 42. An unbalanced shaft is clamped at the outer end of the test piece, and this is driven by a flexible coupling from a motor, the speed of the shaft being in the neighbourhood of 2000 revs. per minute. The unbalancing

Fig. 43. Vibratory Testing Machine, Tinius Olsen, Philadelphia.

of the shaft is arranged by weights on two opposite discs, and the centrifugal action set up thereby, gives a definite vibration to the " free end " of the test piece, alternately stressing each of the width surfaces in tension and compression. The amplitude of this end movement is controlled by the speed of the motor affecting the " out-of-balance " action. As might be expected, arrangements are made for taking with extreme accuracy all figures pertinent to such tests, but to the author's mind two awkward points present themselves in the general connection, the first being the fact that hardened and tempered spring steel plates are not mathematically true, being slightly warped, and also have probably a few thousandths difference between the edge thicknesses, to say nothing of the concave variations. To fix these in a machine desiring high accuracy entails machining the test piece—in which condition it is no longer a spring steel plate. The second point turns on the use of the cantilever—which as previously shown in the spring drawing, Fig. 24, works backwards from the extreme line of support, and considerably militates therefore against very extreme accuracy in the ultimate calculations. The use of a beam, supported at each end, paralleling semi-elliptic spring action, would have been free from these objections.

Another form of machine for vibrating tests made by Tinius Olsen, Philadelphia, is shown in Fig. 43. This has the same objection as the one previously described, namely, that it tests the piece as a cantilever, but as this particular machine is intended primarily for commercial work, the need for high accuracy in the calculations is not as prominent as it is in the case of purely research work. The vibratory movement is imparted to the test piece by an eccentric motion, motor-driven, at a maximum speed of 500 r.p.m. This implies that the piece will be relatively highly stressed at the commencement of the test, so as to bring the number of revolutions down to a practical figure, instead of running into millions.

In the U.S.A., certain firms use, in normal working, a type of transverse testing machine, also made by Tinius Olsen, which is shown in Fig. 44. This is a very efficient little machine, and is fitted with an autographic apparatus, which provides the necessary information regarding the behaviour of the piece. A standard test piece section, say 1 in. × ¼ in., must be employed, so as to maintain uniform the diagram paper necessary, as the weight control is a constant. Curves obtained from various steels of various treatments can be compared, and interesting information obtained, but with the higher class of alloy steels, it is not easy to detect the yield point, as the graph obtained on such steels is practically a straight line.

The above resumé of spring steel testing methods shows the wide range covered by specifications. The buyer must of necessity be the dictator, but it will be observed that the general conclusion of the foregoing remarks is that spring steel for making into springs which can be tested by the purchaser, along the lines of Chapter XI, can be accepted on the analysis only, whereas in cases of spring steel purchased as such, the most satisfactory test is the B.S. "80 times thickness."

FIG. 44. TRANSVERSE TESTING MACHINE, TINIUS OLSEN, PHILADELPHIA.

CHAPTER XIII

"NIP" AND "PULL"

In the process of hand spring-fitting, as ordinarily carried out, the fitter always arranges the curvatures of his plates so that the ends only of each plate bear upon the plate above. He does this chiefly because it facilitates the fabrication of the spring, as it is comparatively easy to make plates touch on the "points," and comparatively difficult to make them fit snugly all along their lengths. A not uncommon idea, also, is that the introduction of plenty of "nip" in a spring increases the load carrying capacity, owing to the increase of friction at the bearing points. The latter theory is, however, fallacious, as in the ordinary way of measurements, it has not yet been shown that any difference in this direction is detectable between springs fitted with "nip" and springs fitted without "nip." Nevertheless, the statement has at times occurred in print that a "nipped" spring is stronger than a "nipless" spring. A further prevalent idea is that the more "nip" included in the manufacture, the more lively is the spring in service. This again is erroneous, as the "liveliness" or otherwise of any spring under shock or bump loads, turns on its static deflection or periodicity, when it receives the shock or bump, which factor is obviously not included in any arrangement of "nip." Furthermore, generally speaking, laminated springs are not desired to be of great "liveliness," but rather the reverse.

Some automobile authorities claim that "nip" is very

necessary on account of the tendency of the spring to become unloaded during the " rebound " which succeeds a bad bump, in which case, if no nip is present, the plates separate and chatter as they re-set themselves. This may occur, with a bump and corresponding rebound of abnormal character, but one wonders what happens to the back plate, harnessed to the chassis, under such condition. As a matter of fact, well placed clips along the spring length will prevent this " separation," and also save the back plate, which is in just as bad a condition under the above circumstance whether the spring be fitted with or without nip.

Mr. Hendrickson (of the Mather Spring Co., Toledo, U.S.A.) puts the whole matter of nip in simple form in the following words :—" If you take two springs and fit them up, one without any tension between the leaves, and the other with considerable tension, the deflection and riding qualities would be practically the same. I have heard it said that the spring that did not have tension between the leaves would be " dead," that is, it would not have any spring qualities. This is a ridiculous argument, however, because a certain number of beams will bend a certain amount under a load, regardless of what the internal stress is in those beams. The reason for putting the pull or tension between the leaves, is that the old-fashioned spring fitter feels it is necessary in order to obtain a good fit. He picks up a few leaves between his tongs, and if he can see daylight between them, he peens the uneven spots until they are in perfect contact. If there were no tension, the points would not lie in good contact when he got through."

The term " nip " may be defined as the " daylight " between successive plates, and " pull " is the amount the top of the back plate travels, as the spring plates are " nipped " together with the tongs. (See Fig. 34, ante.) It is very obvious that all possibility of uniform stressing throughout vanishes when the plates in an individual spring are all formed to substantially different radii, and in a spring of plates of uniform thickness, " nip " results in an initial positive stress being present in the shorter plates, and an initial negative stress being present in the longer plates, one, or perhaps even none (according to the fitting) of the total plates being under " no load." Before elaborating this aspect, it will be well to regard the diagrams Fig. 45 which show four " fitting " possibilities examined below :—

SPRING A. "FLAT-FITTED"

Radii of plates ..	With the back plate as a basis, the radius of each plate increases by one thickness.
Nip	None.
Pull	None.
Initial stresses in working order	None.
Loaded stresses..	Maximum in back plate, minimum in short plate, but the difference between the two is very slight, and due only to the back plate having a smaller radius of (so many) thicknesses than the short plate. Practically, stresses can be taken as identical in each plate.
Test loss.. ..	Negligible—perhaps three per cent. of the test deflection.

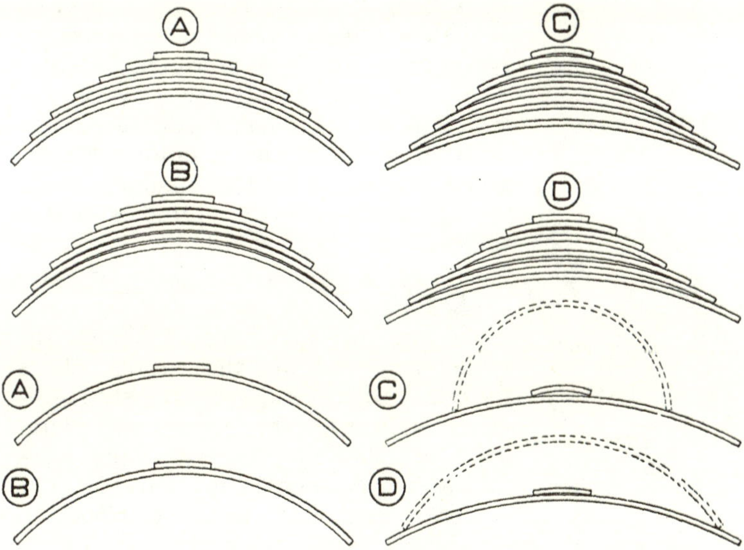

FIG. 45. SPRING FITTING: FOUR EXAMPLES.

SPRING B. "ONE-RADIUS."

Radii of plates .. The same in all cases.

Nip .. .. Very small.

Pull .. .. Very small.

Initial stresses in working order Very small negative on back plate, and equivalent positive on short plate. Negligible, however.

Load stresses .. Identical throughout the plates.

Test loss.. .. Negligible—perhaps three per cent. of the test deflection.

SPRING C. "REGULATOR."

So called because of the regulated "nip" between each plate, which should decrease uniformly from a maximum between the two top plates, to a minimum (or practically nothing) between the two short plates. Springs are often specified to be so made that when the hoop or clamp is removed, a regularly decreasing "nip" is shown between the plates.

Radii of plates .. Decreasing from the back plate in order that the daylight between the succeeding plates may diminish uniformly.

Nip .. .. The aggregate total varies with the spring fitter, as no rule can be stated. To a large extent, the amount of "nip" given varies with the design of the spring, the nearer the spring is to the theoretically perfect shape, the less "nip" being required; and the converse holding good.

Pull .. .. According to the "nip." When the spring is pulled up, the various plates include positive and negative stresses—in other words, the component plates fly apart directly the restraining force (of the tongs or buckle) is removed. The algebraic sum of the negative stresses in the top plates, and the positive stresses in the bottom plates becomes zero, otherwise, these minus and plus stresses balance.

Initial stresses in working order	Maximum—negative—on the top plate. Maximum—positive—on the short plate.
Load stresses ..	It is possible for the " nip " and " pull " to be so arranged that under working load, the back plate returns to its " as made " position, in other words, it carries no static stress. In any case, however, the maximum positive stress in the back will be considerably less than what might be termed the " natural " stress, due to the loading deflection had there been no " pull " in the plate, causing thereby original negative stress. The short plates have a positive stress when light, which reaches its maximum in the shortest plate.
Test loss.. ..	Varies according to the " nip " included in the spring. Will be anywhere, in ordinary spring work, between 5 per cent. and 15 per cent. of the test deflection, with carbon (water) steels. As the test deflection must equal that of the British Standard Specification plus any excess of camber over the required camber, the reason for the permission of an overload of 15 per cent. on the calculated B.S. test, is obvious.

SPRING D. " VERY ORDINARY FITTING "

Radii of plates	Varying without method—the " fitting " being devoted to keeping the points well on.
Nip	Excessive, and not regulated.
Pull	According to the " nip."
Initial stresses in working order	Follow the same lines as the " regulator " (C) but will have no uniform variation.
Load stresses ..	Similar to those in the " regulator " (C).
Test loss ..	Anything up to 25 per cent. of the total test stroke. Many plates would accordingly be badly overstressed.

In the case of springs fitted like the example (D), if the customer's inspector saw the first test stroke—which he does not, and cannot rationally be expected to do, as that would involve constant attendance on the works—such badly fitted springs as these would at once be detected, and probably rejected. The best spring firms, however, insist on the standard of fitting being of such quality, that practically all the testing falls within the 15 per cent. over-tolerance of the B.S.S.

The dotted lines on the sketches, Fig. 45, show the appearance of a plate of full length if made with the radius of the short plate, in the cases under review. These give an idea of the possibilities of overstressing which can occur in springs like (C) and (D) when excessive nip is present.

Fig. 46 presents some further aspects of this problem, in which :—

(A) is a " one-radius " spring, which, when pulled up, remains for all practical purposes the same as when loose. Under the B.S. Test, the stress throughout all plates is constant at 46·5 tons as noted on the stress diagram. (The stresses throughout this diagram are taken as $\frac{4}{5}$ the usual calculated stress due to fibre extension, and $\frac{2}{3}$ the usual calculated stress due to the bending moment and (*f*Z) formula, for reasons which will be suggested in Chapter XIX). When free, all plates have no stress.

(B) shows a spring built up on the lines of a non-stressed back plate, that is, the spring is intended to work straight in service, and the back plate accordingly is made straight, the nip in the spring pulling it down to the other plates so that the desired free camber is obtained. By reference to the stress diagram, it will be seen that when pulled up, the middle plate only is free from initial stress. When loaded to working requirements, the back plate has no stress. In tabulated form, the following appears :—

Back Plate, No. 1.	No load,	— 46·5 tons.	Working load,	0 tons.
Plate No. 2.	,,	— 23·2 ,,	,,	+ 23·2
Plate No. 3.	,,	0 ,,	,,	+ 46·5
Plate No. 4.	,,	+ 23·2 ,,	,,	+ 69·8
Short Plate No. 5.	,,	+ 46·5 ,,	,,	+ 93·0

This indicates the extent to which overstressing can occur with the introduction of considerable nip. On this particular spring, when loaded, with one non-stressed plate, the average

As made - Free
"Pulled up"
Radii (xT) all = 56
T = Thickness of a plate.
Radii (xT) all = 56.
LOAD LINE
Camber of Back :- Light :- 2".
A
Pulled :- 2". Loaded :- 0".
Radii (xT) = ∞. 112. 56. 37. 28.
Radii (xT) all = 56.
LOAD LINE
Camber of Back :- Light :- 0".
B
Pulled :- 2". Loaded :- 0".
Radii (xT) = 56. 37. 32. 26. 25.
Radii (xT) all = 32.
LOAD LINE
Camber of Back :- Light :- 2".
C
Pulled :- 3½". Loaded :- 1½".
TONS STRESS
f as BT²/4
100
90
80
70
60
50
40
30
20
10
0
-10
-20
-30
-40
-50
LOADED A
LOADED C
LOADED B
PULLED UP A
PULLED UP C
PULLED UP B
B LOADED
C LOADED
A LOADED
B PULLED UP
C PULLED UP
A PULLED UP
POSITIVE STRESS
NO STRESS
NEGATIVE STRESS
1 2 3 4 5
TOP OR BACK
EYE PLATE
← PLATE Nos →
SHORT PLATE
Diagram of Stresses in Plates of Springs A-B-C.
A is a Spring 'flat-fitted'
B is a 'non-stressed back'
C is an exaggerated 'regulator'
Springs of uniform thickness.
As made
Pulled up
Stress
POSITIVE +
ZERO 0
NEGATIVE −
Carrying no Load
In all 3 cases the algebraic sum of the stresses on the plates with the spring "pulled up" = 0, and the sum of the stresses on the spring as loaded is in all cases 5 (no. of plates) x 45·6 (natural stress)

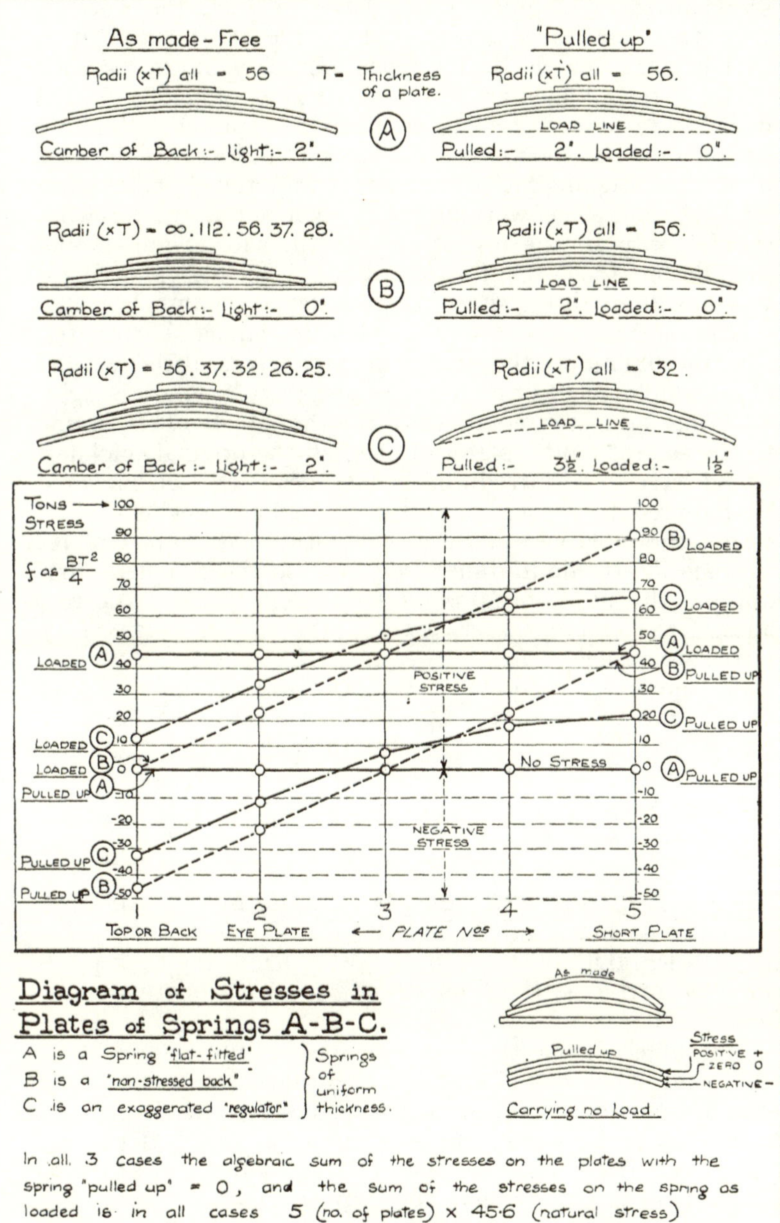

FIG. 46.

stress in the remainder, which have to share the work, is $\frac{5}{4}$ of the "natural" stress, and in the short plate, it is well beyond any elastic limit possibilities of normal carbon or alloy steels, which are heat treated so as to be of sufficient ductility to permit them to be made into springs. Whether such a spring as the above would stand up in service depends upon the designing factor used; if it were three, and the test load $1\frac{1}{2}$ times the working load, such spring would work satisfactorily, as the test stress—on the short plate—would then be 46·5 tons. On the other hand, if the designing factor was less than three, short plates would probably break in service, and then resort would be had to thinner short plates, which would reduce the stresses in these leaves, at the expense of the employment of more plates and therefore more weight. To bring the test stress down in the short plate of (B) spring to the correct figure, plates of one-half the original thickness would have to be introduced, which means that eight would be required to replace one of the original thickness. The second short plate should then be replaced by thinner plates of some intermediate dimensions, and so the complete non-uniform thickness spring is built up, along perfectly correct lines if the assumption is granted that nip is a necessary virtue in spring fitting.

(C) shows a "regulated" spring, along the lines of Fig. 45—D, and the tabulated stresses therefor are as follows:—

Back Plate. No. 1.	No load.	— 33·5 tons.	Working load,	+ 13·0 tons.
Plate No. 2.	,,	— 11·0 ,,	,,	+ 35·5 ,,
Plate No. 3	,,	+ 8·0 ,,	,,	+ 54·5 ,,
Plate No. 4.	,,	+ 18·0 ,,	,,	+ 64·5 ,,
Plate No. 5.	,,	+ 21·5 ,,	,,	+ 68·0 ,,

The average stress over these plates, as all five are working, is 46·5 tons, and it will be noted, that the short plate stress is much less than in (B). On the other hand, the back plate does not arrive at the "no stress" ideal frequently fondly held.

The subject is one that can lend itself to much interesting treatment in the way of diagrams and calculations, but as the author's opinion of the ideal spring resolves itself into the fixed idea that such spring is of uniform thickness plates; and, if not fitted along the lines of "one-radius," at anyrate, fitted with as little practicable nip as possible; it is not worth while dealing much further with the point now that the elements governing the same have been sketched out.

The idea is largely held by automobile people that the perfect spring should be so made that the back plate is of the "no stress" variety, as Fig. 46—B. Thousands of springs have been made in this way, with plenty of nip to obtain the pull necessary to provide the "no-stress" back, and well graded non-uniform thickness plates to provide in turn for the nip wanted. In spite of this, back plates break. The reason that such interest in "no stress" back plates is limited to the automobile trade, is owing to the fact that designers in this trade expect the spring back plate to act as a spring, radius rod, tractor bar, torque rod, and anti-rolling device, separately and in combination. No question arises as to the extreme practical and simple character of the anchoring devices of spring to chassis that necessitate such a series of properties, but it must be admitted that it is rough on the spring back. However, with all these functions to fulfil, clearly the breakage of such a plate is a serious matter, wherefore the special interest displayed therein, to the point of the attempted elimination of stress under static working load. The aspect overlooked, however, is that it is not loading as such which breaks materials, but rather the range of loading, corresponding to range of stress. Now, the range of stress in a back plate will always be the same, whether the initial static stress be zero or infinity, as the range turns on the actual movement received on the road. Assume that the maximum bump load is equal to three times the static load, and that two springs are present, both of which have been tested to B.S.S., only one has been "flat-fitted," that is, with no nip, and the other is arranged with a "no stress" back plate. The following figures then appear :—(Stress figures used are on the same lines as in Fig. 46.)

"Flat-fitted" Spring.—Under load, 15·5 tons, with bump, 46·5. Range 31.

"No stress" Back.—Under load 0 tons, with bump, 31·0. Range 31.

If the highest figure attained is reasonably under the elastic limit, it is the repetition of such stress, through the 31 tons range, that ultimately causes fracture, and this has been verified many times. The idea of the superiority inherent in the "no stress" back plate should therefore be shelved, and eliminated from specifications at present calling for it. Wohler's researches in this direction still hold good, and it might be of value if some of the conclusions are repeated here :—

(1) The range of stress and not the maximum stress, determines within certain limits, the number of stress repetitions before fracture.

(2) The number of repetitions before fracture increases as the range of stress is diminished, and there is a range of stress, "the limiting range," at which the number of repetitions is infinite.

(3) The limiting range of stress diminishes as the maximum stress increases.

Clearly the ideal stress range to be found is the "limiting range," and this point will be returned to later.

Reverting to the nip and pull in ordinary hand fitting, as previously indicated, the main reason thereof is to keep points or plate ends, "on," or touching the plate above. With plenty of nip, and a good powerful testing machine, when the springs come along to the latter and receive their test stroke, the effect is to give all the plates a good "bedding" one to the other, as indicated by the set that takes place. Every spring, whether fitted with or without nip, takes a "set" on the first stroke of the testing machine, varying from the 2 or 3 per cent. (of test stroke) of the "nipless" spring, to the 25 per cent. or thereabouts of the badly fitted spring. This "set" is generally attributed to all, or some, of the plates having been stressed beyond the elastic limit, but it is a moot point as to whether this is really so. Undoubtedly, in the case of springs of high loss, say from 10 per cent. upwards, certain plates have had their external skins really travelled beyond the elastic stage, and the set of the whole spring is largely due to these partially plastic plates retarding the general recovery. A large percentage of set, however, which grows larger with the decrease of loss under test, is doubtless due to what might be termed "molecular locking" or otherwise, the bedding internally of the steel structures. It must be remembered that before the spring plate arrives at the testing machine, it has been heated, quenched in oil or water, re-heated to a tempering point, and then, whilst in this warm state, hammered over the block until shapeable (that is, until twist has been eliminated, and the desired curvature or sweep arrived at). These operations must considerably shake the general structure, and the first test stroke will, by stressing the plate, persuade the molecules into the best position for resistance—or close their ranks—and thereby "set" the plate to an automatic position of constant deflection under repeated loading.

As remarked, every spring fitted in the ordinary way, and being therefore a "regulator" or an attempt thereat, takes a distinct set under the first test stroke, and the greater the "nip" the greater the "set" will be. If a plate is slightly soft, the "points" come off, as Fig. 47, it is returned to the fitter, who in correcting, leaves it harder to raise the elastic limit, and, with the initial stress due to nip, it is probably near the danger point in working—owing to its lacking ductility. A spring with a short plate much longer than it should be, if it is fitted without nip, will always show the points off after the first test stroke—showing thereby bad design. Any spring of superimposed plates has an automatic load distribution from plate to plate (see Fig. 47), and given ordinary workmanship as regards the hardening and tempering, "points off" under the standard test show that the plates are too long. In practice, instead of taking up the question of drawings showing short plates (and consequently intermediate plates) of excessive length, manufacturers work to them, and cause the springs to be fitted with plenty of nip to keep the points on. It must be admitted, however, that the results of discussion with designers of springs are not as a rule encouraging, in fact, the difficulty in argument is generally in exact proportion to the excessive length of the short plate shown.

This desire to keep points on is little more than fetish worship, given the fact that reasonably uniform workmanship

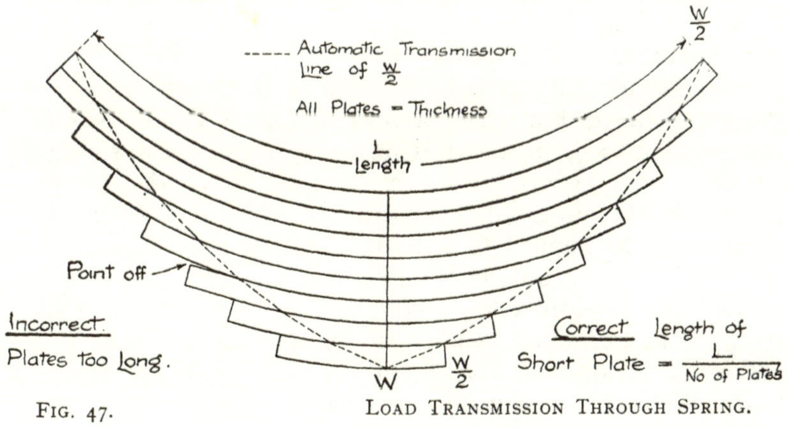

FIG. 47. LOAD TRANSMISSION THROUGH SPRING.

and heat treatment is present. With machine-fitted springs, all plates of one radius, it cannot be guaranteed that after tempering all plates will retain exactly the same curvature, the result being that when such springs are assembled, certain plate ends are slightly " off " the surfaces of superimposed plates. In a light spring, probably this will only be obvious when it is lying on its side, and when in vertical position, the weight of the spring alone will bring the points on. With heavy springs, working at perhaps 5 tons, the worst cases will generally come on with a ton load. Any spring doubtful in this direction can always be repeatedly tested, and if the plate closes down, and gets no worse, the spring can be taken as satisfactory.

Probably Professor Perry in his book on *Applied Mechanics* included more matter on plate springs than any other author of engineering works, and whilst the deflection formula he gives is not above criticism, it is certainly refreshing to come across a commonsense remark of the following nature in his book :—" I think that carriage spring makers and buyers are too particular in their wishing to have all the plates lying very tightly together when bolted up. There is too much of such tightness. Usually, too, in actual use a carriage spring never becomes unloaded, and its plates are therefore always very much more tightly pressing together than when the spring is examined unloaded. If, however, such tightness is found to be necessary, and if to obtain it we must have such great differences in curvature as I have found in springs by the best makers, it would be advisable to make the shorter plates thinner than the rest. But if all the plates are of the same thickness, they ought certainly to have the same curvature before bolting up."

No doubt a spring looks better with all the points dead on the plate above, than with a few gaping off $\frac{1}{32}$ in. or so, but when loaded, unless it is an exceptionally bad spring altogether, the points will touch the plate above, and carry practically their share of the load. If a certain amount of observation be applied to some of the springs on " private owner's " wagons running here, twenty and more years old, which have " weathered " or corroded at the plate ends, the truth of the foregoing remarks will be appreciated.

Certain Continental specifications, including the German State Railways, specifically state that " no space must be left between plates." These authorities, however, always design " one-thickness " springs.

The British Standard Specification states " All the springs are to be fitted and regulated by hand. When required, springs which have satisfactorily withstood the test shall be examined by removing the centre clip or buckle, after which the plates must separate so as to touch each other at the extremities, and each plate must show a proper and uniform curvature."

French and Belgian specifications state that nip must be allowed according to length, and must decrease uniformly from the back plate. A general requirement is that the greatest space between the back and second plate must not exceed 8 mm. (say $\frac{5}{16}$ in.) per metre (say 40 ins.) of span.

The specification of the American Society for Testing Materials includes the paragraph :—" The springs shall have the leaves properly graduated in length, properly bent, and fitted to reasonably true circular arcs." By compliance with the first portion, namely, the proper length graduation of the plates, the remainder is comparatively easy. As American springs are invariably designed with the shortest possible " short plate," and plates of uniform thickness throughout, the amount of " nip," even with hand fitting, can be kept to a minimum.

CHAPTER XIV

PLATE THICKNESSES

THE previous chapter explained, in connection with "nip" and "pull," the need, practically and theoretically, for the employment of plates of varying thicknesses throughout any one spring, and showed that the short plates should be the thinnest, and the grading uniform up to the thick long plates, always assuming the virtue of "nip" fitted springs. In passing, it might be pointed out that even with "nip" fitted springs, a moiety of design retains uniform thickness plates, and general world practice is approximately as follows :

BRITISH

Railway :— Locomotive.	Largely designed of non-uniform thicknesses, but the trend to-day is towards the employment of uniform thicknesses in one spring.
Railway :— Coaching Stock.	Generally of uniform thickness, but there still remain a high percentage using back plates thicker than the remainder of the plates.
Railway :— Freight Stock.	Invariably to-day of uniform thickness.
Tramway.	Mixed practice—about one-half to each type.
Automobile :— Private Cars.	Invariably of non-uniform thickness plates. A few exceptions only of uniform thicknesses—thereby proving the rule.
Commercial Trucks.	Generally the same as the private cars, but with more exceptions, which are increasing.

Continental

Railway :— Locomotive, Coaching Stock. Freight Stock.	With such few exceptions as to be negligible, all classes of rolling stock are on uniform thickness springs.
Tramway.	The same remarks apply.
Automobile :— Private Cars.	Mixed practice—a large percentage of springs of non-uniform plates.
Commercial Trucks.	Generally designed as of uniform thickness.

American

Railway :— Locomotive, Coaching Stock, Freight Stock.	With very few exceptions, all classes of stock on uniform thickness springs. The only class of freight stock on laminated springs is, however, the 4-wheel " caboose " or (British parlance) brake van.
Tramway.	Mixed practice, tending towards the elimination of all except uniform thickness designs.
Automobile :— Private Cars.	Nearly all on uniform thickness springs —a few exceptions only on non-uniform plates.
Commercial Trucks.	With negligible exceptions, all on uniform thickness springs.

American and Continental practice tend to eliminate the need for " nip " by the introduction of the shortest possible " short plates," and in the U.S.A. it is still further eliminated by machine-forming, with little or no succeeding hand work.

The use of uniform thickness plates, it need hardly be emphasised, reduces substantially the number of sections to be stocked, and has other practical manufacturing considerations. If a designer really lets himself go on this matter, he can readily arrange a spring of ten plates, with five different thicknesses, which will be accepted without question by manufacturers in this country, and made up. One awkward aspect of such scientific grading is, however, the fact that in a spring designed with plates of $\frac{1}{4}$ in., $\frac{9}{32}$ in., $\frac{5}{16}$ in., $\frac{11}{32}$ in., and $\frac{3}{8}$ in, the $\frac{1}{32}$ variations prove rather subtle in handling, and sometimes get misplaced, so that the finished spring, whilst being of correct overall thickness, actually grades as perhaps: (from the bottom) $\frac{9}{32}$ in., $\frac{1}{4}$ in., $\frac{1}{4}$ in., $\frac{5}{16}$ in., $\frac{9}{32}$ in., $\frac{11}{32}$in., $\frac{5}{16}$ in., $\frac{11}{32}$ in., $\frac{3}{8}$ in., and $\frac{3}{8}$ in. Needless to say, this rather spoils the original calculations, but nevertheless, such things happen.

The development of the plan of a spring designed with non-uniform thickness plates cannot, of course, be worked on the same simple lines as the corresponding development for uniform thickness plates. The thick plates should be converted into terms of the thinnest plate, and the plan view

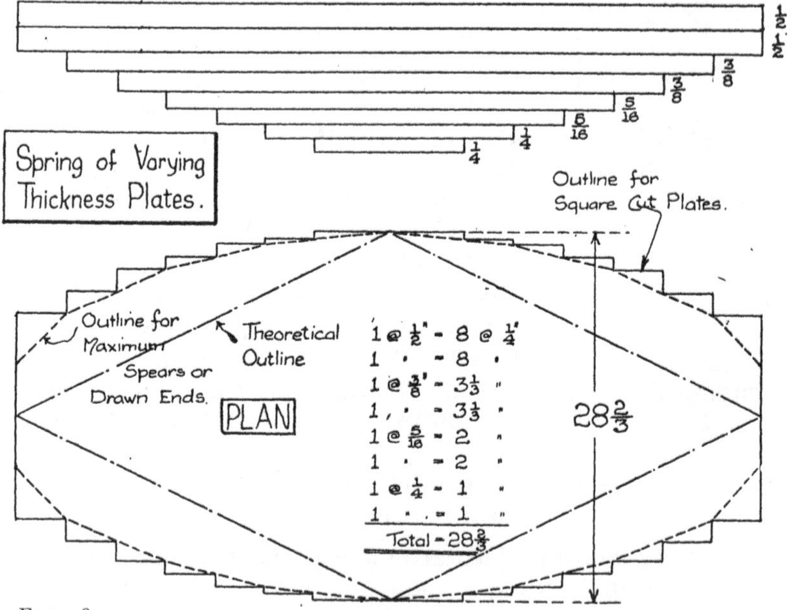

Fig. 48.

is then as shown in Fig. 48. The large amount of excess material over the rhombus will be noticed, and it would appear accordingly that the unit deflection of such springs would give results "stiffer" than those which are obtained by the recommended formula. In practice, however, as such types of springs include nearly always, "drawn" or tapered thickness plate ends, the effect of the weakening caused thereby, is to correct the general spring back towards the standard formula deflection.

At times, in the case of light springs, a spring may need to be made of such thin material that the employment of the same thickness plate throughout would cause the harnessing attachments on the back plate to appear very weak. Consequently, thicker back plates are employed purely from the point of view of obtaining sound chassis connections. (This remark generally is obviously only applicable to light road vehicles). Back plates of such thinness, can however, be effectively stiffened by rolling the second plate completely round the eyes—these springs having invariably rolled-eye attachment means. This has certain disadvantages, as the freedom for longitudinal movement under deflection, of both the plates, is somewhat restricted, but on the whole it would be recommended in preference to the employment of

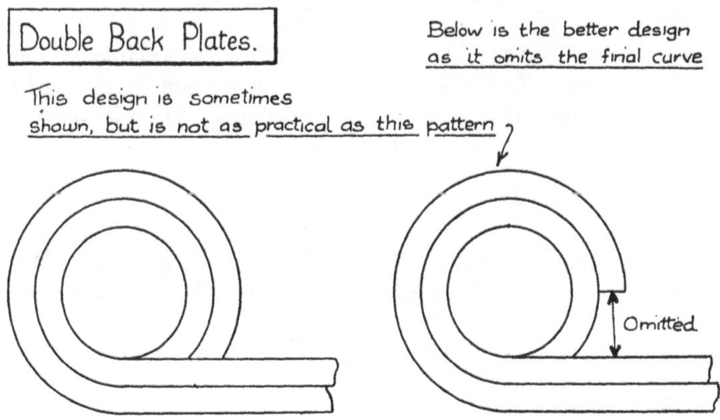

FIG. 49.

a thicker back. If the spring is arranged to run, as it should be, straight under normal loading, little question arises as regards the restriction of end movement. The suggested means is shown in Fig. 49.

The thickness of plates is the most important dimension in the design of the spring, and is generally the chief unknown of the equation. Lengths are more or less fixed, as are also widths, leaving thickness and number of plates to be determined—and the latter factor obviously turns on the thickness desired. This dimension must clearly be maintained within certain limits, although if required, any " laminated " spring could be designed to deflect, and take a load with one plate only. For instance, in Chapter IX, a spring was suggested with 10 plates, 4 ins. × $\frac{1}{2}$ in., and 40 ins. straight length, the calculated deflection being 0.31 in. per ton, and the test-load (B.S.S.) 11·4 tons. With one plate only, such would have a thickness of $1\frac{9}{16}$ in., and would deflect, to B.S.S., 1·13 ins. instead of the 3·55 ins. of the $\frac{1}{2}$ in. plates. Under certain circumstances, it is conceivable that a spring of this nature would be satisfactory, but the use of one plate only nullifies one of the great advantages of the multiple plate spring, namely, that if breakage occurs, the machine or vehicle attached is out of service, whereas breakage of one plate in the multiple plate spring is of comparatively no moment, unless it be the back plate. Practical considerations have to influence plate thicknesses, in maintaining them within certain limits—and these will be reviewed at length in connection with the manufacture. For the present, it is sufficient to state that the range of designing is between $\frac{1}{8}$ in. and $\frac{3}{4}$ in. thick. Railway and tramway springs generally are between $\frac{5}{16}$ in. and $\frac{5}{8}$ in., and automobiles (light) from $\frac{1}{8}$ in. to (heavy) $\frac{1}{2}$ in. The following formula will be found of value to essay thicknesses :—

$$t^2 = \frac{90\text{X (working load) L}}{16bn} \quad \ldots \quad \text{(XIV.—1)}$$

Herein (X) has the usual significance hitherto given, of the "designing factor," in this case of Test Deflection ÷ Working Deflection, as in Chapter X.

With the use of the above, it will be found a relatively easy matter to try out varying thicknesses until the desired one is discovered.

In U.S.A. locomotive practice, the thickness of $\frac{1}{2}$ in. is almost standardized, and under such circumstances, the unknown factor can be regarded as (nb) or (B). The formula for this evolves as below :—

$\frac{1}{2}$ in. Plates. $nb = 0{\cdot}175$ (Working load) L .. (XIV.—2)

On a similar basis, all thicknesses can be adapted for the ready determination of (nb), which can then be factorized as desired. The symbol (X) herein has been taken as 2, and the (nb) obtained accordingly gives a spring with a test load of twice the working load.

CHAPTER XV

LENGTHS OF PLATES

THE determination of plate thicknesses in spring design presents itself as a definite calculations factor, but the question of "offsets" or as it is sometimes termed "grading" of plates is quite another matter. (In this connection, it might be pointed out that as the word "graded" is used in American books with reference both to offsets in the lengths of plates, and also to varying thicknesses of plates, and tends thereby to become confusing, it has been generally omitted by the author in regard to these points.) Certain formulæ purport to provide correct offset lengths, and others to give correct unit deflections having given the offset lengths. It would not appear, however, to be of great value to touch on these, as the only correct offset is the uniform step, from the end of the short plate to the bearing centres of the back plate, in other words, the offset lengths at each end should be :—

Length of offset, or step at each end of semi-elliptic,

$$J = \frac{L - L^s}{2(n - 1)} \quad \ldots \quad \ldots \quad \ldots \quad (XV.—1)$$

In a perfectly designed spring, with (L^s) as $L \div n$, this amends itself to the following :—

$$J = \frac{L}{2n} \quad \ldots \quad \ldots \quad \ldots \quad (XV.—2)$$

In connection with the use of Formula XV.—1, let it be pointed out that if the result of (J) is 2·932 ins. or some similar figure, it is not intended that slavish adherence to rule should induce a designer to include this offset on his drawing, as he will probably not get it. It is quite good practice to omit offsets as such from the drawing, and figure the short plate length only, leaving the manufacturer to divide the distance. If the short plate can be so adjusted for length as to bring offsets to "even" dimensions it assists matters in the cutting-up shop.

As is evident, both from design and the Formula XV.—1, offsets in any spring are determined by the governing features of the length of the back plate and the length of the short plate. The former is a fixed item of design, the latter is not, and many otherwise good designs are completely spoiled by lack of appreciation of the features determining short plate lengths. The theoretical length should be at all times L ÷ n, but this cannot often be adopted in practice—possible examples being chiefly long springs with few plates. In

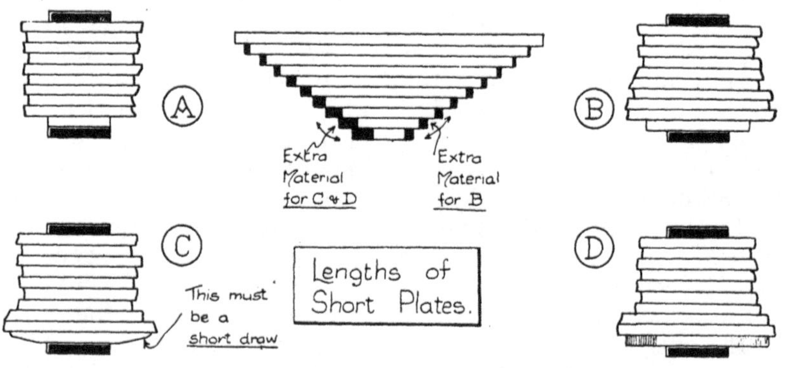

FIG. 50.

relatively short springs, with comparatively many plates, if L ÷ n were used, the short plate would be inside the hoop or bearing pad, and whilst not useless under such conditions, would give rise to many practical objections. Another point which precludes the use of a theoretical short plate is its possible shortness of 3 ins. or thereabouts, which renders it very difficult to handle throughout the manufacturing operations.

In the case of springs fitted with hoops, it is necessary that the short plate should extend to at least 1 in. from each edge of the hoop in order to allow the grip of the vice when placing into position the buckle. The expression for short plate length (L^s) then (H being width of hoop) becomes :—

For springs with square ends :

$$L^s = \frac{L}{n} \text{ or } H + 2 \text{ ins., whichever is greater} \quad (XV.—3)$$

For springs with " drawn " or " speared " ends :

$$L^s = \frac{L}{n} + H \text{ or } H + 4 \text{ ins., whichever is greater} \quad (XV.—4)$$

In Fig. 50 are shown respectively at (A) a hoop and short plate of theoretical length, at (B) a hoop and square points, Formula XV.—3, at (C) a hoop and " drawn " points, and at (D) a hoop and " speared " points, the two latter to Formula XV.—4, which drawings will assist in the appreciation of the above reasoning.

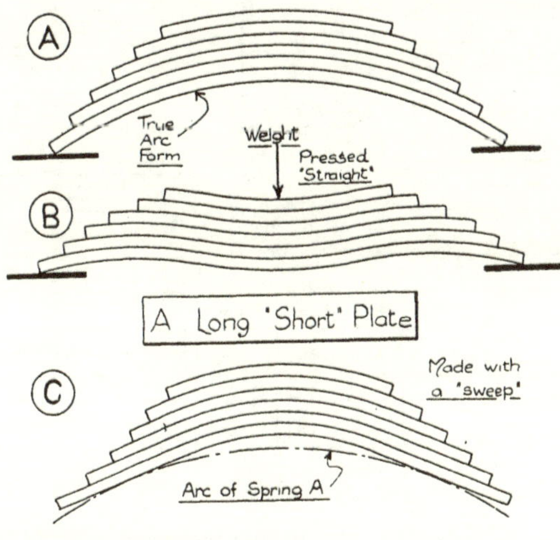

FIG. 51.

The length of the short plate determines the relationship of the spring in general to the theoretical rhombus shape, and therefore influences to some degree the deflection. As a rule, however, with ordinary practical measurements, there is no need to take this into consideration in the standard formulæ, and the main influence it has is on the fitting shape of the spring, and on the amount of useless weight which is introduced, and which has to be paid for. Instances are known of springs, say 40 ins. long, with short plates of 20 ins., the spring having perhaps 6 plates. Springs of this sort are stiff to the deflection formula, and are not formed to a true arc, being made to a sweep. If fitted to a true arc, the resulting effect when pressed "straight" is shown in the sketch in Fig. 51. The sweep indicated in the drawing makes no difference to the stresses along the plates, but admits of the spring being really and visibly "straight" when all the camber is taken out, whereas, if it were fitted to a true arc, it would have depressions in the lengths from the middle to the bearings. Springs of this type a hand-fitter calls "too weak at the centre"—the converse being that they are "too strong at the ends"—and the fact being that they have too much material in the ends, or too little material in the centre, to permit of uniform deflection, and consequent uniform stressing, throughout the plates.

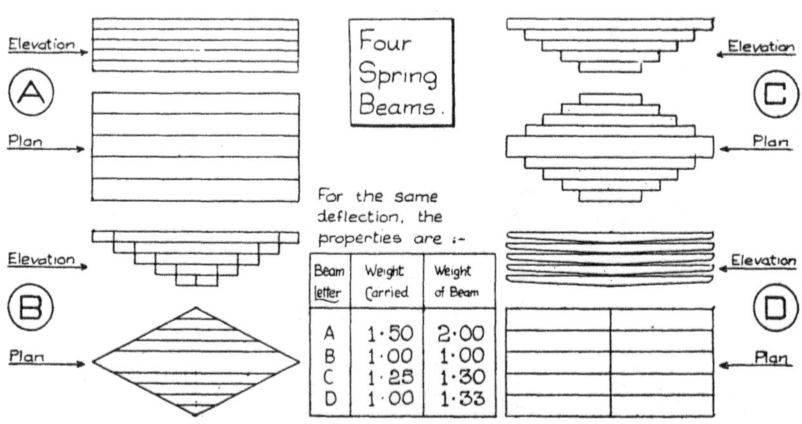

Beam letter	Weight Carried	Weight of Beam
A	1·50	2·00
B	1·00	1·00
C	1·25	1·30
D	1·00	1·33

FIG. 52.

The diagrams, Fig. 52, show spring beams made up along four different lines, as follows :—

(A) represents the multiple plate beam, with all plates of the same length, and corresponding to the "rectangle plan." It also represents the extremity of plate lengths ; and the type of maximum weight and lowest efficiency.

(B) represents the purely theoretical multiple plate beam, of the exact "rhombus plan," plates being cut to their minimum lengths. This type of the minimum weight and highest efficiency, but owing to practical exigencies, cannot be employed in its perfect form.

(C) represents the generally practical spring beam, with minimum possible plate lengths, and with the back plate of full width from end to end. This approximates closely to (B) and has nearly the same efficiency.

(D) represents a multiple plate beam, with units of parabolic elevation, that is, with each plate varying in thickness. There is no use for this in ordinary spring work, owing to manufacturing considerations. The efficiency is the same as (A) and lower than (B) or (C).

The relative efficiencies, on theoretical figures, of the above, are as follows—efficiency being measured on the basis of the weight of the beam compared with the load safely carried—A : B : C : D : : 0·75 : 1·00 : 0·96 : 0·75. The obvious logic is therefore, to maintain the practical shape (C) as closely as possible, by careful judgment of the plate lengths.

In most springs composed of non-uniform thickness plates, the designer figures his short plate to the same length as would have been used had the plates been all of the same thickness. It must be admitted that in the majority of cases, these lengths could be none other, owing to the need for the short plate coming beyond the width of the hoop or axlepad. The effect in general is, however, that they are 10 or 12 times the correct length, whereas the most that is usually attempted with uniform thickness plates, only results in the short plate being perhaps three times the correct length. Fig. 53 makes this clear, and shows that in springs of varying thickness plates, the offset or step from plate to plate should increase from a minimum at the thin short plate to a maximum at the thick back plate. This would appear simple logic, as clearly, the thicker the plate the stronger it is, and with equal weight can be permitted a longer overhang with a safe deflection than the thinner short plates. In practice, however, many designers, not content with the varying thickness plates, vary also the offsets, but always in the reverse direction to that suggested, namely, having long offsets at the thin plates, and short offsets at the thick plates. Some hazy idea would appear to be present in this connection that a parabolic outline is produced—and if applied to uniform thickness plates, this might be admitted. As applied to varying thicknesses, however, the "parabola" obtained can be regarded as doubtful. A spring of non-uniform thickness plates, with ordinary (and probably needful) lengths from short plate upwards, is equivalent to a spring with much superfluity of material, and as such, has a plan as shown in Fig. 48, where the comparisons are indicated. It must be made clear that the author is not advocating theoretical short plate lengths for non-uniform thickness springs, as they

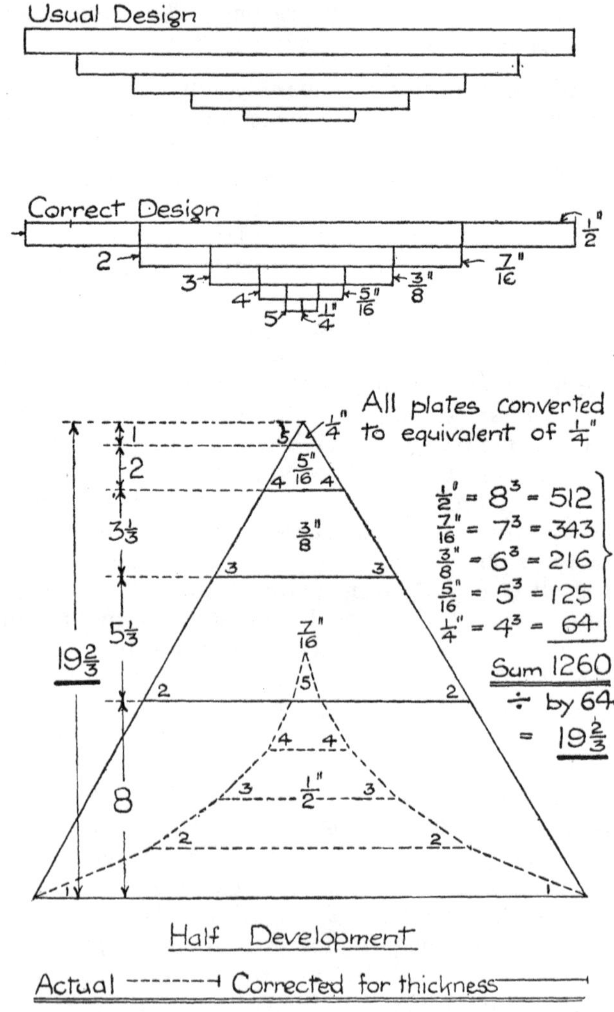

FIG. 53. PLATE LENGTHS—VARYING THICKNESSES.

are impossible, and the fact that the average best that can be done for practical spring design is to make such plates at least 8 times their true length, would appear to be a further excellent argument for the elimination of the varying thickness spring.

In this connection, particulars might be quoted of two springs, of varying thickness plates—length of back plate 30 ins., width 4 ins. The following were the remaining dimensions—

Back Plate.	Spring A.	30 ins. × $\frac{1}{2}$ in.	Spring B.	30 ins. × $\frac{1}{2}$ in.
2nd ,,	,,	24 ins. × $\frac{7}{16}$ in.	,,	$17\frac{13}{16}$ ins. × $\frac{7}{16}$ in.
3rd ,,	,,	18 ins. × $\frac{3}{8}$ in.	,,	$9\frac{11}{16}$ ins. × $\frac{3}{8}$ in.
4th ,,	,,	12 ins. × $\frac{5}{16}$ in.	,,	$4\frac{1}{2}$ ins. × $\frac{5}{16}$ in.
5th ,,	,,	6 ins. × $\frac{1}{4}$ in.	,,	$1\frac{1}{2}$ ins. × $\frac{1}{4}$ in.

The plate lengths for Spring B were calculated out to accord with the general plan of Fig. 53, and the proof of the accuracy of design was clearly shown when these two springs were tested straight, Spring A having the characteristic " humps " of the design with superfluous material, the Spring B going really flat. Both these springs were fitted to a true radius, and without nip. Had these been for actual working, Spring A would not have been made to a circular arc, but to a sweep the centre radius of which would be less than that of the true arc, and the stress at this position, with the spring straight, accordingly greater.

The effect of too many full length back plates is, of course, the same as commencing the design from a too-long short plate—as it increases non-uniformity of stressing throughout the spring beam. In springs with a large number of plates, however, four full length plates (a usual number) make little difference, as indicated in the view of a typical American pattern, Fig. 54. Springs of 20 plates are by no means uncommon in the U.S.A. for locomotives and heavy road trucks, and the four full-length plates specified amount to only 20 per cent. of the total, which possible effect is included in the unit deflection formula—as a 5-plate spring, with its necessarily full width back plate, is a similar equivalent of 20 per cent.

A few words might be said on the subject of the parabolic elevation outline (Fig. 55) favoured by certain designers—chiefly in this country. They probably obtained the idea from text books on bridge building, as no authority of note hints at such procedure for springs. Plate girder bridges are frequently built with numerous flange plates, paralleling, in some senses, the multiple plate spring, and such flange plates are of such lengths as to form overall a parabolic outline, but this is owing to the fact that the bridge, as such, is designed for a distributed load, whereas the spring maintains always, as an entity, a concentrated load ; the distribution being a function of the spring plates, in other words, taking place within the spring. The only effect of such outline shape for the plate lengths is, of course, to slightly stiffen the deflection per unit, and to add weight and cost well out of proportion to the slightly extra stronger spring. The effect of such outline can be counteracted by taking (K) as 0·095 in the deflection formula, if a normal short plate be used.

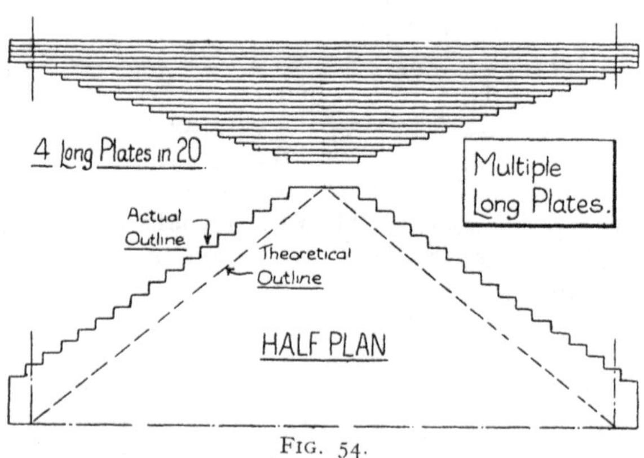

FIG. 54.

In the perfectly designed spring, it will be found that the speared short plate, taken by itself, and tested through the same relative deflection as the whole spring, will, in itself, carry the same weight as that carried by the complete spring. Expressed differently, if a perfect 10-plate spring, when tested straight, carries 8 tons, the short plate removed, and by itself tested straight, will also carry 8 tons. As indicated in Fig. 47, the weight carried at the short plate (or hoop) has to be transmitted through the spring in entirety, to the bearing or suspension points, or *vice versa*, according to the direction of the load application, whether at the ends or at the centre. In normal springs, with the cantilever arms of equal length, otherwise, with the load or reaction in the centre of the span, the weight or reaction at each end is, of course, one half of the middle load. This line of load transmission is automatic, and regardless of plate lengths. With plates of uniform thickness, and the spring straight, it follows a line drawn from the centre of the short plate (if the load is really applied there) to the bearing point at each side. The points at which this line cuts the succeeding plates are generally maintained

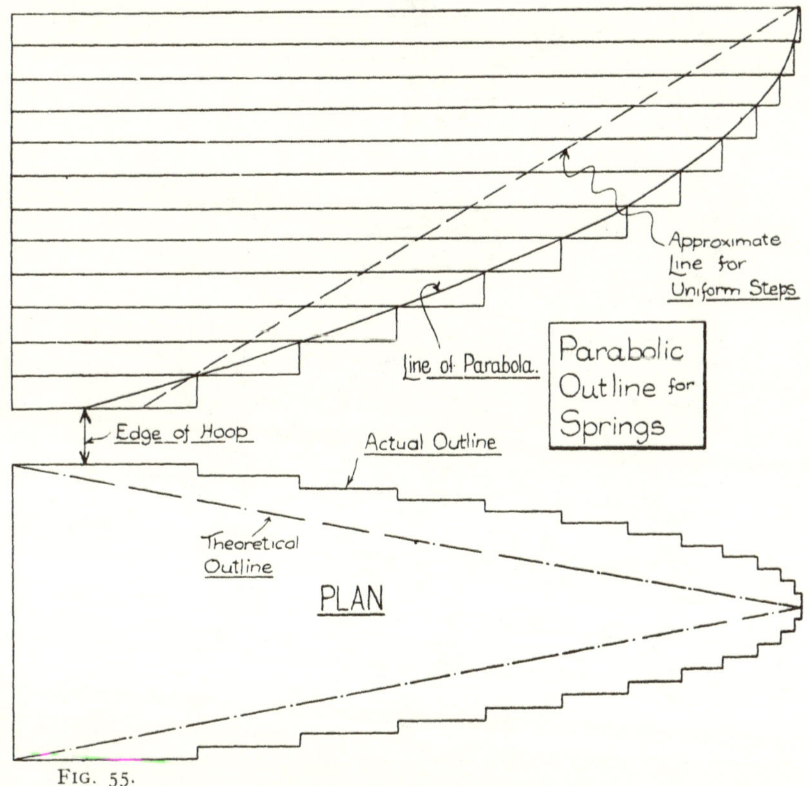

FIG. 55.

during flexure of the spring, only varying owing to the extension and compression of the plate surfaces due to the flexure. Fig. 56 shows some varying aspects of this load line.

If the short plate is too long, the points will distribute less than the transmitted load, and intermediate plates will also distribute less—the full load coming through the spring on lines between the plate ends and the centre. As the whole object of correct spring design is to transmit a given load with a minimum of material, nothing is gained by having plates of excess length, and whilst it is generally impossible to arrange for the short plate to be so designed that it will carry the whole spring loading for a proportionate deflection, nevertheless it can be made to approximate thereto.

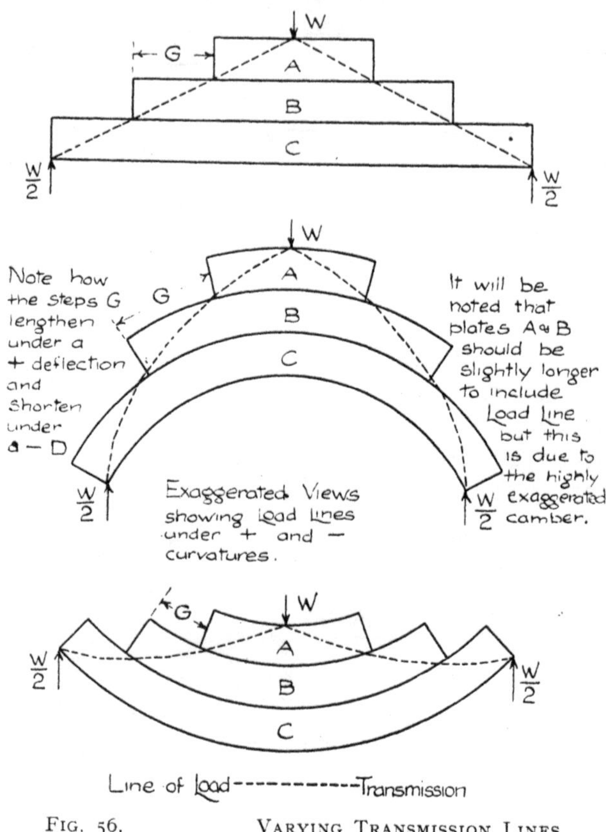

FIG. 56. VARYING TRANSMISSION LINES.

CHAPTER XVI

FINISH OF PLATE ENDS

VARIATIONS in plate thickness can theoretically be infinite, but in practice are very finite, at any rate, so far as the designer is concerned, being limited to certain well-known dimensions. Variations in plate lengths can also be infinite, but as this only involves the cutting up of long bars, is not specially worrying from the manufacturers' point of view. Variation in plate end finishes, however, which can be, and are also infinite, present more substantial and serious aspects, as relatively expensive tools are required, followed by considerable expense in setting such tools—these questions both apart from the employment of special machines, as distinct from the tools used in such machines.

The basis types of plate end finishes are shown in Fig. 57 in which (1) shows a square cut end, which is by far the most practical pattern, requiring only the necessary shearing operation which is involved by cutting the plate to length. Ends such as this are sometimes more or less ground, but this is largely a matter of shop practice. (2) shows at one end the perfect "spear" which is obviously impracticable, owing to the sharp and dangerous point. The other end shows a workable spear, "square spear" shape, which is largely employed, and which is standardized in U.S.A. automobile practice. This involves the use of specially shaped cutting tools, under a powerful machine. (3) shows a "drawn point" (American "rolled end") one end showing the perfect shape, which is never made, and the other end showing the practical shape.

In examining the "spear point" finish, with relationship to perfect rhombus spring plans, it is clear that to attain theoretical accuracy, the spear angle should vary with every design of spring. As, however, at the best it cannot practically be used when coming to a sharp point, which is the theoretical requirement, it would appear that meticulous adherence to theory might be waived in connection with varying the spears for each pattern. Unfortunately, however, the fact that the cost and setting of spear tools is expensive, does not always seem to be within the knowledge of designers, and a set of springs for the same engine, which necessitate perhaps, three different widths, will have three different spears. This shows a praiseworthy study of theoretical aspects, but involves

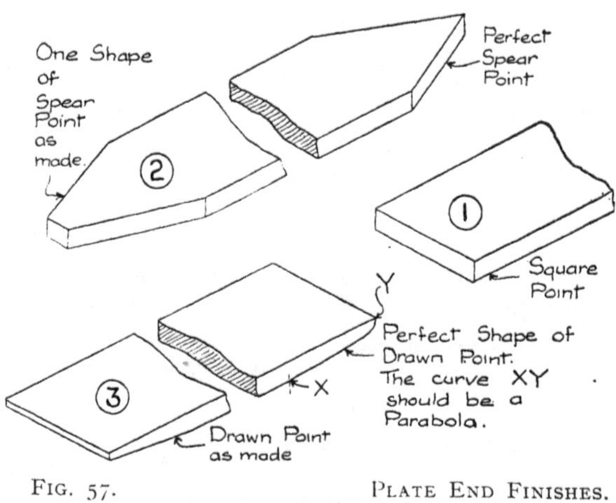

FIG. 57. PLATE END FINISHES.

substantial trouble and cost, and the value received therefor is precisely nil. Anyway, the subject of standard spears will be treated at length at a later moment, and meanwhile, the chief point to emphasise is, that as a general rule, shapes or sizes of spears have not the slightest effect on the unit deflection of springs—at anyrate, when practical measurements are taken. This is generally agreed to by all spring experts, and is a strong argument for the square cut plate—since usually the spearing is mainly for appearances. In long springs, of few plates, long spears, involving probably special forming, will make a certain small difference to deflection, but such springs are not common. A few spear shapes are shown in Fig. 58, which includes remarks in connection therewith. Railway and tramway designers, British and Continental, generally favour the square and round-nose spears, whereas automobile designers, of the same schools, allow themselves to run riot with numerous shapes of circulars, rounds, and gothics. In American practice, most railway and tramway springs are square-cut, and automobile springs square-speared to a recognised standard. With offsets at each plate end of less than the plate width, spearing is certainly a luxury that might be omitted with no disadvantage to the spring, and possibly a small balancing consideration in price.

The chief point, of course, of spearing plate ends, should be to obtain uniform stress over the ends, which are really small

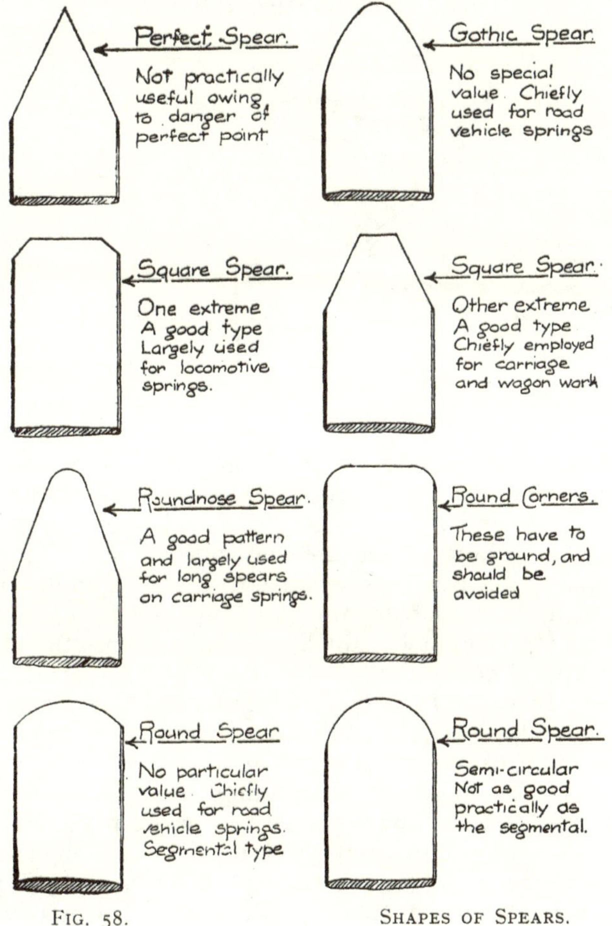

FIG. 58. SHAPES OF SPEARS.

cantilevers, but as pointed out, in the absence of being able to attain in practice the perfect spring design, it would seem of little value to trouble over such a detail as the uniform stress in the offsets. An alternative to spearing has been suggested and patented by David Landau, as " stress equalizing slots." These are reversed spears, in other words, instead of removing material from the exterior width of the plate, it is removed from the interior width, as shown in Fig. 59. The value of such method is debatable, as the cost of the necessary tools is little different to that of spearing tools, and the effect is about the same, as it is just as impossible to make a theoretical " stress equalizing slot " as it is to make a theoretical " spear point."

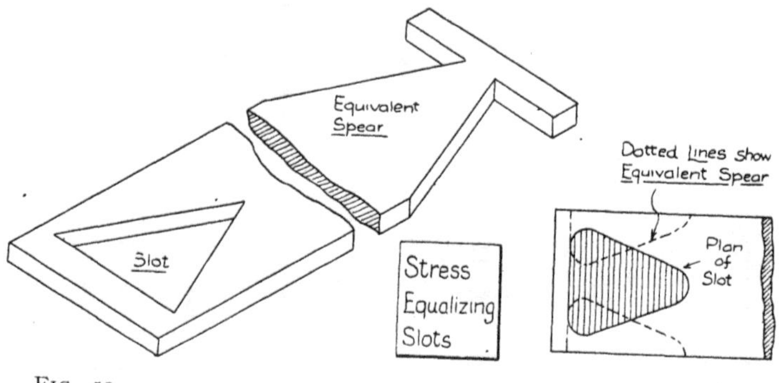

FIG. 59.

Mainly, the spearing of plates turns upon the designers' ideas, some spearing because it " has always been done " within their limited range of vision, others because they have a knowledge of underlying reasons and endeavour to put it into practice, and others because they consider it a " finish " to the spring. No serious evidence has been adduced in practice of its being detrimental or advantageous.

This last fact cannot be equally stated in connection with the " drawing " or " rolling " of plate ends—the tapering in thickness. In order to obtain the uniformly stressed offset by this method, the small cantilever represented by the offset should be in side elevation a parabola—which shape can only be made by stamping, or grinding to template, neither of which methods find employment in normal manufacture. Actually, the ends are invariably thinned in " tapering rolls," which gives a uniform graduation of thickness. When offsets are less than the plate width, the " drawn point " becomes quite useless, particularly on the plates of $\frac{3}{8}$ in. or over, as the length of the " draw " is so limited, that the finishing thickness remains fairly substantial. Under $\frac{3}{8}$ in., and with springs of narrow plates, other aspects present themselves, as comparatively long draws are obtained, with thin ends. Such finish will materially increase the deflection of a spring under unit load, particularly if it be made of a few plates, and have a relatively long length. Generally speaking, however, the bulk of drawn point springs have also varying thickness plates, designed as usual, that is, with uniform offsets ; and in such cases, the deflection obtained will be in practical agreement with the standard formula. Some authorities have used considerable labour in dealing with aspects of drawn points, of which there are many, but it is the author's opinion that the process should become obsolete, and certainly the trend of practical opinion is in this direction. It must be remembered that the plates of " drawn point " springs cannot be properly cambered by either hot or cold 3-roller processes. In America, the " rolled end " is nearly abandoned for automobile work, largely because springs for this trade are machine made, and great difficulties present themselves in obtaining a good result without considerable following hand work. Strangely enough, however, quantities of railway springs are supplied there with plates having an end finish which has been put on in the tapering rolls. As many of the offsets of such springs are less than 3 ins., on plate widths of 5 ins. and 6 ins., it would clearly be a mis-description to call such ends " drawn points," and they are known here as " dubbed " ends. Needless to say, no possible explanation can be forthcoming of this quaint practice, unless it be taken on the double basis of old custom, and providing an end finish which displaces the grinding which is sometimes carried out.

Not infrequently, designers add to the beauty of their work on the drafting of plate ends by adding some form of spear to a " drawn point." Probably it is only their ignorance of " stress equalizing slots " which prevents them from also adding this arrangement. The perfect (offset) cantilever can be obtained only by varying uniformly in width, or parabolically in thickness, and an arrangement by which the thickness and width can both be tapered so as to provide a uniformly stressed member for this offset is obviously too subtle to be gone into, particularly as the result would be valueless. When this double process is involved, it not infrequently will be found that the spring is weaker than anticipated, if the standard formula has been used for calculation purposes, but such result is determined to such a large extent by other features of the design, that here again, it is of no value to endeavour to obtain a correcting constant. In due course, characteristics of this latter type will doubtless

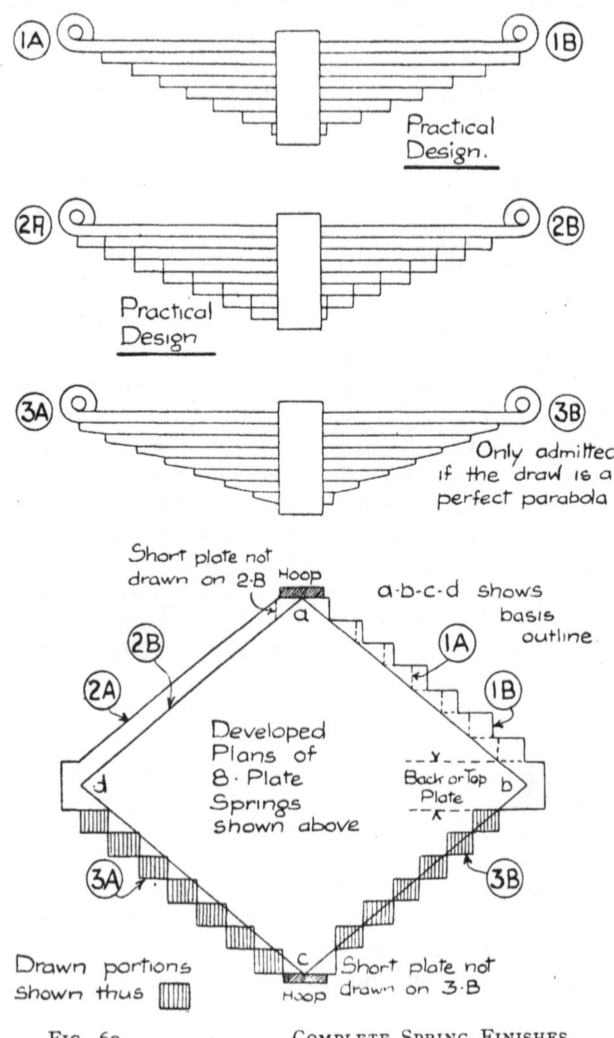

FIG. 60. COMPLETE SPRING FINISHES.

die a natural death, as they revert back to the earliest days of road vehicles, and are carried on very largely as a matter of ancient faith.

Theoretical and practical spring designs with the square, spear, and drawn, plate-end finishes are shown in Fig. 60, which includes the necessary amendments to the rhombus plan. In the first place, the spring cannot be otherwise than the " full plate " section at all points throughout the hoop, or between axle-pad clamp bolts. This means that no spearing or drawing of the ends can be carried out until the edge of the buckle, or other fastening, is reached. The back plate must be full width throughout (certain exceptions occur as " diminished " ends but the reduction leaves still parallel ends on the top plates) to carry the suspension or bearing arrangements, and the second plate should go either to the bearing centre or to the extreme end of the back from the point of view of support being given to the harnessing attachment. This is particularly necessary in welded back plates, and in automobile springs, the second plate is frequently wrapped round the rolled eye so as to hold the fastening arrangements in case the eye breaks off.

CHAPTER XVII

THE STRESS EFFECT OF CENTRE FASTENINGS

ALL classes of semi-elliptic springs include some type of "centre-fastening," which will not, of course, be geometrically central with the length in the case of asymmetrical springs. This fastening is necessary for two main reasons: (1) it acts as a register to prevent the plates moving longitudinally, and (2) it acts as a securing means for the hoop and spring, or alternatively, as a registering means into an axle-pad.

Clearly, the provision of any means for centre fastening for the above purposes entails the plate being pierced or deformed in some way at the plane which is taking the greatest bending moment. At this stage, only the broad distinctions between the two patterns of fastening will be touched upon, namely, the type that includes the removal of material necessary for the introduction of a rivet, cotter, or bolt; and the type which only causes deformation of the material, in order to emboss nibs on the plates. These are shown in Fig. 61.

The weakening of the plates, on their plane of maximum stress, is obviously a very serious feature, and one of the prime causes of failure. On the surface, it would appear that a plate pierced with a hole, as compared with a plate not so pierced, would result in the former giving a higher unit deflection result than the latter, due to the removal of material. As a matter of fact, no cognisance need be taken of the hole or deformation, as the unit deflection result remains the same as it would if the plate were plain. Naturally, the larger the hole, the quicker is reached the limiting and breaking, or permanent deforming, point; but the deflection per unit is unaffected. The explanation would seem to be given by the stress diagram, Fig. 62. In this, it will be noted that whilst a high increment over normal stress is caused by the presence of the hole, the area due to this, which counts in the general diagram, is so small in proportion, that its effect is negligible, showing that the unit deflection is unchanged.

Only within recent years would it appear that any practical research attention has been given to the weakening effect

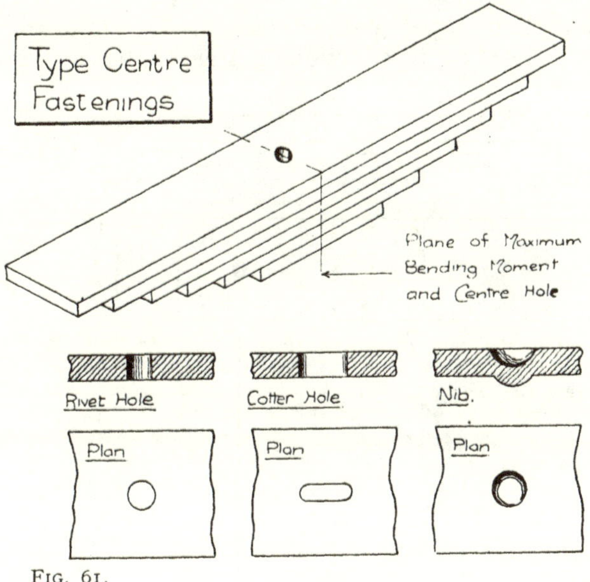

FIG. 61.

of holes on plates, and that in two directions, (1) by Professor E. G. Coker in his fascinating studies with polarized light, and (2) by the author in an extensive series of tests on various centre fastenings. In the latter experiments, it was found that the presence of a centre hole, $\frac{1}{2}$ in. diameter in a 4 in. plate, equivalent to the removal of $12\frac{1}{2}$ per cent. of the material, weakened the plate to the extent of 25 per cent., or twice as much as that which might have been anticipated. Empirical figures were obtained, for comparative purposes, by adding together the deflection and the tonnage of each test piece at the moment of fracture, in the following way :—

Plain Plate. Fractured at 9·9 tons, deflection 1·46 ins., index 11·36.

Holed Plate. Fractured at 7·5 tons, deflection 1·08 ins., index 8·58.

The above are typical of a number of results on hardened and tempered material.

It will be noted that the actual tonnage reduction is nearly 25 per cent., which points to the possibility of unequal stresses, embracing some abnormally high stress and consequently breaking point, across the plane of the hole, or plane of fracture. This of course, could not be proved, and all that was known was that $12\frac{1}{2}$ per cent., or similar reduction of area, did not mean $12\frac{1}{2}$ per cent. reduction of strength, but 25 per cent.

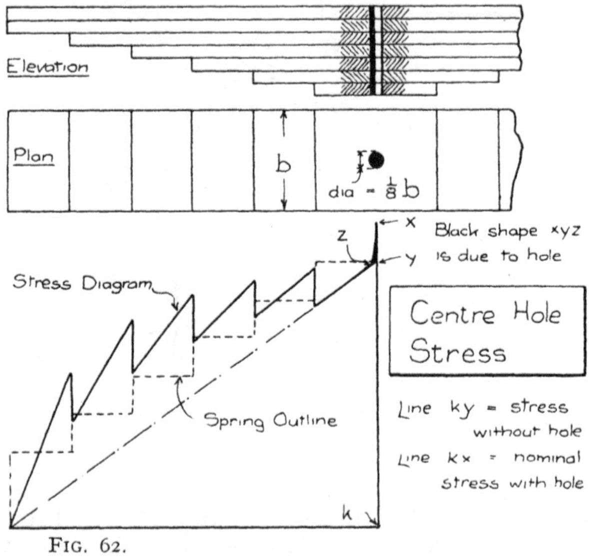

FIG. 62.

Similar results were obtained with forms of notches and cotterholes, and the true explanation of the phenomenon is now available due to recent stress researches on nitro-cellulose by means of polarized light. There is no occasion to detail the arrangements by which the ultimate results are arrived at—the matter of importance from the spring makers' and designers' points of view are the results themselves.

Professor Coker in a recent paper (1921) includes some experiments on bars with holes, and bars with notches, both of which are applicable to spring manufacture, and Fig. 63 shows the stress curves obtained with these examples. It will be noticed that at the edges of the hole or notch, the stress rises to a high value, and then rapidly falls until it becomes a minimum value at the edges or middle of the plate, as the case may be. These specimens were taken under pure tension stress, representing the upper surfaces of a spring plate under load—and it can be safely assumed that the lower surfaces under compression include stresses of similar magnitude. The result of these experiments clearly shows why the breakdown of holed or notched plates occurs at a lower figure than might be expected. The fact of the stress at the edge of the hole or notch being substantially higher than the finishing stress, indicates that the material at these edges will arrive at the plastic stage in advance of the bulk of the section, and by contraction, will throw the load on to the decreased effective section until rupture occurs.

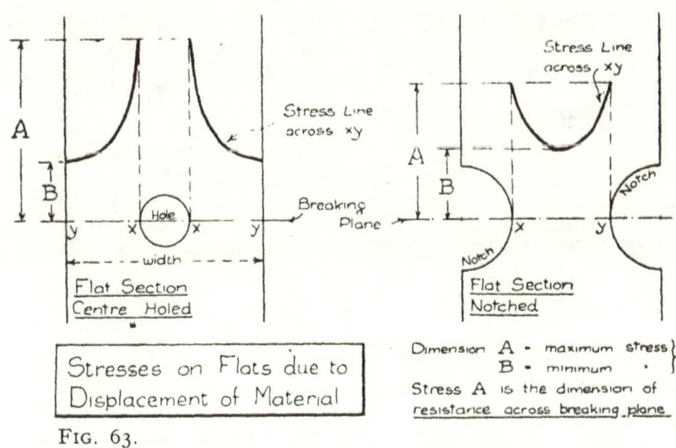

FIG. 63.

Judging from these results, and regarding the matter from another point of view, it might be said that the stress curves adjacent to holes are determined by the intensity of the stress lines through the bar. This aspect is shown in Fig. 64. It is imagined that certain stress lines are present in a loaded specimen, and at the central hole these divert around. There is shown (1) the example instanced in Professor Coker's Paper, of a hole one-quarter the width of the test bar, (2) a further example from the same of a hole nearly the size of the test bar, (3) a hole of small size useless for spring work, (4) a fairly normally dimensioned hole for a spring plate, and (5) a plain plate, with no hole. It will be fairly clear from these why, if holes cannot be avoided, they should be kept as small as possible.

Fortunately, considerations occur which will, up to a certain point, maintain the plane of the plates which contain the hole out of the line of maximum stress, but when, owing to slackness of fastenings, or other reasons, this maintenance ceases, fracture is the result.

So far, published research by polarized light does not appear to have included a parallel case to that of the deformed spring plate, by the introduction of "nibs." These can be of many and various forms, good, bad, and indifferent, but

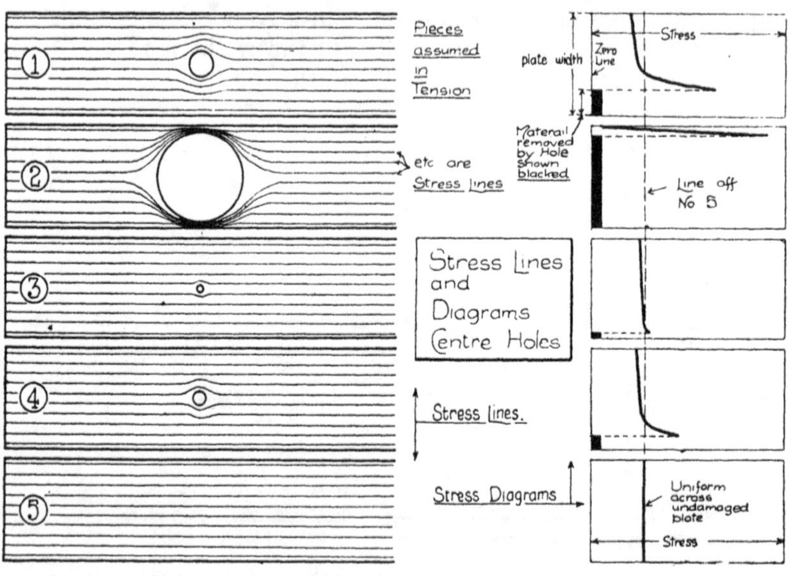

FIG. 64.

when used, the depression should always be on the tension side of the plate, and the embossment on the compression side of the plate. (The reasons for this are discussed fully in the section dealing with "Manufacture.") So formed, the resulting register is known as a "downward-nib." The author's experiments on a good form of these, typical results being shown, were as follows :—

Plain Plate. Fractured at 9·9 tons, deflection 1·46 ins., index 11·36.

Nibbed Plate. Fractured at 9·0 tons, deflection 1·20 ins., index 10·20.

The comparative index figures for the three instances then stand :—

Plain Plate ..	11·36,	100 per cent.		
Plate with hole	8·58,	75 per cent.	Reduction	25 per cent.
Plate nibbed	10·20,	90 per cent.	,,	10 per cent.

From the above, and from the bulk experiments, there is no doubt whatever that the downward nib form of fastening is the most superior of any definite registering fastening, which has to be common to all types of springs. This statement is without regard to practical conditions of manufacture, in which the hole is frequently preferred. Automobile designers generally embody a hole, as most springs for this trade are unhooped, and the bolt through the spring will retain the plates. This effect, however, can be equally contrived by clipping the short plate, and these clips will then hold the spring as an entity, by which means the employment of a centre downward nib is rendered possible.

CHAPTER XVIII

A STUDY IN SKIN STRESSES

Few articles of commercial use are stressed in service as highly as the ordinary laminated spring. Reasons for this are undoubtedly the fact that (1) it is a comparatively simple structure ; (2) practice justifies the use of such high working stresses, equal to one-third to one-half the elastic limit ; and (3) to carry present-day weights, a low stressed spring would have to be of great length, or contain many thin plates ; either designing method leading to great weight. An examination of the skin stresses of springs is of great interest, but before touching on instances, it would be well to refer as briefly as possible to the discrepancy which has been before remarked upon, namely, the great difference obtained from calculation based upon fibre extension, and that based upon bending moments.

Select for example, a plate, 4 ins. $\times \frac{1}{2}$ in. $\times$ 30 ins. long, cambered to B.S. test, equal to 2 ins. If this be pressed flat, and then the skin stress worked out from Formula IV.—4, it will be found that the result (as shown adjacent to this formula) is 58 tons per square inch. If this plate be centrally loaded, the deflection per ton will be practically 1·10 ins., and it will carry therefore, for 2 ins. deflection, 1·80 tons, or 1·90 corrected for nominal sizes. The standard bending moment formula for skin stresses (No. I.—4) then gives the following figures :—

$$f = \frac{WL}{4} \times \frac{6}{BT^2} = \frac{1{\cdot}90 \times 30 \times 6 \times 4}{4 \times 4 \times 1} = 85 \text{ tons.}$$

The discrepancy between 85 tons and 58 tons is hardly negligible, and that such should occur in the ordinary way of calculation with standard text-book formulæ, without explanation, does not assist the belief of the man generally unfamiliar with such things, whom one is endeavouring to educate to put some trust in "figures."

Let an examination now be made of the skin stresses on the R. C. H. Buffing spring illustrated by Fig. 65. In service, this spring is put on the wagon and held in normal position with a nip of $3\frac{1}{4}$ tons, central load, representing 5 ins. deflection. The stroke following this, when wagon strikes wagon, travels it a further 5 ins., this latter dimension being the full possible movement of the buffer heads. The total deflection is therefore 10 ins., and it is specified to have a resistance of $6\frac{1}{2}$ tons at this deflection, which corrects to 6·8 tons for calculating with nominal sizes. Working on the standard bending moment formula, I.—4, the following figures appear :

$$f = \frac{6{\cdot}8 \times 70 \times 6 \times 4}{4 \times 42 \times 1} = 68 \text{ tons.}$$

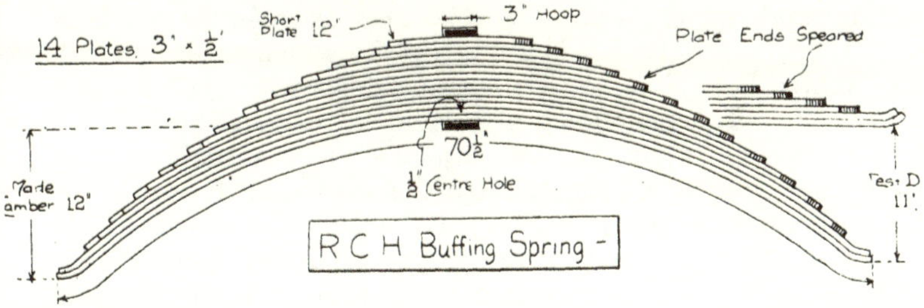

Fig. 65.

The test deflection absorbs 7 tons, which makes the skin stress 73 tons per square inch. Anyhow, the 68 tons is an actual working stress on the material, based on standard calculations. This 68 tons would only be correct, however, if the spring were fitted without nip, and if the plates were undamaged by holes or nibs. Actually, a hole $\frac{1}{2}$ in. diameter is introduced, and the plates are fitted with nip. In consequence of the latter, the stresses of the shorter plates will be at least 4 tons over the above, or 72 tons. Also, the presence of the hole, equal to one-sixth of the width, increases still further the stresses of the short plates, and on a basis of pure proportion, sends them up to 86 tons. Actually, however, as found by the experiments reported in Chapter XVII, at the edge of the hole this is substantially greater, probably equal to 120 tons, and certainly up to 105 tons. The paradox then appears of a spring, made from ordinary water-hardening carbon steel, and tempered from back plate to short plate, through a range of 65 to 75 tons, with elastic limits about 50 to 60 tons—standing up in continuous service, year after year, receiving thousands of " live " shocks as the buffers bump home, and yet having its component plates working (according to standard calculations) at skin stresses of 60 to 70 tons throughout, and up to 105 tons locally.

The British Standard Test load is for all practical purposes, equal to 70 tons per sq. inch (exactly 68·5) fibre stress, when based on bending moment calculations—and this is far in excess of the elastic limit of the steel of the bulk of the springs which are tested to the B.S. requirement—yet the springs remain elastic, and can be overtested 15 per cent., with, in many cases, very small " sets," particularly if little or no nip has been included in their manufacture.

The explanation, to which the author has given considerable attention, would appear to lie in two directions as regards the normal spring, and one direction as regards a single uniform-section beam. The first line to which attention will be directed is applicable to both, and deals with the modulus of section. The resistance of a solid beam, such as that of flat, round, or square section, appears to be actually higher than that which is obtained by the standard formulæ for these sections, which in turn is derived by the graphic construction shown in Fig. 3. In the case of shaped beams, such as H-sections (joists), the modulus (Z) obtained by the graphic construction is probably correct, as the reason for the shape of such beams is the elimination of superfluous material such as obtains in the solid beam. Accordingly, there is relatively no spare material in such shapes to offer increased resistance to loading, whereas in the rectangular solid beams, it can be assumed that one-half of the section is inert. Put in another way, the maximum stress is on the surfaces only, and this limits the carrying capacity, whereas, if it were arranged for the section to be uniformly stressed throughout, one-half the material, properly arranged, would give the same carrying power. Clearly, in the case of a thin rectangular beam of spring plate form, this arrangement is not a practical possibility, and, in spite of the apparent waste of material, the approximate rectangular section has to be maintained.

Moduli of Four Sections.
1'
7½'
FLAT.
3"
ROUND.
2¾"
SQUARE.
JOIST.
8'
4'
Comparative Particulars of Sections shown, as regards 'normal" Z and "solid" Z.
Flat
Round
Square
Joist
Curve A, Wgt lbs per ft.
Lbs.
26
25
24
20
15
Inches
10
5
0
Curve B Z, inches, normal
Curve C Z, inches, solid
Curve C
Curve B
Curves ABC Base Line Curve D
0
50
%
100
150
200
Curve D C/B %
Base Line Curves EF
0
Square Inches
5
10
Curve F. Graphic Area
Curve E Full Area
Base Line Curve G
0
Curve G = F/E %
50
100

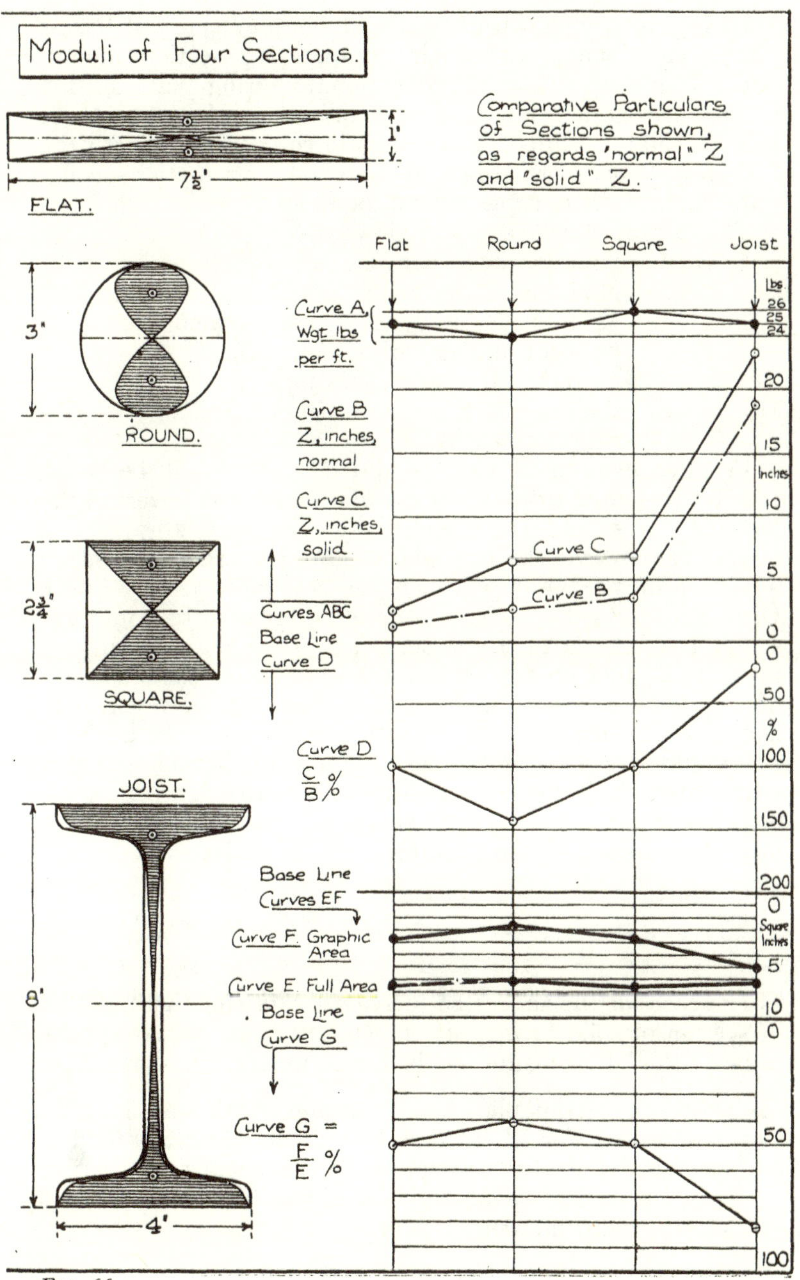

FIG. 66.

In Fig. 66 are shown instances of the four types of beams above mentioned, with their comparative section moduli and relative weights. A proportionate spring plate section is shown, together with a round and a square—these three being "solid" beams; and the main comparison is with the "shaped" joist, or I-beam section, all four sections being approximately the same weight per foot.

The curves shown are as follows:—

A .. pounds weight per foot run of each section.
B .. Modulus (Z) in inches, normal calculation.
C .. Modulus (Z) in inches, "solid" calculation.
D .. per cent. increase of (C) over (B).
E .. geometrical sectional area.
F .. geometrical "uniform stress" area.
G .. ratio percentage of F : E.

From an examination of the above curves it will be noted that on the standard basis of calculating (Z) the lowest is the flat with 1·25, and the highest the joist with 18·77. The latter has, however, been specially shaped in order to produce a section of maximum efficiency, as best indicated by Curve G, which shows that, with the material reduced to a uniform stress area, the latter is 82 per cent. of the actual sectional area, as against 50 per cent. for the rectangle sections, and only 41 per cent. for the circular section. If, however, it is conceivable that the whole section is enduring a uniform maximum stress at the moment of rupture, the low-stressed material of the solid beams—adjacent to the neutral axis—is brought into action, the centres of resistance remaining locked, and the following comparisons are arrived at:—

Flat.	Standard Z,	— 1·25.	Maximum Z,	— 2·50.	Increase	100 per cent.
Round.	,,	— 2·65.	,,	— 6·40.	,,	141 per cent.
Square.	,,	— 3·46.	,,	— 6·92.	,,	100 per cent.
Joist.	,,	— 18·77.	,,	— 22·80.	,,	22 per cent.

The case for the suggested increased value for the modulus of section of solid beams under bending stress would appear to be on the following lines—illustrated by Fig. 67, in which:—

(1) Shows a rectangular beam section, with the graphic construction for (Z) of orthodox form. It is contended that this is correct as a purely geometrical construction for a theoretical beam, which is of the abstract type, and has therefore no weight.

$$(Z) \text{ here is } BT^2 \div 6,$$

(2) Shows the same rectangular beam section, with amended construction assumed on the basis of the actual (Z) under load, that is, with any definite and visible beam, which has weight, and has therefore to carry its own weight when placed on two supports, and sustain external weight as required.

(Z) here is $BT^2 \div 4$.

(3) Shows the same section, with the bending continued until the beam is at the point of rupture, with all its interior construction resisting to the utmost limit.

(Z) here is $BT^2 \div 3$.

(4) Shows the known (Z) in direct tension.

(Z) here is BT.

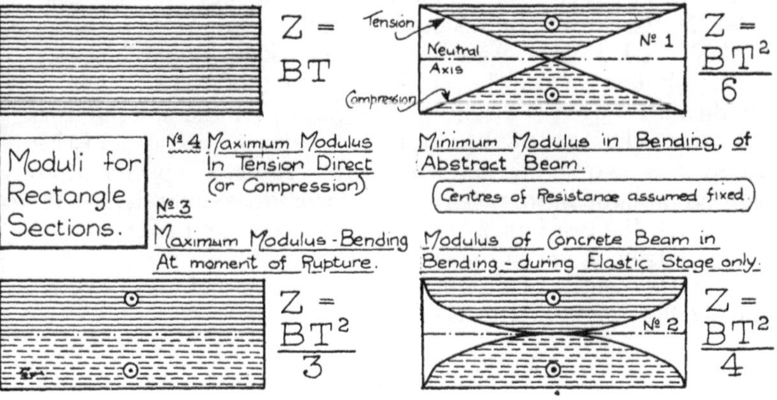

FIG. 67.

The relative values of the above, on a plate, say 4 ins. × ½ in., are as follows :—

(1)	.. .. ..	$Z = \frac{1}{6}$, or 0·166.	$8\frac{1}{4}$	per cent.
(2)	.. .. ..	$Z = \frac{1}{4}$, or 0·250.	$12\frac{1}{2}$	per cent.
(3)	.. .. ..	$Z = \frac{1}{3}$, or 0·333.	$16\frac{1}{2}$	per cent.
(4)	.. .. ..	$Z = 2$, or 2·000.	100	per cent.

It is not necessarily suggested that the divisor for BT^2 in Case 2 is exactly 4. It may be 3·9 or 4·1, but certainly it would appear to be very closely approximate to 4.

Occasionally standard text books will refer to certain doubts as to the accuracy of the accepted beam theories; which are in the main correct. Usual calculations based on standard formulæ however, are, it has long been recognised, quite useless when the beam arrives at breaking point. For instance, experiment has shown that a piece of 4 in. × ½ in. spring steel, hardened and tempered, and resting on supports at 10 ins. centres, will sustain a load of 10 tons when just at the point of fracture. Based on the usual formula, the following skin stress is arrived at :—

$$f = \frac{10 \times 10 \times 6 \times 4}{4 \times 4 \times 1} = 150 \text{ tons.}$$

Actually, on a true rectangle section, this would be about 158 tons. To suggest that ordinary carbon steel, 0·60 per cent. content, with a breaking stress in the as-rolled state of about 46 tons, and in the hardened and tempered state as tested, of about 80 tons—can withstand a skin stress of 150 tons, clearly causes one furiously to think, and the knowledge of this has led certain authors to remark on possibilities of the breakdown of the theory under such conditions. Lest it should be thought that the plastic nature of the steel at the later stages has some subtle influence, take the instance of the transverse testing of cast iron bars. A usual specification requirement for hard cast iron for locomotive cylinders is as follows :—The test bar, 2 ins. × 1 in. section, and 14 ins. long, placed edgewise on supports 12 ins. apart, must sustain a load in the centre of not less than 90 cwts. before fracture. Calculating this out, the skin stress is :—

$$f = \frac{4{\cdot}5 \times 12 \times 6}{4 \times 1 \times 4} = 20{\cdot}3 \text{ tons.}$$

The actual weights supported would, on sound bars, be between 95 and 105 cwts. for material to this specification, which would give skin stresses of 21·4 and 23·7 tons. The maximum deflection tolerated is only $\frac{1}{16}$ in. so little argument can be made regarding plasticity. The tensile stress obtained from such metal on testing would be found to be about 11 tons per sq. inch.

In both these cases, of the 4 in. × ½ in. spring steel, and the 2 in. × 1 in. cast iron, if the modulus of section for rupture be taken as $BT^2 \div 3$ as suggested, the skin stresses at fracture will be found to be 75 tons and 11 tons respectively, which are the same as the material strength as found by actual tensile testing.

Professor Dalby (Paper on " Laminated Bearing Springs," Institution of Civil Engineers, London, 1911) after calculating out the skin stress due to the application of the B.S. test for Spring Steel, Chapter XII. (80 times thickness) and finding it to be equivalent to 82 tons, on material which he has found by tensile tests averages 69·9 tons ultimate stress, and 60·5 tons yield point, thinks that the recommended camber test is too severe, and makes the following statement :—

"This calculated stress is based upon an imperfect theory of solid beams, and it is well known from experimental evidence that the actual stress caused by bending a beam of rectangular section is less than the calculated stress. But allowing for this, the bending test prescribed (B.S.S. No. 6—b, 80 times thickness) would produce a stress in the laminæ near the surface of the plate considerably above 60 tons per square inch if the material remained elastic up to a higher limit. What probably takes place is that as soon as the stress reaches the region of the elastic limit, the laminae near the surface yield, and consequently near the surface of the plate the stress does not increase beyond the elastic limit but remains approximately constant in value, the area of the cross section over which this condition of constant stress prevails extending gradually towards the neutral axis of the section as the bending proceeds. In this way, a complete re-distribution of stress over the section is produced, a stress partly elastic and partly non-elastic, so that the ordinary theory of beams no longer applies. Possibly the more accurate way to state the effect of the prescribed test on the plate would be to say that it produces the maximum stress at the surface laminæ corresponding with the elastic limit of the material, but the area over which this stress is constant is not known."

It will be noted that the trend of the first part of these very pertinent remarks of Professor Dalby's, is to indicate a possible modification in the resisting power of the solid beam or spring plate under consideration ; or otherwise, to render necessary a revised section modulus (Z). This is a considerable step towards the solution of the problem, but the author cannot agree with the suggestion that such modulus would alter according to the high stressing of the plate— as if this were the case, the deflection would be greatest for the first unit load, and very appreciably smaller for the unit load which was causing the steel of any spring to approach

M

its elastic limit, owing to the increasing resistance. Furthermore, if it is admitted that the section resistance is only increased by stressing up to the elastic limit, the result of a second loading of a new spring would be very different to the first. The general reasoning regarding this modification of resistance during stressing is probably best disproved by the fact that " soft " (as-rolled) plates, which are not uncommonly taken straight from the stock bank to essay deflection results, give the same unit deflections as cut off and put under the machine, as the made-up spring which has been hardened and tempered, and fully scragged (tested) many times up to the B.S. test, of $L^2 \div 900$ T.

The author is in agreement with the idea that appears to be put forward in the last sentence of the remarks of Professor Dalby, namely, that the elastic limit is automatically maintained on the outer surfaces, and a graph is given at a later stage, which it is believed, indicates the true elastic limit of the material by the study of the " set " of a camber test piece.

This most able Paper concludes by suggesting the limitation of (f) to 55 tons " that is within 5 tons of the limit of elasticity," and states that with (E) suggested as 13200 tons, the test deflection should be

$$D = \frac{L^2}{960T}. \qquad \ldots \qquad \ldots \text{(XVIII.—1)}$$

Professor Henry Adams, in his book *Building Construction*, makes the following remarks :—

" A misapprehension of the relationship between the maximum fibre stress and the modulus of rupture is very general. The common equation for a rectangular beam is :—Bending Moment $= C\frac{BT^2}{6}$, where (C) is the co-efficient of transverse strength, which is generally supposed to be the same as the maximum fibre strength, but is not so. This is a very important matter, and it is well to know what other authorities say about it.

" (1) Humber's *Handy Book for the Calculation of Strains in Girders, and Similar Structures*, says :—' Modulus of Rupture.—The theoretical value of (C) is the resistance of the material to direct compression or tension, but it is found from experiments on cross breaking that this value is not sufficiently high. Amongst the reasons that have been assigned for this are (1) That in addition to the resistances of the particles of the beam to a direct strain, there is another resistance arising from the lateral adhesion of the fibres to each other, termed the "resistance of flexure" (see Barlow on the *Strength of Materials*, 6th Edition); and (2) that in most metallic beams (especially when cast), the outer skin, which is strained more than any other part of the section, is very much stronger (from many well-known causes) than the average section; whereas if the direct tensile or compressive resistance of the same beam, in the direction of its length, were being experimentally ascertained, it would be the average section at least, and perhaps the centre (weaker) portion especially, from which the strength would be determined. However, there is evidently a necessity to employ a higher value than that for the direct resistance.'

"(2) Professor Fidler's *Bridge Construction* says, with regard to the common theory:—'The simplicity of this theory would be very satisfactory if it could be regarded as a true and complete statement of the facts; for nothing could be easier than to calculate by the formula the weight required to produce any given tensile stress; and if we know the ultimate tensile strength of the material, it would seem that we ought to be able, by this means, to find exactly the load that will break the beam. But if we take a rectangular beam of cast iron, and put the calculated breaking load upon it, the beam will show no symptoms of tearing at the stretched fibres, and no inclination to yield in any way; and as a matter of fact, it will not break until we have increased the load to about $2\frac{1}{4}$ times the amount calculated.'

"(3) R. H. Cousins says:—'Experiments have shown that the compressive and tensile strength do not possess equal values as factors in determining the transverse load that a beam will bear, and that the influence of the tensile strength predominates.'"

After quoting the above various authorities, Professor Adams concludes by stating "The whole subject is one of considerable difficulty, although from want of sufficient knowledge it is generally considered extremely simple."

Professor Goodman, *Mechanics Applied to Engineering*, suggests a possibility of the modification of the standard modulus, after the elastic limit has been passed. This is debatable, and in the author's opinion is not so, but it is of interest to have this suggestion from such an authority, who points out additionally, that such modified modulus is of greater proportionate value in the case of solid beams than in the case of shaped beams—owing, of course, to there being present more superfluous material capable of being included for additional resistance. It will have been noted that both Professor Goodman and the author agree on the idea of a modification of (Z) after the elastic limit has been passed, the difference of opinion being that the former authority would then modify it from the standard, whereas the author modifies it from an already amended value.

The author's suggestion is that the modulus of section for solid rectangular beams, under bending stress, is $BT^2 \div 4$, or a value 50 per cent. higher than that usually adopted—which means that proportionately higher resistance is presented to the bending action, the maximum surface stresses being backed up by higher internal layer stresses than are derived from the graphic constructions. The most reasonable explanation of these is the assumption that this additional strength is provided by the lateral cohesion of the metal structure tending to resist the deformation caused by flexion, as indicated by Humber's remark (1) previously quoted. With this suggested modification of (Z) an examination will be made of the actually obtained results quoted at length in Chapter XII, dealing with the testing of spring steel to the French Railways' specification. The metric dimensions therein translate into English units as follows :—

bT = 100 mm. × 15 mm.	= 3·94 ins. × 0·59 ins.
L = 1000 mm.	= 39·40 ins.
W = 1980 kg.	= 1·950 tons,
W corrected for section	= 2·06 tons.
D = 83 mm.	= 3·27 ins.

The correction for (W) is made owing to the section used being normal round edged and concave spring steel, and light accordingly to the nominal section. Either the weight or section could have been corrected, as the nominal dimensions will be used in the calculation.

To obtain the fibre stress, maximum, under the central load :—

$$f = \frac{2{\cdot}06 \times 39{\cdot}4 \times 4}{4 \times 3{\cdot}94 \times 0{\cdot}348} = 59 \text{ tons/sq. inch.}$$

The above is with (4) as divisor for (Z) instead of the usual (6). Taking the latter figure the skin stress arrives at 89 tons. According to the specification, adherence to the formulæ given therein should produce a stress of 150 kg. per sq. mm. or 95 tons per sq. inch. These latter have been worked along standard lines, and the discrepancy between 89 and 95 is merely due to the steel being probably on the light side to a small degree. The material employed could not give under any treatment rendering it suitable for springs, an ultimate breaking stress of more than 80 tons. How then can it remain well elastic at 89 tons ? Furthermore, this particular piece took no set until 2180 kg. actual, was applied. Corrected this is 2300 kg. or 2·26 tons. Taking this loading, on orthodox lines, the skin stress arrives at 96 tons, and the piece was still in the elastic stage. With the divisor as (4), the 98 translates into 65, which is reasonable.

There is no object in introducing further figures along these lines—continuously will it be found that with $Z = BT^2 \div 4$ results are obtained which correctly agree with known material, whereas with $Z = BT^2 \div 6$ the results bear no relation to anything concrete, and can only be taken as a series based on a standard, such as " It is known that material of 70 tons ultimate stress and 55 tons elastic limit will endure skin stresses in working or testing of 80 tons. Therefore, material of 90 tons ultimate stress and 75 tons limit, will endure elastic stresses in working or testing of 110 tons," and so on, in proportion.

It might be wondered why the modification of (Z) as suggested, to 50 per cent. above its generally accepted value (for SOLID beams) has not been hitherto seriously noticed. In this connection, however, it must be pointed out that the standard (Z) as $BT^2 \div 6$ is on the side of increased factor of safety, albeit more weight is included. The result is the same whether (Z) with divisor (6) be worked to with a factor of safety of 4, or whether (Z) with divisor (4) be worked to with a factor of safety of 6. At the same time, it appears possible that the extra resistance of solid beams above the usually accepted value has been appreciated on the Continent and in America, in view of the lightness of some of their constructions compared with British designs for similar purposes.

In the conclusion of this aspect, Fig. 68 shows three graphic resisting sections of beams in bending, (1) being the orthodox conception; (2) being after the ideas of Dalby and Goodman, and (3) the suggested revision—there being certain similarities between (2) and (3). Curves are also shown representing the stresses in laminæ at varying positions from the neutral axis.

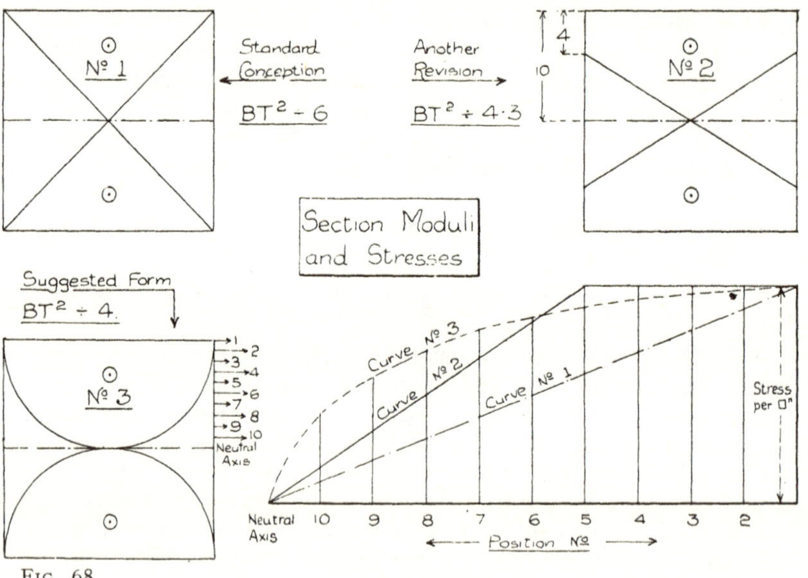

FIG. 68.

It now remains to be seen to what extent this modified (Z) will affect the deduction of the standard deflection formula obtained in Chapter IV. At the stage where (Z) is introduced, as $BT^2 \div 6$, the divisor (6) appears in the numerator (Formula IV.—6) and the resultant unit deflection is shown finally, as the consequence of using (6), as :—

Formula IV.—7 $$d = \frac{0{\cdot}118L^3}{Bt^3}.$$

With the use of (4) instead of (6) the unit deflection would be reduced 50 per cent. For the derivation of the above formula, however, a beam of varying section, uniform stress, was assumed, and herein arrives the second modifying factor in the general scheme, namely, that relative to (E). With the Formula IV.—7 altered by the inclusion of the (Z) factor as (4) instead of (6) the numerator constant becomes 0·079. The point now to be derived is that dealing with the correction of the modulus of elasticity (E) for spring shapes, or approximate rhombus plan beams.

With the (Z) of a beam increased by 50 per cent. as above, the formula unit deflection has decreased 50 per cent. With the constant 0·118, however, the deflection results are known to agree with experiment. In the examination of other formulæ relative to laminated springs—Chapter IX—it was shown that many authors had cheerfully amended the (E) factor in order to bring their formulæ results in accordance with known results, and it was also indicated that such procedure, in the absence of sound explanation, must be regarded as an illegitimate means of obtaining corrections, as it gives to such expressions an air of "first-principles" in order to attract confidence. An attempt will now be made, however, to indicate a legitimate modification of (E) for rhombus beams.

In Fig. 69 are shown two shapes, one (A) of ordinary test form and the other (B) the rhombus (uniform thickness). The elastic properties of steel in bending are parallel to those of steel in tension or compression, and it will be assumed that each of these shapes is subjected to tensile stress of sufficient intensity to double its length, the piece remaining elastic meanwhile (the usual hypothesis for modulus of elasticity). A measure of (E) is thereby obtained. The sectional unit dimensions at the middle of the length are, of course, equivalent. In the case of piece (A) the centres of resistance before pulling are distance apart equal to (L/2). After pulling to double length they are apart a distance (L). If the original section had been one square inch, the necessary pull would have been 13000 tons. Now regard the piece (B). The original centres of resistance are apart a distance equal to (L/3). As stretched to double length they are (2L/3). All the material of this rhombus-shaped piece has been elongated to twice its original length, but owing to the shape, the stress needed is less because of the smaller movement of the resisting centres. (It is obvious that such a piece could not be so stretched even to a small degree, as the ends would tear off before the middle was much influenced, and the case is as hypothetical as that of the rectangle shape A. The same result would have been arrived at by assuming compression instead of tension.) With both pieces made from similar material, and with the same central section, taken as one square inch, the stress required will therefore be in proportion to the travel of the centres of resistance, or :—

13000 tons has moved the centres of A a distance L/2
∴ 8700 ,, ,, ,, B ,, L/3

This gives, therefore, an (E) for rhombus shapes of $\frac{2}{3}$ the normal (E), or 8700 tons. For those who prefer working in

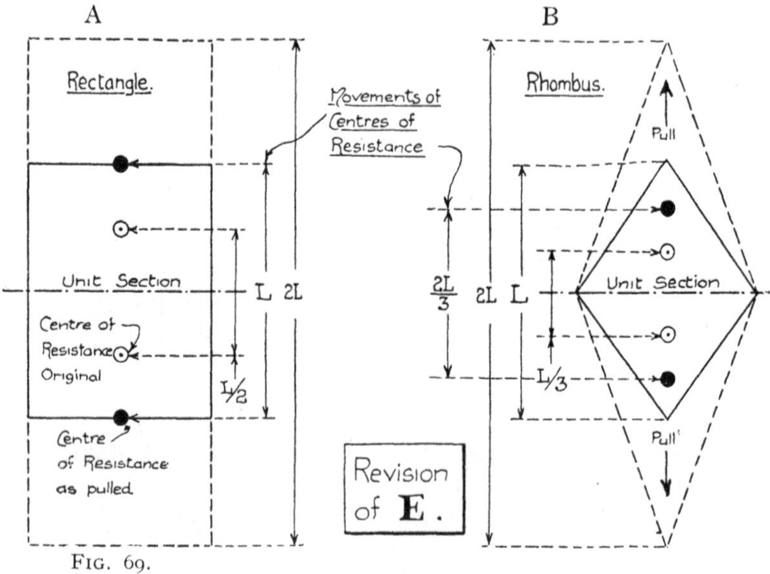

FIG. 69.

pounds, it becomes $\frac{2}{3}$ of 30,000,000 lbs. = 20,000,000 lbs. With this new value in the Formula IV.—6, matters correct themselves, and No. IV.—7 stands as before with constant 0·118.

An alternative method of calculation can be taken off a standard formula, of the ration E : *f* : : L : *i*. With two beams of the same central section and length, and deflected to the same degree, one being of rhombus, and one of rectangle plan, it is known that the weight (W) required for the given deflection will be 50 per cent. greater for the rectangle than for the rhombus. The stress (f) will accordingly be 50 per cent. greater, and to maintain the ratio E : f, the (E) will be proportionately less, thus :—E(1) : f(1) : : E(2) : f(2). If [f(1)] be taken as the rectangle beam, then :—

13000 : 1·50 : : X : 1·00 or 13000 : 1·50 : : 8700 : 1·00

The above method of reasoning, it is claimed, rationalise the use of an amended modulus of elasticity (E).

The details of the amendments—for rhombus shapes—are as follows :—

Formula IV.—6a :— $d = \frac{16^2 \times 6L^3}{13000Bt^3} = \frac{0{\cdot}118L^3}{Bt^3}$.

As amended :— $d = \frac{16^2 \times 4L^3}{8700Bt^3} = \frac{0{\cdot}118L^3}{Bt^3}$

The above is the key formula of all, as it is based on a uniform fibre stretching. The rectangle plan beam is modified from this on the basis of the known curve to which such non-uniformly stressed beam arrives on loading. Alternatively, either can be arrived at by the " stress diagram " methods of Figs. 20 and 21. Alternatively the formula for the rectangle plan beam can be derived by introducing the full (E) of the shape into the denominator. Under such conditions, (E) for the constant (K) of 0·10 which has been taken as a practical (K) for normal springs, will be on the proportional basis as under :—

(E) for rectangle plans	..	13000 tons, K = 0·083
(E) for perfect rhombus plans		8700 tons, K = 0·124
(E) for practical spring plans		10800 tons, K = 0·100

It must always be remembered, particularly for those who desire meticulous accuracy as against practical accuracy, that the value of (E) is rather indeterminate. It is generally agreed to be between 28,000,000 and 32,000,000 lbs. per sq. inch, and for memorising and calculating purposes, is best retained as 30,000,000 lbs. (13,400 tons) or 13000 tons (29,100,000 lbs.).

A reversion will now be made to the R.C.H. buffing spring referred to in the earlier part of the chapter. According to ordinary calculations, it was shown as having a skin stress under test load of 73 tons, calculated on the weight carried, whereas according to B.S. test, based on fibre extension, this should be 58 tons. With reference, first, to the stress due to load carrying, with the (Z) divisor as (4) instead of (6) the skin stress reduces to 46·5 tons. Secondly, take out the stress according to fibre extension, on the basis of Formula IV.—4, which was arrived at before the introduction of any question of (Z).

Formula IV.—4 :

$$f = \frac{E(4DT)}{L^2} = \frac{10800 \times 4 \times 10{\cdot}8 \times 1}{70 \times 70 \times 2} = 47{\cdot}6.$$

The two results, 46·5 tons per bending moment calculations, and 47·6 tons per fibre stress calculations, would appear to be in such close agreement as to generally confirm the foregoing hypotheses dealing with the amendments of (Z) for all solid beams, and of (E) for varying section beams.

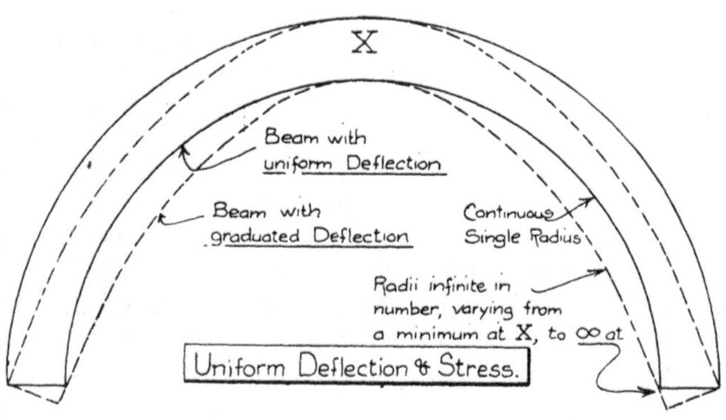

FIG. 70.

It will probably be a matter of surprise that whereas the B.S. test on the single spring plate has been given as equal to 58 tons, when applied to the completed spring, it reduces to about 47 tons. This is in one sense, due to the altered aspects of surface stressing on centrally loaded beams of the rectangle and rhombus plans. For an equal deflection, the maximum stress on the former type is higher than that of the latter type, as will be seen from a glance at Fig. 70, which shows a uniform stress beam, and a non-uniform stress beam, and clearly shows why the latter has the smaller central radius, which is due to the stiff ends reducing deflection along the bulk of the length. The effect of obtaining the same deflection, either by forming the beam to rhombus plan, or distributing load in such a way as to obtain the same result—

of uniform stressing throughout—is to immediately increase the central radius, and reduce thereby the stress. It might, however, be pointed out that by first principle calculations, any solid section beam pressed flat to B.S. test for instance, should have throughout a stress of 58 tons, in accordance with Formula IV.—4. By very accurate measurement, it should be possible to determine the strain over the surfaces, plate by plate, of various designs of springs, and this would assist in the general elucidation of desired facts. With the suggestions given, however, it is evident that calculations of skin stress (Z as $BT^2 \div 4$) will vary according to every design of spring ; as each has its own particular (K), varying, it is true, for normal design, not greatly from the 0.100 standard, but nevertheless, indicating a varying stress for the same deflection. It is contended that for " normal design " springs the B.S.S. test can be taken as equal to 47 tons per square inch, maximum—which maximum is always reached along the plane of greatest bending moment, which, in ordinary semi-elliptic designs, is in the geometrical longitudinal middle. No practical spring can be made to a perfect arc, but very close approximations can be obtained, and it is only by such manufacture that the central stress can be reduced. The further the departure of the design from this perfection the higher the central maximum stress. It is not a question of the fitting, as no matter what amendments are therein made to conserve a tolerable " straight " appearance, the stresses will be just the same under a given load as if the spring were made of uncambered plates, or made to a true arc. In Fig. 71 are illustrated three " spring sweeps." The first, if fitted to an arc, will " come down " practically flat. The second, and third, if fitted to an arc, will not be flat when the camber is pushed out, but will be shaped as shown. In manufacture, they will therefore be made with a curve which is not a true arc, and which corrects the imperfect shape to a straight shape. Such manufacture, however, cannot correct the stresses, which are a resultant of the load, length, and modulus of section.

The whole art of spring design is to obtain uniform stressing throughout the spring entity, as non-uniform stressing causes the equivalent of the beam shown in Fig. 70, which has " the action crowded into the centre," similarly to the second and third springs of Fig. 71. The more uniform the general stress, the less pronounced is the maximum central stress. The design, therefore, becomes a question of " proper distribution" —which expression is even a shop term—the spring fitters themselves seeing at a glance from the plates they collect, whether or no the load is properly distributed, and gauging the amount of fitting they will have to perform in accordance therewith.

In view of the foregoing difficulties in determining maximum skin stress, it will be appreciated that not the least clever feature of the B.S. test deflection formula is the fact that it carefully misses this very debatable point, and adheres to a deflection, without reference to any possible or probable " tons per square inch " obtained by the application of the pressure necessary for the deflection.

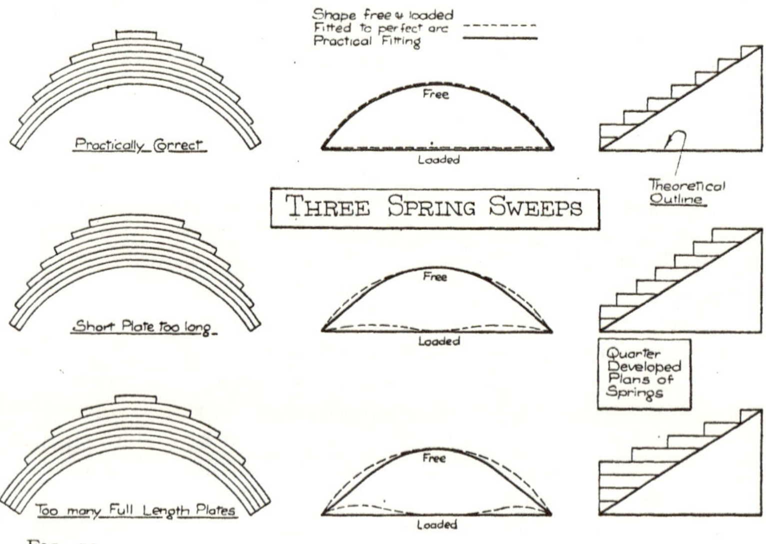

FIG. 71.

With the foregoing arguments in mind, it should be now relatively easy to understand the reason why a spring of the R.C.H. 14-plate buffing type, stands up in service. Its maximum working stress, on a no-nip spring, based on the arguments which have been raised, is 43 tons (the working stress being less than the test stress, working deflection being 10 ins., and test deflection 11 ins.). With 4 tons extra (a low figure) for the short plates of a nip-fitted spring, the figure arrives at 47 tons. Nominal allowance for the $\frac{1}{2}$ in. hole should be 16 per cent. of the area, but allowance for maximum stress at the edges of the hole increases this to 33 per cent. The stress addition caused thereby brings the absolute maximum to 71 tons, which is somewhat above the elastic limit of the short plates (generally left somewhat harder than the longer plates). However, so far as these are concerned, the situation is saved by the hoop bearing causing the hole to be " dead," as the buckle will be flat on the spring for about 1 in. at this point. Fig. 72 shows probable stresses in this design.

The conclusions arrived at with the whole of the foregoing arguments amount to :—

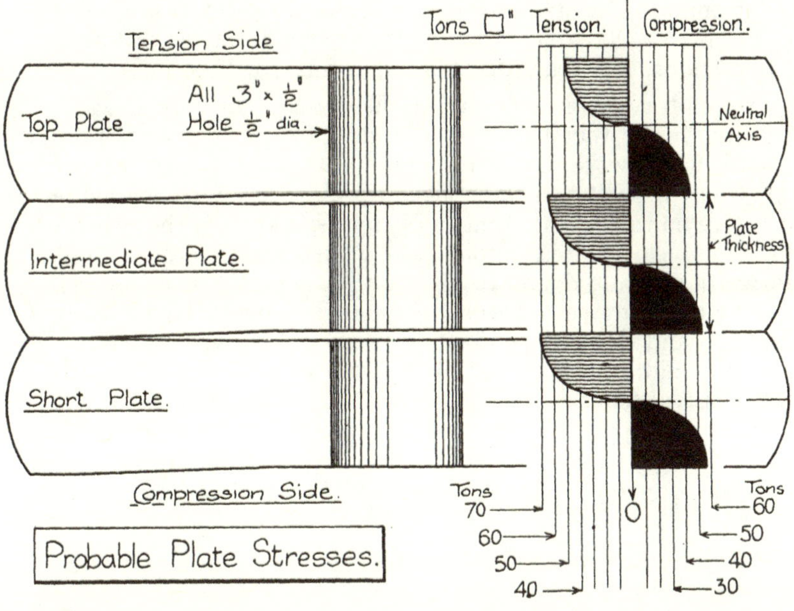

FIG. 72.

(1) That the modulus of section (Z) of a solid rectangular beam, during the elastic stage of deflection, is approximately $BT^2 \div 4$.

(2) That the modulus of section (Z) of a solid rectangular beam, at the instant of rupture, is $BT^2 \div 3$.

(3) That the modulus of elasticity (E) of solid beams having a non-uniform section between the point of loading and the point of encastrement, or points of support, is

proportional to the total movement of the centres of resistance of such beams, when the complete beam is elongated to double its original length, or compressed to one-half its original length—the movement of corresponding centres of resistance of a piece of uniform section to double its original length (or one-half its original length) whilst in the elastic stage, requiring a stress of 13000 tons per square inch, in accordance with well-recognised standards.

(4) That the modulus of elasticity (E) of solid beams of non-uniform section as described above, is in inverse proportion to the constant (K) employed in the unit deflection formula—the standard of the uniform section beam being (E) as 13000 tons, with (K) as 0·079 for true rectangle sections, and 0·083 for normal spring steel sections.

An endeavour has been seriously made in this chapter to illuminate the apparent paradox of springs remaining elastic when working up to (nominally) 80 and 90 tons per square inch, whilst the steel from which they have been made is under 60 tons elastic limit. So far as is known, the subject has not been treated before, and whilst the hypotheses put forward are not dogmatized upon as correct, nevertheless, the figures obtained thereby seem to be confirmed by their mutual and inter-dependent agreement. Considerable research could still, however, be applied to this problem, with much advantage.

CHAPTER XIX

STRESS AND BENDING MOMENT DIAGRAMS

In previous chapters, stress diagrams have been frequently referred to, and it may be of interest to follow out the working of one relative to some well-known standard spring. The example selected, and illustrated in Fig. 73, is the 5-plate wagon bearing spring of the R.C.H., of which many scores of thousands have been made for the "private owners" 10 and 12-ton wagons.

To plot the diagram, commence to any convenient scale, with the lines AB (base) representing the spring length, and BC (ordinate) representing a stress, which may be any stress —either that due to working load, or test load, or intermediates, as under all conditions of deflection, the stresses throughout the plates remain proportional. The height BC should be divided into 5 parts—representing the number of plates, and if the corresponding lengths then be set off from BC, a concentrated plan of the spring is shown. Join points A and C, forming thus the triangle ABC, which is the stress diagram for a spring of all full-length plates, that is, of 5 plates (in this case) 44 ins. long—otherwise, of a beam of rectangle plan, in which the stress decreases uniformly from a maximum (BC) to zero, at A. The spring under consideration is speared, but the spear is small in comparison with the offset (see sketch) and will not here be taken into account.

From the point C, the stress decreases uniformly along line AC, until the end of the short plate is reached at F. Here, owing to the retirement of this short plate, the stress increases to J, and is represented by PJ. The method of obtaining this ordinate height is merely one of simple proportion. For instance, if the spring had remained 20 ins.

Construction for Stress Diagram.
Span 42"
'Straight' Length between Bearing Points is 44"
SPRING
5 Plates, 4" × 5/8"
5/8" dia hole
Centre of Spring
Offset 6"
1½"
2½"
PLAN of SPEAR.
Spring Outline (Fine Lines)
12"
No. 5
No. 4
No. 3
No 2
Back
8"
Back = 44"
2nd = 44"
3rd = 36"
4th. = 24"
5th = 12"
Plate Lengths
STRESS DIAGRAM (Thick Lines)
Stress through Centre of Spring
Final Plane of Centre of Resistance
Z
A
B
C
D
E
F
G
H
J
K
L
M
N
P
Q
R
S
T
10
6
15
27
29·4
37
Area = × Centre of Resistance
ABC 550 × (29·4 + 10) = 21600
AGD 28 × (6 + 10) = 450
KDEH 54 × (15 + 10) = 1350
LEJF 40 × (27 + 10) = 1480
672 24880
24880/672 = 37 from Z
or 27 from A
Stress Ranges
Maximum Stress
1·2
3
4
5
Stress from end to end of each Plate if the Spring was a solid Flat Beam.

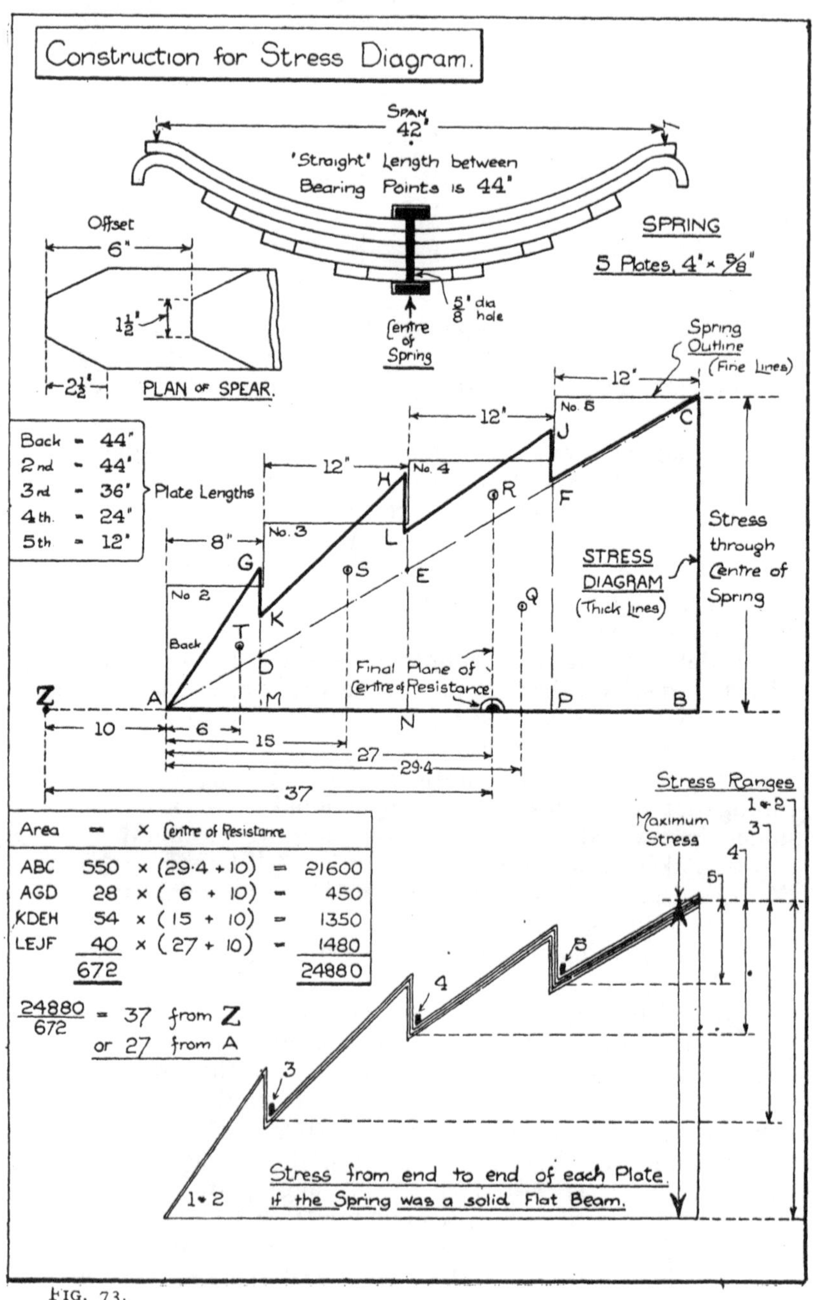

FIG. 73.

(or 5 plates) wide, the ordinate would be PF. It has reduced, however to 16 ins. (or 4 plates) wide, and the stress is therefore PF × 5/4. Alternatively, had the spring been of width PF (the true rhombus outline) the stress would have remained as BC, which also represents the full width. Then BC (or 5 plates) : 4/5 BC (or 4 plates) : : PJ : PF. Similar calculation is needed at every step, where two stresses appear, one at the absolute end of the plate retiring (the lower stress) and the other immediately above this on the next longer plate (the higher stress). The full diagram is as shown, and the ordinates BC — PF — PJ — NL — NH — MK — MG represent respectively stresses at vital points along the spring beam.

The resistance of this diagram is the total area multiplied by the distance of the centre of resistance from A. The area can be arrived at by planimeter, but without this instrument, it can be considered as so many separate areas, the centres of each found, and then the composite centre discovered. In this case, areas have been taken as shown on the drawing, which also indicates the position of the centres of resistance, and the total moment from a point Z, assumed 10 units removed. The result of these moments shows that the resisting centre of the whole is 27 units from the point of support A, and the area is 672 (square) units. Total resistance is therefore 18150. Had the spring beam been of full length plates, the area ABC × distance AQ would have been the total resistance ; which equals 16150. With such a spring beam, (K) = 0·083, therefore the following ratio presents itself :—18150 : 16150 : : x : 0·083, giving (x), or the (K) for this particular spring as 0·0935. The deflection per ton with this constant calculates at 0·397 ins., and in practice, it is between 0·37 ins. and 0·40 ins., according to the thickness of steel, and variations of bearing points.

If the spring were a solid beam, the commencing and finishing stresses at the points corresponding to the retirement of plates, would be as shown on the bottom diagram of Fig. 73. Actually, however, as the spring beam is made up of independent superimposed pieces, the stresses along each separate plate cannot be derived direct from the stress diagram, and recourse must be had to a study of the bending moment and distribution diagrams, to be now shown.

A completed spring represents a series of beams and cantilevers, and if it is fitted without nip, and perfectly designed, when under load the stresses are uniform throughout the whole of the plates. It has been sufficiently shown, however, that springs cannot be theoretically perfectly designed owing to the intrusion of necessary practical features, and therefore the happy result of absolute uniform stressing cannot be attained. It may be of interest, however, to study certain springs which are tolerably near perfection, as compared

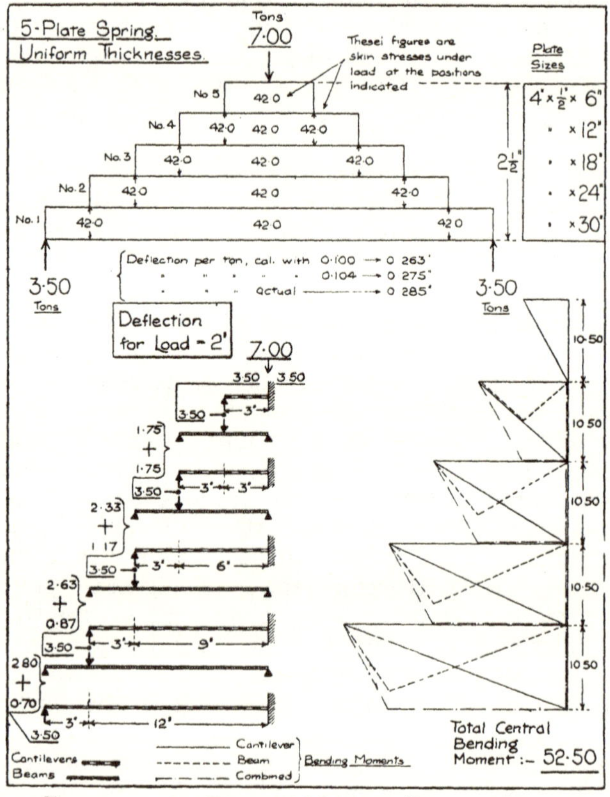

FIG. 74. BENDING MOMENT, DIAGRAM No. 1.

with similar springs which are obviously not so. All the plates of the actual springs under consideration were square cut.

Fig. 74 shows a 5-plate spring, short plate length L/5, and it will be noted that the calculated deflection closely agrees with the actual deflection. All plates are of uniform thickness, offsets all 3 ins., and the spring is fitted without nip.

For the calculations, the whole spring has been treated as a beam of width (bn) and the central bending moment thereby determined—which totals 52·50 or, 10·50 inch-tons per plate width. Working from the back plate, the total re-action (in this case 3·50 tons per end) does not go beyond the first "point"—that is, the end of the superimposed plate—where it commences to distribute. The full load has only

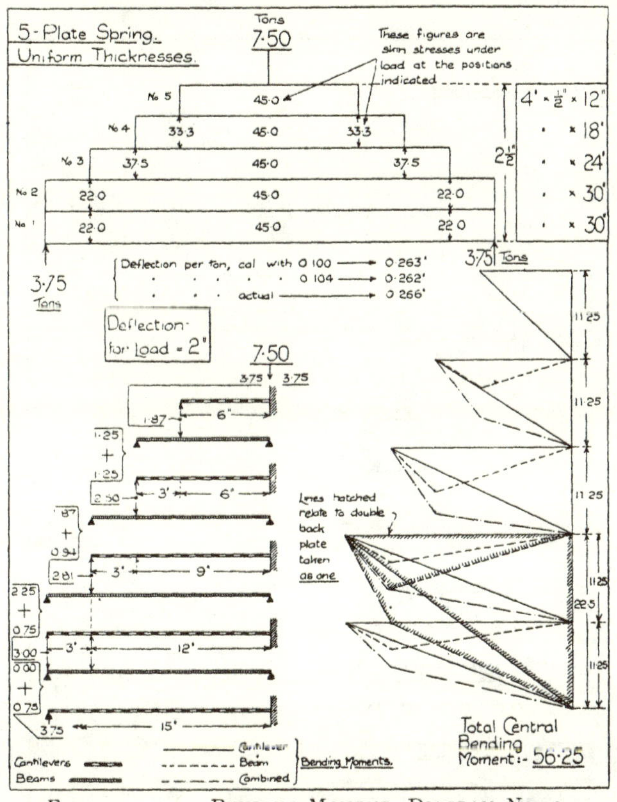

FIG. 75. BENDING MOMENT, DIAGRAM No. 2.

therefore to be dealt with, so far as this plate is concerned, over the distance or offset of 3 ins. The load on the bearing point, 3·50 tons on a leverage of 3 ins., gives a bending moment in the back plate, under the end of the second plate, of 10·50 inch-tons—the plate end acting as a cantilever. The reaction due to the plate itself, taken separately, is 0·70 tons (load of 7 tons ÷ 5 plates ÷ 2 as only one end is being dealt with) which, acting at 15 ins., half the spring length, gives the central bending moment of 10·50 inch-tons. Each plate is dealt with in the same way, and the diagram illustrates the matter better than any description. The bending moments for each plate, taken separately, with the plate considered consecutively as a beam, and as a cantilever, when combined, give the full bending moment diagram on the basis of each plate sharing 1/5 of the total.

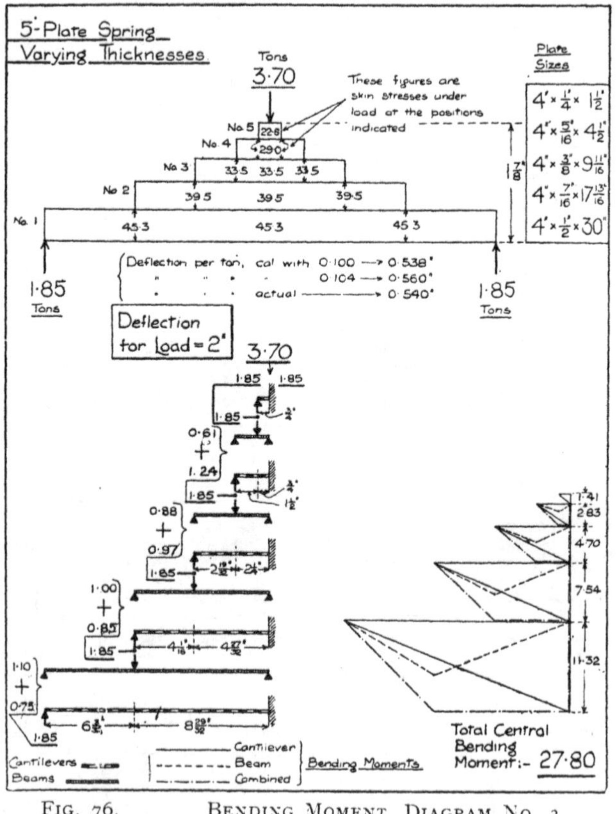

FIG. 76. BENDING MOMENT, DIAGRAM NO. 3.

In Fig. 75 is shown a similar spring, differing in having two top plates, and a short plate twice the length of the perfect length. In this case, the two top plates can be conceived as one plate of double width, and treated accordingly, as indicated on the diagram, where the moments are obtained for the wide plate, and divided for each separate plate. The combined moments for intermediate plates indicate the non-uniform stress. It is interesting to note, however, that if the offsets, counting from the spring middle, with half the short plate length as equal to one offset, are identical, as in Fig. 74. uniform stressing is obtained over the intermediate plates up to the "retiring" points of superimposed plates. To achieve this, however, if the short plate is 6 ins. (one-half

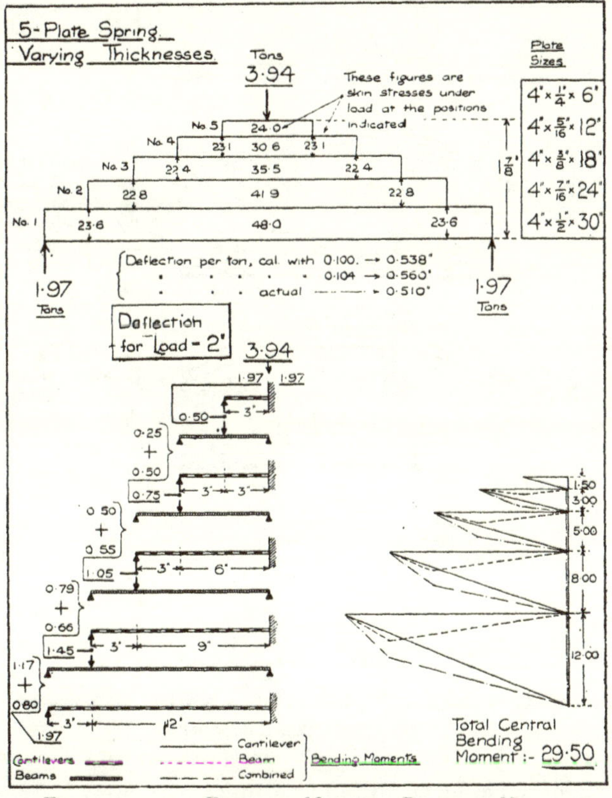

FIG. 77. BENDING MOMENT, DIAGRAM NO. 4.

equals 3 ins.), all offsets must be 3 ins., in other words, each plate must be a multiple of the short plate.

Fig. 76 illustrates a spring made of plates of non-uniform thickness, each plate having a different thickness. In this case, the plate lengths have been calculated out on the basis of the diagram in Fig. 53, with the result that the short plate is 1½ ins. long, and uniform stressing throughout results.

A similar spring to this is shown in Fig. 77, with the difference that the plates have uniform offsets of 3 ins. Consequently more material than required is present, and the bending moment diagrams clearly indicate this.

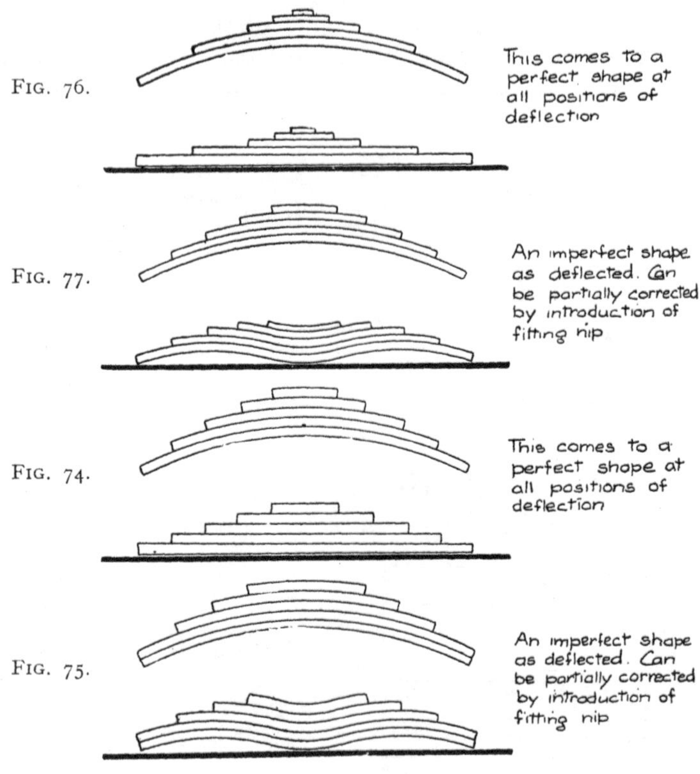

FIG. 78.—DEFLECTED SHAPES.

The effect when large and indeterminate amounts of "nip" are introduced can be imagined, as combined diagrams would be needed to indicate the un-loaded spring, and these would then have to be re-combined with the loading diagrams shown.

In this last spring, it will be noted that the deflection obtained is less per unit weight than calculated, which is owing to what might be called, the proportion of "top plates." The total spring strength as (t^3), is 1260; of which the ½ in. back plate accounts for 512, or 40 per cent. With this proportion, the constant was suggested as 0·093, (Chapter VII) which corrects to very nearly the actual loading figure.

These four springs give characteristic shapes when tested straight, as indicated in Fig. 78. Clearly, the two patterns illustrated in Figs. 75 and 77, are not admissible as shown, particularly if the loaded working is approaching straight, and to correct the shapes, the fitter employs the joint means of setting to a " sweep " instead of an arc, and introducing nip as required—according to his ideas.

In Fig. 79 is included a bending moment diagram similar to the foregoing, but relative to the 5-plate wagon spring illustrated in Fig. 73. Successive plates of the spring are multiples of 12 ins. the short plate length, and as a result, the stresses are uniform along these until the plate above ceases. Diagram 79—C shows the complete bending moment diagram for the centrally loaded spring beam, and if the ordinates representing combined bending moments on the separate plates be transferred from 79—B to 79—C, it will be found that they add up to the full ordinates at corresponding points of 79—C.

To obtain the stresses along each separate plate, it is now necessary to study the stress and bending moment diagrams together. Fig. 80 includes the stress diagram off Fig. 73 with ordinate AB representing the maximum unit stress of the whole spring beam, in other words, the unit stress at the centre under the load. Short plate No. 5 acts as a cantilever only, and as it is full width, the unit stress in this ranges from the maximum to zero, along line AC. Plate No. 4 is uniformly stressed to maximum until No. 5 leaves it, wherefore JD shows the stress drop, as the plate end is assumed square cut for the time being. Plate No. 3 behaves similarly, and line HE is the dropping range. The two top plates, being of the same length, with a unit stress represented by ordinate EG, obviously remain as the stress diagram, and as the bending moment diagram showed a uniform drop from maximum to this point, line AG is a representation thereof. Taken away from the left-hand side, the distribution of unit stress throughout each plate is as shown on the right-hand side. These outlines, however, only generally parallel the actual spring, as certain considerations, small in themselves, but of vital importance in practice, occur to modify the purely

5-Plate R.C.H. Wagon Bending Moments.
A Distribution Diagram
B Bending Moment on each Plate, from Distribution Diagram
C Complete Diagram built up from Distribution Diagram, B
Maximum Bending Moment on Spring
Cantilever
Beam
Combined
Test D = 3·4″
Test W = 8·5″
d per ton = 0·4″
with K as 0·094″
8·50
4·25
3·11
1·55 + 1·56
2·08 + 1·03
2·55 + 0·85
3·40 + 0·85
12″
24″
36″
44″
6″
18″
4″

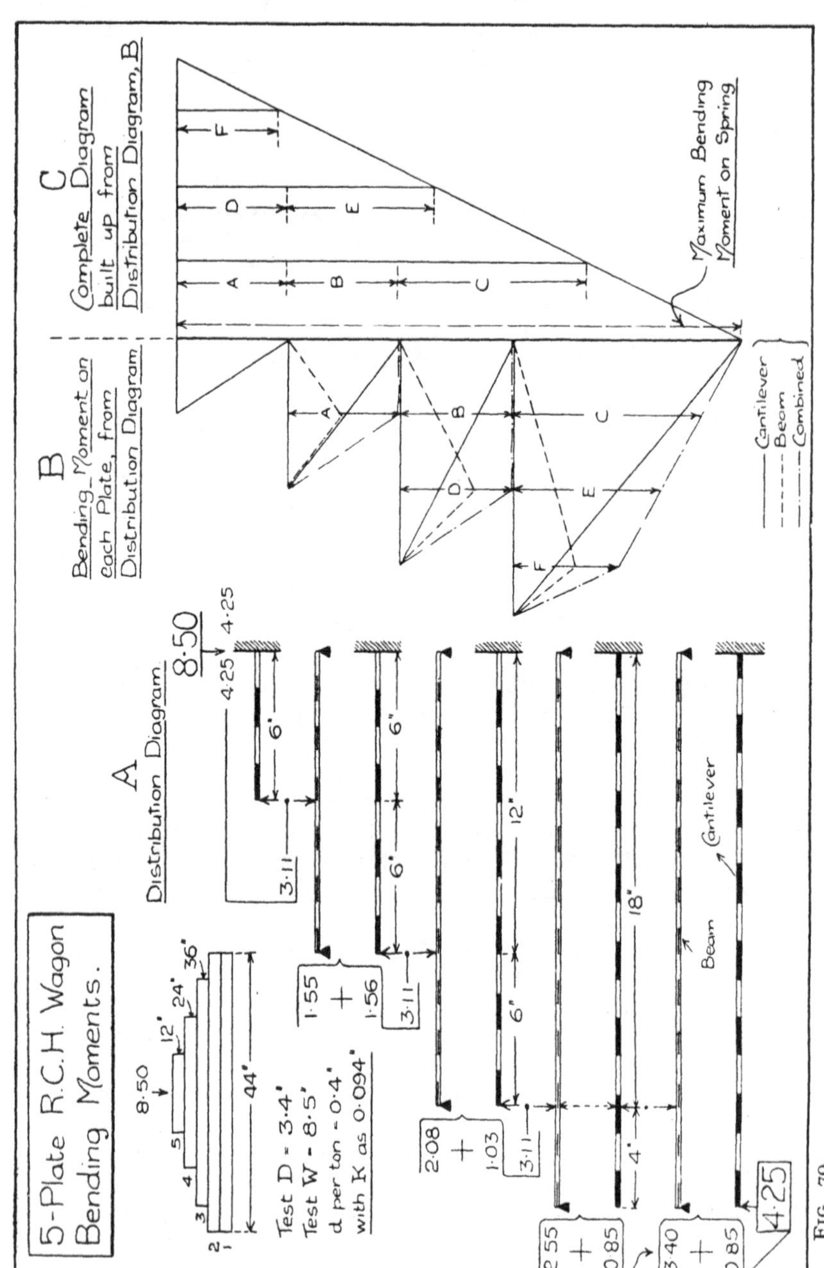

FIG. 79.

5-Plate R C.H Wagon.
Stresses on Plates. Plain Spring
Stress per Unit Section along the length of each plate, is represented by outlines as follows:-
Plate No 1 — ABGF
" 2 — ABGF
" 3 — ABEH
" 4 — ABDJ
" 5 — ABC
AB - Maximum Stress
Heavy Line is outline of Stress Diagram (per unit section) of Complete Spring Beam
Maximum Stress
Zero

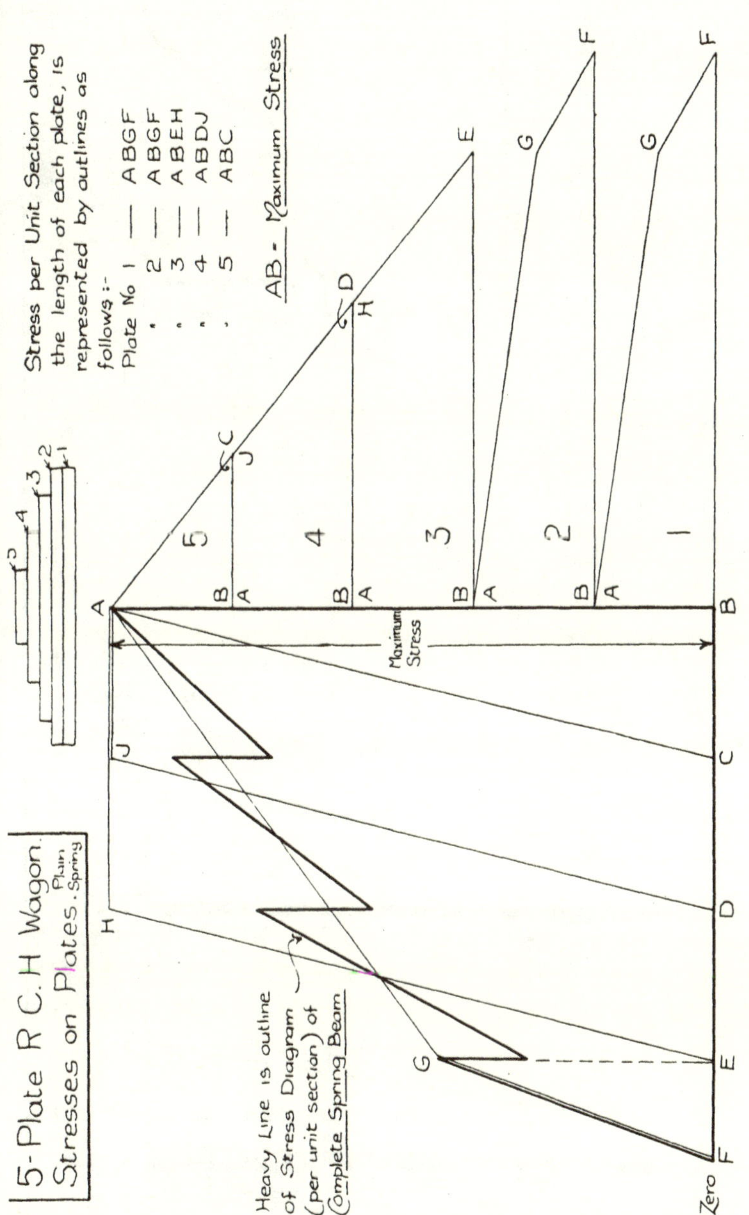

FIG. 80.

5-Plate R.C.H. Wagon Spring Stresses on Actual Plates.
PLAN
Centre Holes
Studs
Slits
Black Area is Positive Stress
Hole
Plus
Nº 5
Nº 4
Black Area is Positive Stress
Slit
Stud
Centre Hole
Modification of YY due to Spear
Nº 3
Z - Test Range of Stress.
Outlines thus represent initial stress in each plate, when the spring is held by the hoop, and is carrying no load. This has always to be added to, or deducted from, ordinate S.
Black Area is Negative Stress
Nº 2
Minus
Black Area is Negative Stress
Nº 1
Minus

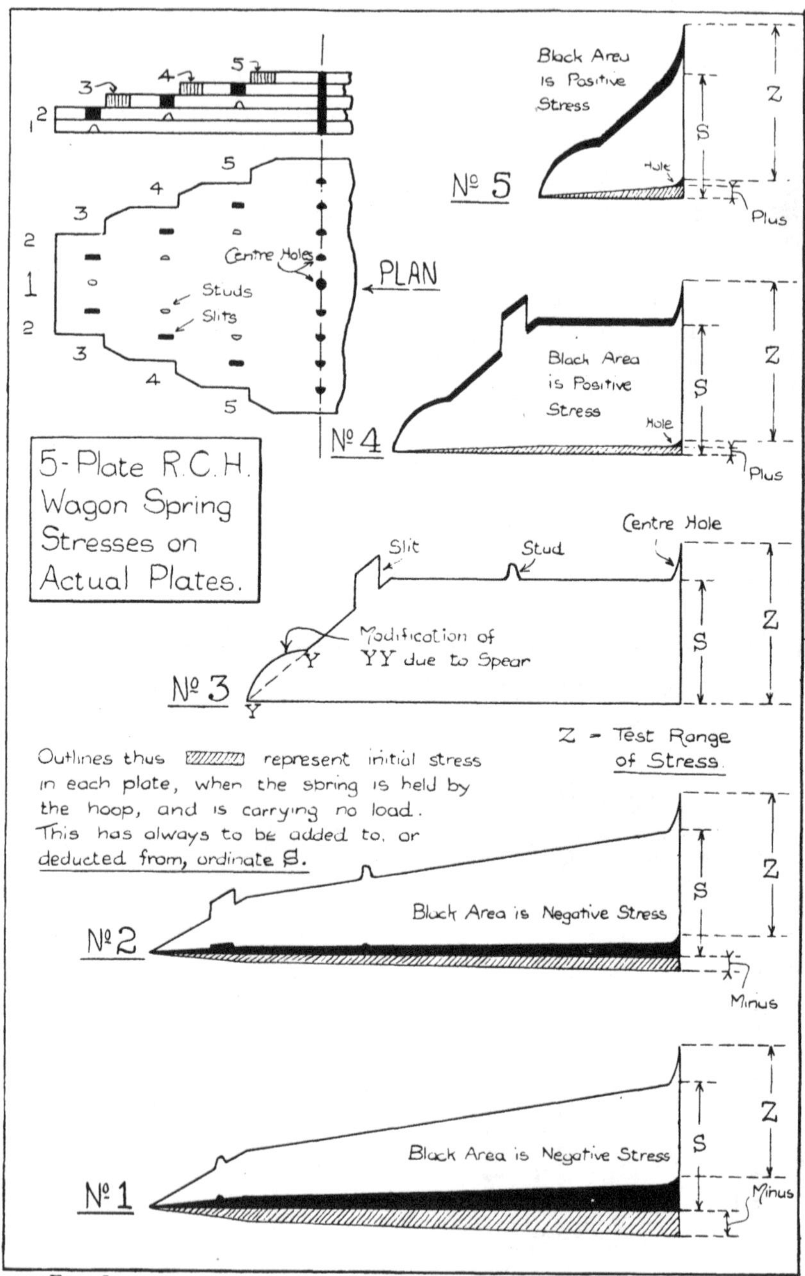

FIG. 81.

theoretical aspect. These include the more or less unknown and individual aspects of "nip," the provision of centre holes for hoop attachment, and the provision of studs and slits to check lateral play. Additionally, the specified spearing alters slightly the unit stressing at the plate ends. Fig. 81 includes views of the spring as actually made, with the positions of spears, centre-holes, studs, and slits. The unit stress diagrams for each separate plate then have to be amended in accordance therewith, and also should include the allowance for an average "nip." On Fig. 81 therefore, are shown such diagrams. A survey of these shows that, as is known, the highest working or test range, and also the highest stress from the basis line, occurs with the short plate, at the position of the centre hole. At all these holes are necessarily planes of high stress, but the short plate hole is assisted to a small degree by the hoop, which, if tight, will cause it to remain outside a working plane. This feature probably balances the extra stress due to nip. The next danger points are clearly the slits in Plates Nos. 3 and 4, and the stud in Plate No. 3; and, as is well-known, frequent failures occur across these planes, particularly across the slit planes. The highest stresses in the top plates are across the centre hole, and as these holes, in this pattern of spring, are always within the working area, failures can be expected here. Such are, however, partially obviated by the top plates being somewhat softer than the following plates, and tending therefore to "set" some little before breakage.

Similar diagrams can be produced for all types of springs, and will probably assist in the explanation of many failures. If, however, studs and slits—when required—are kept at the extreme ends of plates, and centre holes are replaced where possible, by downward nibs;—it will be found that fewer failures require explanation, and it can be appreciated that designing along such lines, and obtaining thereby a reasonably good spring pattern, is a very much quicker and more satisfactory process than evolving and co-relating stress and bending diagrams to discover reasons for broken plates.

CHAPTER XX

RESILIENCY, AND SPRING WEIGHTS

THE store of energy in a spring is called its "resilience." Energy is the capability of doing "work," or absorbing "work," which in its turn, represents the product of the weight or force exerted or absorbed and the distance through which such force travels. An important difference presents itself between a weight, falling under gravity control, and a spring, inasmuch as the "weight" of the weight is constant, whereas the "weight" or resistance of the spring varies from zero to a maximum. For instance, a weight of 1000 lbs. moving through a distance of 10 ins., represents a power of doing work of 10,000 inch-pounds; whereas a spring, which has required a weight of 1000 lbs. to force it through a deflection of 10 ins. has a "work" capacity of only one-half this, or $1000 \times 10 \div 2$, as the commencing pressure is zero, which rises to the finishing pressure of 1000, making the average force exerted as 500 lbs. over 10 ins. or 5,000 inch-pounds. This average only holds good, of course, in the case of springs of the laminated, or "cylindrical helical" pattern, where the load-deflection curve is a straight line. The diagram Fig. 82 shows what is meant in this direction, The resilience therefore of a laminated spring is represented by the formula :—

Resilience, $$r = \frac{W \times D}{2} \quad \ldots \quad \ldots \quad (XX.—1)$$

The measure of efficiency of any spring design is its resilience, but this must be referred to some definite unit of the spring. Professor Perry, *Applied Mechanics,* states :—" The average resilience per cubic inch of a spring is the whole resilience divided by the volume." This definition it will be observed, takes a unit volume as the measure of resilience, but this is not a practical unit, as no one regards as such the number of cubic inches in a given spring. A better factor is the co-related unit of weight of the spring, and this then gives the formula in the following :—

$$r = \frac{W \times D}{2w} \quad .. \quad .. \quad .. \quad (XX.—2)$$

Units in inches-pounds or millimetres-kilogrammes as required.

This formula, however, only gives inch-lbs. or mm.-kg. as a result, and there is still nothing to refer to as the final efficiency. Viewing the exterior aspect of things, it would appear that the true theoretical rhombus beam would be

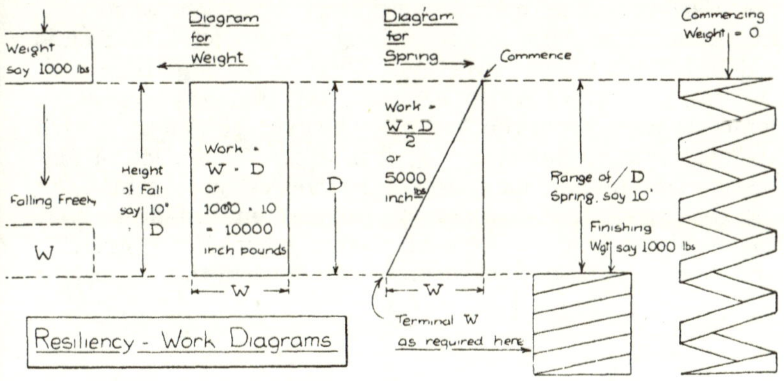

FIG. 82.

that of highest efficiency, and this gives 340 inch-lbs. of " work " per pound weight of spring or beam. A rectangle plan, uniform section beam, will absorb 255 inch-lbs. per pound weight. In each case, the stress imposed is taken as that due to the B.S. deflection test of $L^2 \div 900$ T. Certain known springs, however, such as buffing springs which are generally of considerable length, with numerous plates, will actually give a figure somewhat higher than 340 inch-lbs. as instanced by a particular spring, which absorbs 7·0 tons in a deflection of 11 ins., equivalent to 86000 inch-lbs. The weight is 235 lbs. and the resilience per pound weight is therefore 367. In an endeavour to obtain a " percentage " standard of efficiency, instances of springs being 110 per cent. cannot be tolerated, and in order to obviate this possibility, such as might be discovered in practical working, where test deflections are not gauged to hundredths, thicknesses not taken in thousandths, and loading weights not measured in pounds, an empirical standard of a maximum of 400 lbs. has been taken as representing 100 per cent. The efficiency formula, referred to this standard in its simplest form, is as follows :—

$$\text{Resiliency efficiency, } r \text{ per cent.} = \frac{1000\ WD}{3 \cdot 5w}. \quad \text{(XX.—3)}$$

An example of this might be given, which presents itself as probably the highest efficiency spring in use, namely, the (British) Railway Clearing House approved design for the 14-plate buffing spring required for "private owners" wagons, see Fig. 65. The main requirement of such wagons is minimum prime cost, and accordingly, every detail thereon is reduced down to the lowest possible weight compatible with running efficiency, which accounts for the very creditable figure obtained from the spring. The average weight of such springs (without hoop) is 235 lbs., B.S. test deflection and test weight (giving maximum safe stresses) are 11 ins. and 7·0 tons respectively ($d = 1 \cdot 57$ ins. per ton, which is about the usual obtained). The back plate is 70½ ins. long, straight, and can be reckoned as 70 ins. working length; plates 14 in number, 3 ins. × ½ in. section. Then :—

$$r \text{ per cent.} = \frac{1000 \times 7 \cdot 0 \times 11}{3 \cdot 5 \times 235} = 94.$$

Criticisms could, however, be made of this design, as follows :—(1) The short plate of 12 ins. could be reduced to 8 ins, and (2) the plates might as well have been left square cut, as have been "round-speared" as shown. The weight would remain the same with these amendments, as the saving on the short plate and intermediate lengths would balance off the waste of the spears. However, in general, and as designs go, one must admit this particular instance to be perfection. In the dimensioned drawing, Fig. 65, it is clearly shown how one reduction in weight is obtained—namely, by the lack of full length and full width top plates to take hanger connections. Arrangements like this are possible with buffing springs, and certain other types, but are not practicable with the bulk of plate springs. Consequently, the added weight necessary for such attachments, reduces the resiliency efficiency, and it is not easy to design locomotive springs for instance, with a better factor than 75 per cent. Clearly, plates of varying thicknesses will substantially reduce the percentage, as the permissible range of stress must be based on the thickest, or top plate, which means that the lower plates are working below their potentialities, and the efficiency naturally turns on all portions of the spring being as equally stressed as is practical. Furthermore, all "short plates" of excess length reduce the efficiency, owing to the slight increase in stiffness of the unit deflection being many times counterbalanced in proportion by the extra weight involved in intermediate plates.

A few examples of further actual springs are given herewith, indicating the extent of low factor that can be obtained.

(1) *Locomotive Spring.*

33 ins. straight length, c. to c. 3½ ins. wide. 9 ins. short plate. 1 top plate, ½ in. thick, 13 plates, $\frac{5}{16}$ in. thick.

B.S.S. Test	2·42 ins.
Actual Deflection per ton	0·43 ins.
Actual test weight—tons	5·50
Weight of spring-lbs. (calculated) ..	112
Resiliency efficiency per cent. ..	33

The very low result obtained here, is, of course, entirely due to the use of a ½ in. back plate, with following plates of $\frac{5}{16}$ in., in all respects a thoroughly bad design.

(2) *Locomotive Spring.*

39 ins. straight length, c. to c. 5 ins. wide. 9 ins. short plate. 13 plates, ½ in. thick.

B.S.S. test	3·37 ins.
Actual deflection per ton	0·17 ins.
Actual test weight—tons	20
Weight of spring—lbs. (calculated) ..	260
Resiliency efficiency, per cent. ..	74

The above represents an excellent design, and about the best that can be done with locomotive springs. The difference of efficiency of these two locomotive examples should be specially noted, as they represent extremes of this type of spring. Clearly, both cannot be right, and if the test of commercial and mathematical efficiency be applied, equally clearly Example No. 1 is completely wrong. Such instances are, however, still designed and made.

(3) *Automobile (Heavy) Rear Spring.*

61 ins. straight length, c. to c. $3\frac{1}{2}$ ins. wide. $12\frac{1}{2}$ ins short plate. 8 plates $\frac{7}{16}$ in. thick, 5 plates $\frac{3}{8}$ in. thick.

B.S.S. test	9·45 ins.
Actual deflection per ton	1·64 ins.
Actual test weight—tons	5·75
Weight of spring—lbs. (actual) ..	222
Resiliency efficiency, per cent. ..	70

This is not a bad result, because 8 plates out of 13 are full thickness, and the thickness variation is only $\frac{1}{16}$ in. Had this been of all $\frac{7}{16}$ in. plates, the efficiency would have been 78 per cent. (11 plates at $\frac{7}{16}$ in. are practically equivalent to the combination given above.)

(4) *Automobile—Trailer.*

$43\frac{3}{4}$ ins. straight length, c. to c. $2\frac{1}{2}$ ins. wide. 28 ins. short plate. 6 plates $\frac{3}{8}$ in. thick.

B.S.S. test	5·67 ins.
Actual deflection per ton	2·30 ins.
Actual test weight—tons	2·45 ins.
Weight of springs—lbs. (actual) ..	62
Resiliency efficiency—per cent. ..	64

It will be noticed that even the long short plate has not had a serious effect on the resiliency as might have been expected owing to the lesser unit deflection obtained thereby.

(5) *Railway Carriage Spring.*

63 ins. straight length. c. to c. 3 ins. wide. 12 ins. short plate. 2 top plates $\frac{1}{2}$ in. thick, 12 plates, $\frac{3}{8}$ in. thick.

B.S.S. test	8·80 ins.
Actual deflection per ton	2·20 ins.
Actual test weight—tons	4·00
Weight of springs—lbs. (calculated)	206
Resiliency efficiency	49

Here again is a case where the thinner bottom plates reduce the efficiency, and, generally, such examples of designing prove very potent in the obtaining of a low resiliency figure —which, from another point of view, can be taken as partly representing the relative costs of springs. Such are not proportionate, but the general fact obtains, namely, the higher the resiliency figure the lower the comparative cost.

In Fig. 83 is shown a curve, plotted from resiliencies of rectangular section beams, in inch-lbs., against the " constant" used in the standard unit deflection formula, which has been obtained from corresponding stress diagrams. This curve shows that for all practical purposes, the theoretical resiliency is uniform from " constant " 0·124, representing the maximum deflection, minimum weight, and purely theoretical spring ;

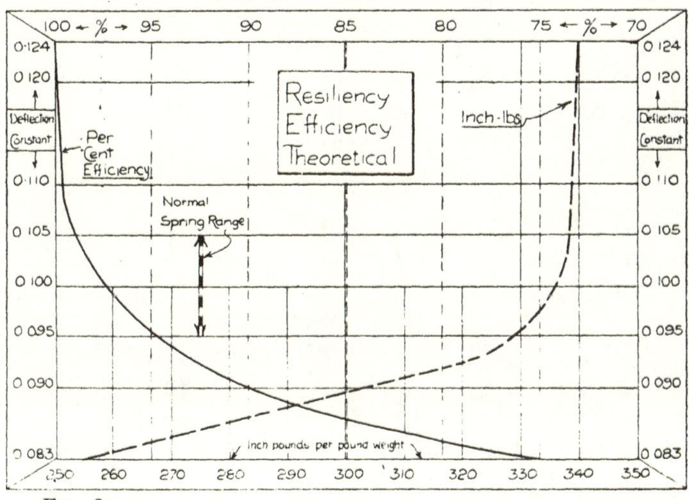

FIG. 83.

to " constant " 0·098, from whence it drops smartly down to 0·083, which represents the minimum deflection, maximum weight spring, of rectangle plan, or having all plates the same length. The " constant," 0·100 as recommended, and as found correct for normal design, gives therefore a high efficiency, if all material is uniformly stressed, otherwise, and chiefly, if all plates are of the same thickness. The necessity of the use of a constant below 0·100, due to a large percentage of full length top plates, or abnormally long short plates, causes the efficiency to drop, but these features of design cannot reduce the resiliency percentage to the same extent as the use of non-uniform thickness plates.

For estimating purposes, in the drawing office of the vehicle builders, and in the works of the spring manufacturer, it is necessary to know the more or less accurate weight of the springs designed and to be used, which is also needed if any study is going to be made of the resiliency factor. Of course, this knowledge can be obtained with great exactness by finding the length of each plate, adding together, and deducting allowances for spears, drawn points, studs and slits, centre holes, etc., but on the whole, the time expended is not worth the result obtained. The total spring weight of any vehicle rarely exceeds 10 per cent., and this figure is only approached on light automobiles. The usual amount on various services is approximately as follows :—

Railway.

Locomotives.	Light, 3 per cent.	Loaded, 3 per cent.
Coaching Stock.	,, 3 per cent.	,, 2½ per cent.
Freight Stock.	,, 3 per cent.	,, 1¼ per cent.

Tramway.

Cars with all springs of laminated form.	Light 4 per cent.	Loaded, 2 per cent.

Automobile.

Small Cars.	Light 10 per cent.	Loaded 6 per cent.
Large Cars.	,, 5 per cent.	,, 4½ per cent.
Heavy Trucks.	,, 5 per cent.	,, 3 per cent.

Spring steel of normal section—round edge, and slightly concave—does not, of course, weigh equal to the nominal section, and the weight of any one nominal section is continually varying as the rolls wear. A good average figure to take for these weights is 95 per cent. of the full flat, and a table drawn up on such lines, is of value. Apart from the actual length of flat plate in the spring as drawn, allowances are necessary for varying shapes of back plates, and deductions for spears, etc. However, to simplify matters, the following formula will be found as accurate as is necessary for all estimating purposes :—

$$\text{Weight of springs (lbs.) } w = \frac{(L + L^s) \times b \times TT}{k} \qquad \text{(XX.—4)}$$

The constant (k) varies as follows :—

For springs for use with shoes such as British type buffing and bearing, certain heavy automobile rear, etc.	$k = 7$
Light automobile type, with drawn points ..	$k = 6\frac{1}{2}$
All other springs	$k = 6$

In practice, it will be found that the results obtained by the use of the above, will give satisfaction.

The calculation of the weights of hoops, which are also necessary to the estimating departments concerned, cannot be reduced to such simplicity. Plain hoops, that is, hoops made from rolled flats, and welded up, can have their weights obtained as follows :—

Weight of hoop (lbs)

$$w^h = \frac{4(b + TT + 2T^h) \times H \times T^h}{7}. \qquad (XX.—5)$$

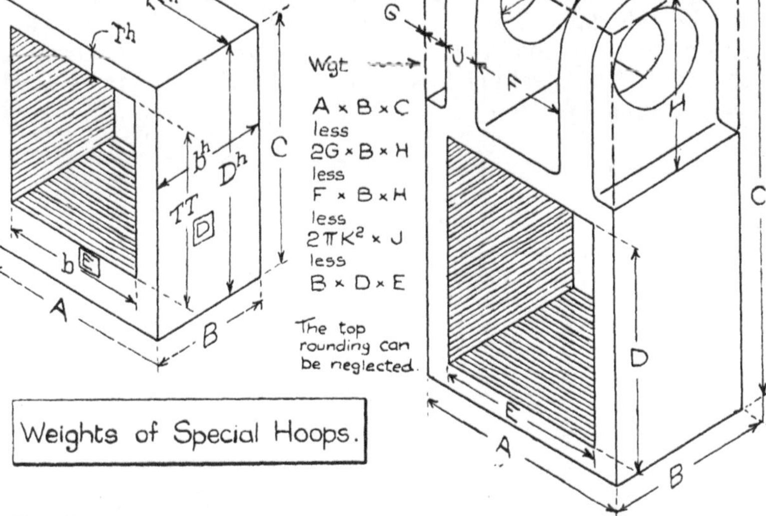

FIG. 84.

In this expression (H) is the breadth of the hoop, and (T^h) the thickness. Alternatively the length of material required can be calculated :—

$$\text{Length in inches} = 2b + 2TT + 4T^h. \qquad (XX.—6)$$

All types of hoops can be dealt with so far as the boxes are concerned, by taking the external size and deducting therefrom the spring size, the result being multiplied by 0·28 which gives the weight in lbs. The sketch, Fig. 84, generally illustrates this method. Included also in this drawing is an example of a lugged hoop, with the easiest calculation means of obtaining weight—all based on the assumption of the solid enclosed block, and deductions thereform.

Simple expressions for readily obtaining the weights per foot of any sections in use in the trade are as follows :—

Weight per foot—concave spring steel	=	$3{\cdot}25\ b\mathrm{T}$
,, flat steel	=	$3{\cdot}40\ b\mathrm{T}$
,, flat iron	=	$3{\cdot}30\ b\mathrm{T}$.

As will be observed from the above, a still simpler approximate calculation, which is not outside practical ranges of accuracy for estimating purposes, can be arranged by taking $10(b\mathrm{T})$ and dividing by 3. This gives the weight of a 3 in. by $\frac{1}{2}$ in. section as 5 lbs. per foot, thus $10 \times 3 \times \frac{1}{2} \div 3 = 5$. Metric sizes to be worked out in shops using British units, can be calculated thus :—Millimetres (b) $\times$ millimetres (T) and divided by 200 gives almost the exact weight of normal concave spring steel per foot in pounds, thus $75 \times 12 \div 200 = 4.5$ pounds per foot.

CHAPTER XXI

PERIODICITY

THE term "periodicity" is best defined as that rate of oscillation (inch-minute units) acquired by a spring consequent upon the sudden removal of the load, before it attains rest. The dimension of "periodicity" is a measure therefore of the easy riding of the spring, as it is clear that the lower the rate of oscillation—otherwise, the longer the period for one oscillation—the more smooth riding will be the vehicle to which it is attached. A very stiff, solid, spring, of corresponding low static deflection, requires no demonstration or proof that it will be a very bad riding spring, and a spring of this type has a high periodicity, with a low oscillation timing.

Considerations of periodicity are not included in railway or tramway work, owing to necessarily standardized limits for the heights of rolling stock bumpers, and to the use of steel track, which at its worst, is no more than equivalent to a tolerably good road. With the growth of the mechanically propelled road vehicle, however, the important part played by the initial static deflection of the springs or "set" of the car, has had to be recognised, and is generally regarded as a leading feature in the design. Periodicity varies inversely as the square root of the static deflection, and all springs with the same static deflection have the same periodicity.

The rates of oscillations of all systems of spring suspension are necessarily dependent on gravity, and the formulæ relative thereto follow accordingly those for pendulums and the like. It must be understood that "periodicity" implies "oscillations" as distinct from "vibrations," the true definition of the term "vibration" being the movement of a pendulum from one end to the other end of its swing, a uni-directional movement; whereas an "oscillation" embraces this movement plus the reverse movement to the starting point; which complete travel requires twice the time of a "vibration." For instance, a standard length clock pendulum makes 60 vibrations per minute, but 30 oscillations per minute, and this latter figure, 30, is its periodicity.

The pendulum formula is as follows:—

$$\text{Time of vibration, (seconds) } v = \pi\sqrt{\frac{\text{L(in feet)}}{\text{gravity}}} \quad \text{(XXI.—1)}$$

The obtaining of periodicity, which is oscillations per minute, needs a re-formation of this into the following terms:—

$$\text{Periodicity, } \quad P = 60 \div 2\pi\sqrt{\frac{\text{L(in feet)}}{\text{gravity}}} \quad \text{(XXI.—2)}$$

With gravity (a constant) as 32·2 feet per second, per second, and (L) in inches, and changed into (D) as representing static deflection, the following is obtained:—

$$P = 60 \div 2 \times 3{\cdot}14\sqrt{\frac{D}{12 \times 32{\cdot}2}}$$

therefore

$$P^2 = \frac{3600 \times 12 \times 32{\cdot}2}{39{\cdot}4\ D}$$

or

$$P = \sqrt{\frac{35300}{D}} \quad \text{(XXI.—3)}$$

Alternatively, and working from a known periodicity,

$$D = \frac{35300}{P^2}, \text{ say } \frac{35000}{P^2} \quad \text{(XXI.—4)}$$

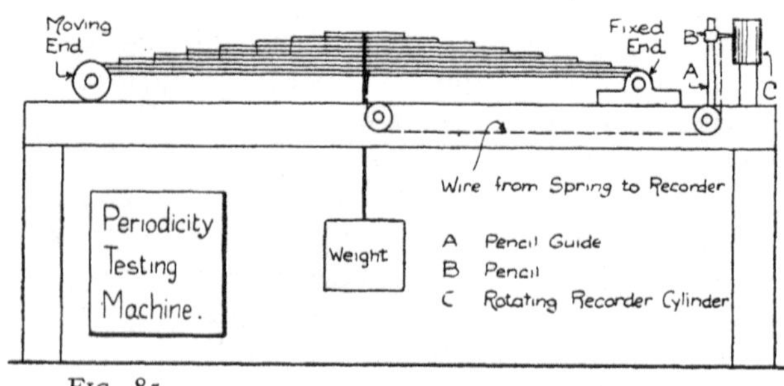

FIG. 85.

The numerical factor here (35000) is a practical one, for easy recollection, and a sufficiently accurate approximation.

A simple machine for obtaining spring periodicities by experiment as devised by Mr. A. A. Remington, and illustrated in his 1922 Paper, read before the Institution of Automobile Engineers, (London), is illustrated in Fig. 85. Herein is a

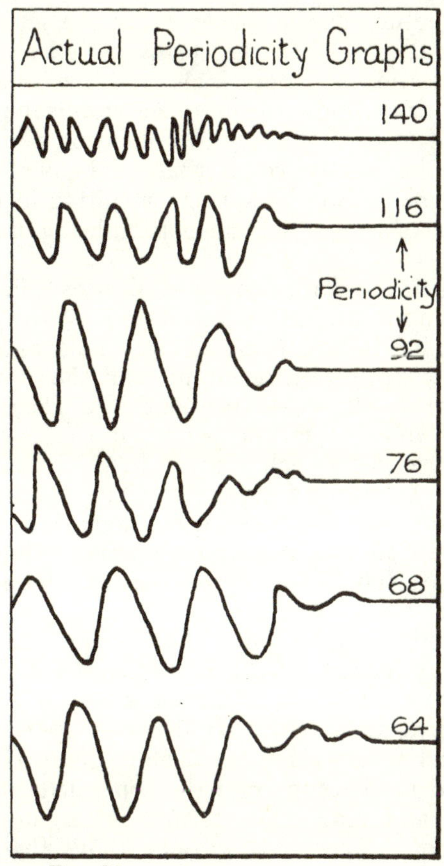

FIG. 86.

predetermined weight, attached to the spring, thereby giving it a definite deflection. The spring middle is also connected to a recording cylinder, moved by timing gear. The sudden removal of the weight causes the spring to " bounce " which action is shown on the cylinder in the form of the graphs of of Fig. 86. It is stated that the results obtained by this means are generally in accordance with calculations. It is also stated that the slowing down to rest of the spring is caused by friction, the implication being that it is interplate friction. This can, however, be doubted, as conversely, a frictionless spring would never come to rest—such as a spring of one plate only. The " friction " that brakes the spring down to " rest " can only be regarded as the internal friction of the steel structure, as the movement of the surface remote from the neutral axis when the spring is deflected from its " no load " position is appreciable and measurable. The higher the static deflection, the more material is in movement, which will tend to cause a reduced oscillation period. This, however, will be partly counterbalanced by the greater inertia of the greater weight spring—resulting in the figures obtained from experiment and calculation being in reasonable agreement.

The remarks in this Paper on the subject of " Periodicity " are worth quoting. Mr. Remington stated : " The periodicity of a suspension has to be considered as a whole, as all portions act in unison to produce a resultant, and the separate components cannot oscillate independently of each other. A suspension has different natural periods in the several planes, normally one period for each plane. In the case of a motor car, these can be divided into :—(1) The front-suspension vertical period ; (2) the rear-suspension vertical period ; (3) the rolling period ; (4) the pitching period. Each of these will be different, and furthermore, it will be found that the rear and front springs interact, and that, for example, the rear vertical period will be slower if the front springs are free than if they are locked, as the movement of the free front springs alters the amplitude of the mass, and so modifies the time of swing to an extent that depends upon the weight distribution and other variables. The rolling period and the pitching period both result from the combined action of the springs, front and rear. The former is largely influenced by the height of the centre of gravity above the suspension level, and is usually considerably slower than either of the vertical periods, while the latter depends upon wheel base and weight distribution longitudinally. The difference between the rolling and the rear vertical periods, and the small practical range of periodicity, constitute the principal disadvantages of normal suspension systems, and indicate the most promising directions in which to seek improvement."

The limitations of static deflections of automobile springs are determined chiefly by clearances between mechanism and body, mudguards and tyres, etc. A universal practice is to include loaded front springs of higher period than loaded rear springs, as this prevents synchronous action, and gives the load over the rear springs a smoother passage. The entire automobile range as regards periodicity is :—

Front Springs, 1 in. to 4 ins. deflection, periods 190 to 94.
Rear springs, 3 ins. to 8 ins. ,, ,, 109 to 67.

If this factor be taken into account for railway springs, it shows up as :—

Locomotives,	from	250 to 150.	All when fully loaded.
Coaching Stock,	,,	130 to 90.	
Freight Stock,	,,	220 to 150.	
Tramcars,	,,	220 to 150.	

In the case of roads, of varying surfaces, it is occasionally found that certain corrugated stretches coincide with the periodicity of front or rear springs at the speed of the moment, which fact provides the obvious reason for designing front and rear springs of different periods. In railway service, with 30 ft. rails, and a train running at 40 miles per hour, the number of pairs of opposite joints travelled over per minute is 117, so that even with indifferent permanent way, the train will have a fairly smooth movement, as the spring periods of modern bogie coaching stock are not often less than 130. Staggered joints—or non-opposite joints—clearly give twice as many shocks, and are further removed accordingly from the possibilities of synchronization so far as the train is concerned, but may agree with that of the locomotive. Such agreement, which is possible in the U.S.A. for instance, where staggered joints are universal, is neutralized by the standard practice of coupling up adjacent springs by equalizing rigging.

In tramway work the problem is complicated with certain designs of trucks, as whilst the body is well spring-borne, the main springs carrying the load on to the axle journals are small stiff, coil springs—necessarily small and stiff because they are cramped between the top of the axle-box and the underside of the truck frame. Occasionally small laminated springs, about 10 ins. long, have been perpetrated for this function, and these are about as useful as a solid cast iron block, the periodicity of which is infinity. Such springs

Graph for Periodicity.
Static Deflection of Spring
7"
6"
5"
4"
3"
2"
70
75
80
85
90
95
100
110
120
Periodicity

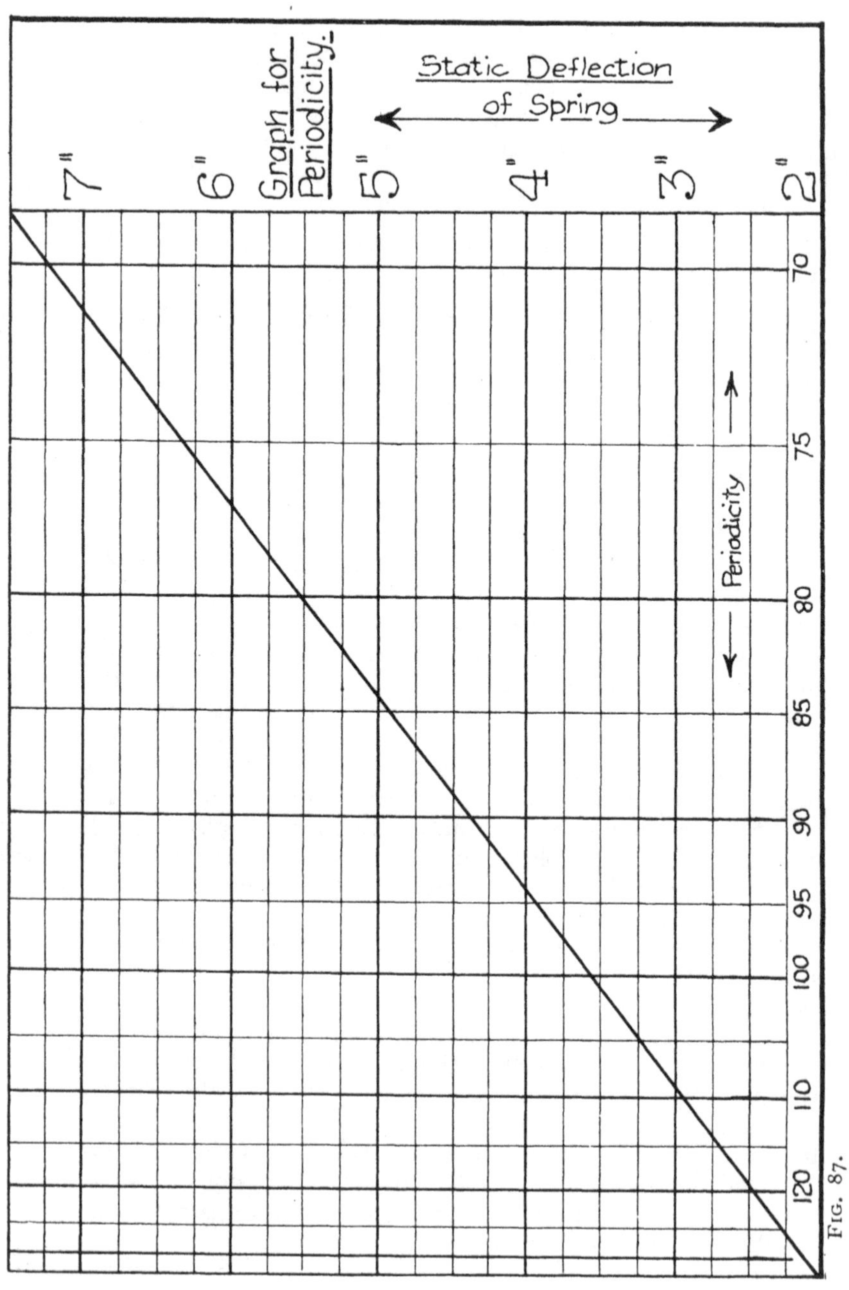

Fig. 87.

as these when fully loaded have (P) in the neighbourhood of 400. A tramcar running 10 m.p.h. will travel over 30 opposite joints per minute—but probably between each joint there will be a number of humps of rail corrugations, which can coincide with the period of the main axlebox springs. Corrugations of all pitches from 2 ins. to 12 ins. are not uncommon, and at varying speeds, will coincide with the spring period for the time being. It would not, however, appear any too easy to obtain a true periodicity of any vehicle of the orthodox tramway motor-car type, or electric railway motor-coach, as such a large amount of weight—in the form of the axle-attached motors—is unspringborne, or otherwise supported by springs of infinitely small deflection, with a corresponding periodicity of infinity. Considering only the known spring period, if this be 400,

At 4 m.p.h. corrugations of 10 in. pitch coincide.
At 6 ,, ,, ,, 16 in. ,, ,,
At 10 ,, ,, ,, 21 in. ,, ,,

With the car running light, the periodicity of the springs will increase, and the coincident corrugating pitches become accordingly shorter. Probably no little rail corrugation can be traced to the springing of tramway trucks, as regards the main journal box springs, and it is a matter that certainly merits attention—as the vertical rigidity of tramway permanent way is unique to this class of railed traction, and provides excellent opportunities for spring reactions. In the case of the orthodox railway tracks, the permanent way is relatively elastic, and if the spring is too stiff to yield at shock, the rail can deflect to a small degree—a condition which is not possible with tramway tracks laid on, and bedded in, substantial concrete.

In the event of any designer preferring to quote deflections in terms of "periodicity," rather than direct, the graph, Fig: 87, will probably be of assistance.

CHAPTER XXII

EXAMPLES OF DEFLECTION CALCULATIONS

THE purpose of this chapter is to include a number of instances of actually tested springs, giving unit deflections as taken from the loading machine, and the same deflections calculated, with a view to inspiring a certain amount of confidence in the use of the standard unit deflection formula which has been stated.

The first instance is of two patterns of springs made up for the commencing test with 8 full-length plates, these being afterwards cut down to varying lengths so that a constant middle strength was assured, which might not have been the case had other plates been used, owing to the great influence on deflection of a small thickness variation of plate. Figs. 88, 89 and 90 show the two springs in their three forms, with stress diagrams relative to each. In these drawings :—

Fig. 88 is the spring with full length plates, all 40 ins.

Fig. 89 is what might be termed " normal design," with plates of lengths 40 ins. (back)—36 ins—32 ins.—28 ins.—24 ins.—20 ins.—16 ins.—12 ins.

Fig. 90 is the perfect design, with short plate length as L/*n*. Plates, 40 ins. (back)—35 ins.—30 ins.—25 ins.—20 ins.—15 ins.—10 ins.—5 ins.

One spring was made " straight " and tested with reverse camber accordingly, and the other was made with a high camber, (7 ins., approximately 1/6 L) with the object of judging, under ordinary shop testing conditions, the effects which could be traced owing to the variation between deflections as measured on the ordinates, and the true deflection paths, referred to at length in Chapter II.

Throughout these three sets of springs, it will be noted that the deflections per ton (the unit load) vary according to the camber. The variation is not serious, and would be comparatively not noticeable had the deflections been taken in the usual fractions of $\frac{1}{16}$ in. This trait is generally attributed to the lengthening of the span, in the case of a spring with camber, but this feature is only incidental, the real reason being due to the difference between the true deflections, and the ordinate deflections measured—coincidence between the two only arriving when the springs are at or approaching the straight. This point is clearly shown in the small diagrams marked "Deflections for each Ton."

In the case of springs A—1 and A—2, Fig. 88, as the make-up constituted a beam of rectangular plan, the (K) for the deflection formula is 0·083. It will be observed that this gives a slightly higher deflection than was obtained, taken over the whole range. On the loadings that approached the straight, however, namely 1 ton on Spring A—1, and 4 tons on spring A—2, the results were very close. For springs B and C, constants have been calculated from the stress diagram, which are in one case less, and the other more, than the recommended (K) of 0.1. In neither case, however, is the result seriously affected. A summary of the general results, with deflections also shown in $\frac{1}{16}$ in. follows :—

Spring No. :	A.1	A.2	B.1	B.2	C.1	C.2
1 ton	1·00 1 in.	0·80 $\frac{13}{16}$ in.	1·26 $1\frac{1}{4}$ in.	1·04 $1\frac{1}{16}$ in.	1·40 $1\frac{3}{8}$ in.	1·25 $1\frac{1}{4}$ in.
2 tons	1·97 2 in.	1·64 $1\frac{5}{8}$ in.	2·39 $2\frac{3}{8}$ in.	2·14 $2\frac{1}{8}$ in.	2·72 $2\frac{3}{4}$ in.	2·55 $2\frac{9}{16}$ in.
3 tons	2·89 $2\frac{7}{8}$ in.	2·58 $2\frac{9}{16}$ in.	3·30 $3\frac{5}{16}$ in.	3·37 $3\frac{3}{8}$ in.	3·95 $3\frac{15}{16}$ in.	3·87 $3\frac{7}{8}$ in.
4 tons	3·75 $3\frac{3}{4}$ in.	3·60 $3\frac{5}{8}$ in.	—	—	—	—
Average per ton ...	0·94	0·90	1·10	1·12	1·32	1·29
Deflection for ton nearest straight	1·00	1·02	1·26	1·23	1·40	1·40
Calculated Deflection :						
K = 0·083	1·02	1·02	—	—	—	—
K = 0·096	—	—	1·18	1·18	—	—
K = 0·100	—	—	1·23	1·23	1·23	1·23
K = 0·105	—	—	—	—	1·29	1·29

Actual Test Results; A-1 & 2.
As Made
A·1
Line of Test
All 40"
As Made
A·2
Line of Test
Outline of Springs
Stress Diagram
K = 0·083
Maximum Stress
Deflections for each Ton.
K as 0·083
A·1
A·2
1·02
Tons
Load Test Graphs.
Collected Heights
K
A 1
A·2
Line for K = 0·083
A·1
A·2
K = 0.083
Deflection in inches
Load — in Tons

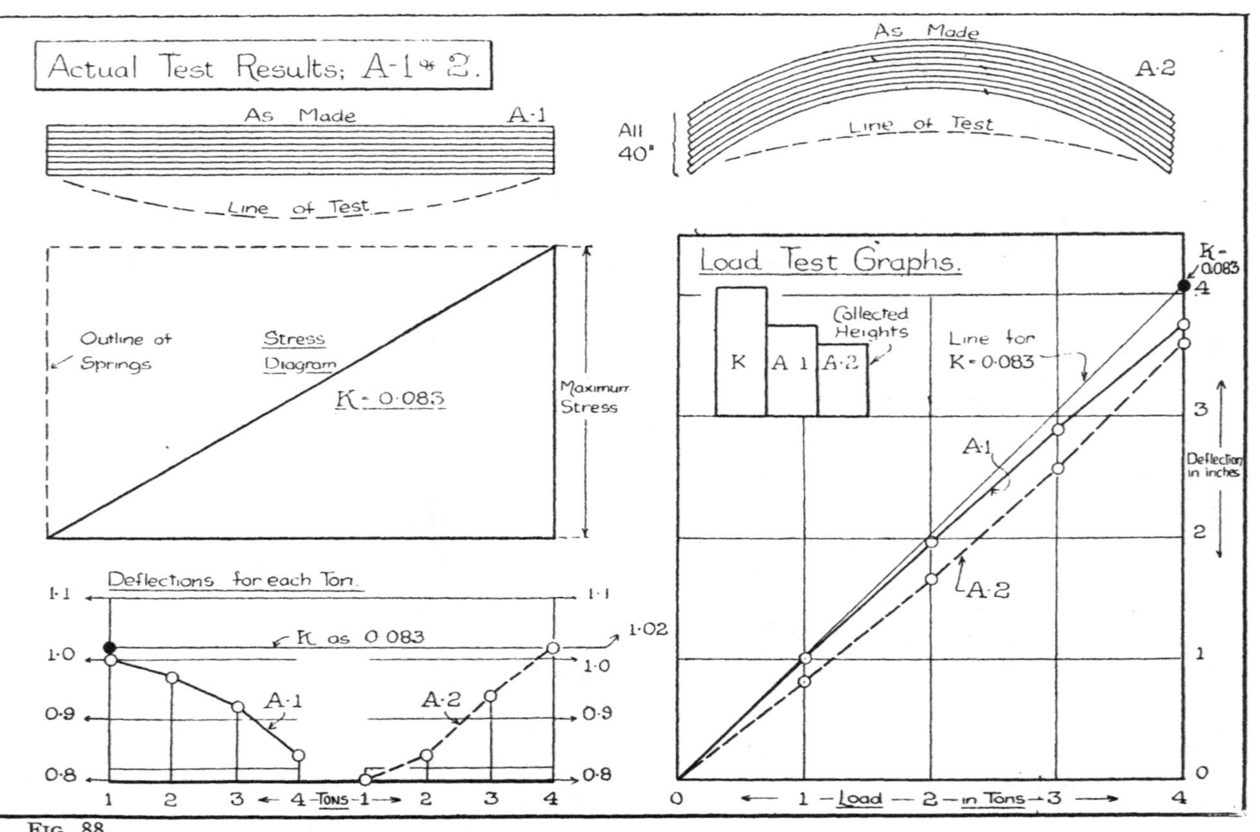

FIG. 88.

Actual Test Results; B-1 & 2.
As Made
B·1
Line of Test
Lengths
12" 16" 20" 24" 28" 32" 36" 40"
As Made
B·2
Line of Test
Outline of Springs
Actual Stress Diagram
Average Stress Diagram
K = 0·096
Maximum Stress
Deflections for Each Ton
K as 0·100
K as 0·096
B·1
B·2
1·23
1·18
1·2
1·1
1·0
0·9
Tons
Load Test Graphs.
Collected Heights
0·096 K
0·100 K
B·1
B·2
Line for K=0·1
K=0·100
K=0·096
Deflection in inches
Line for K=0·096
Load in Tons

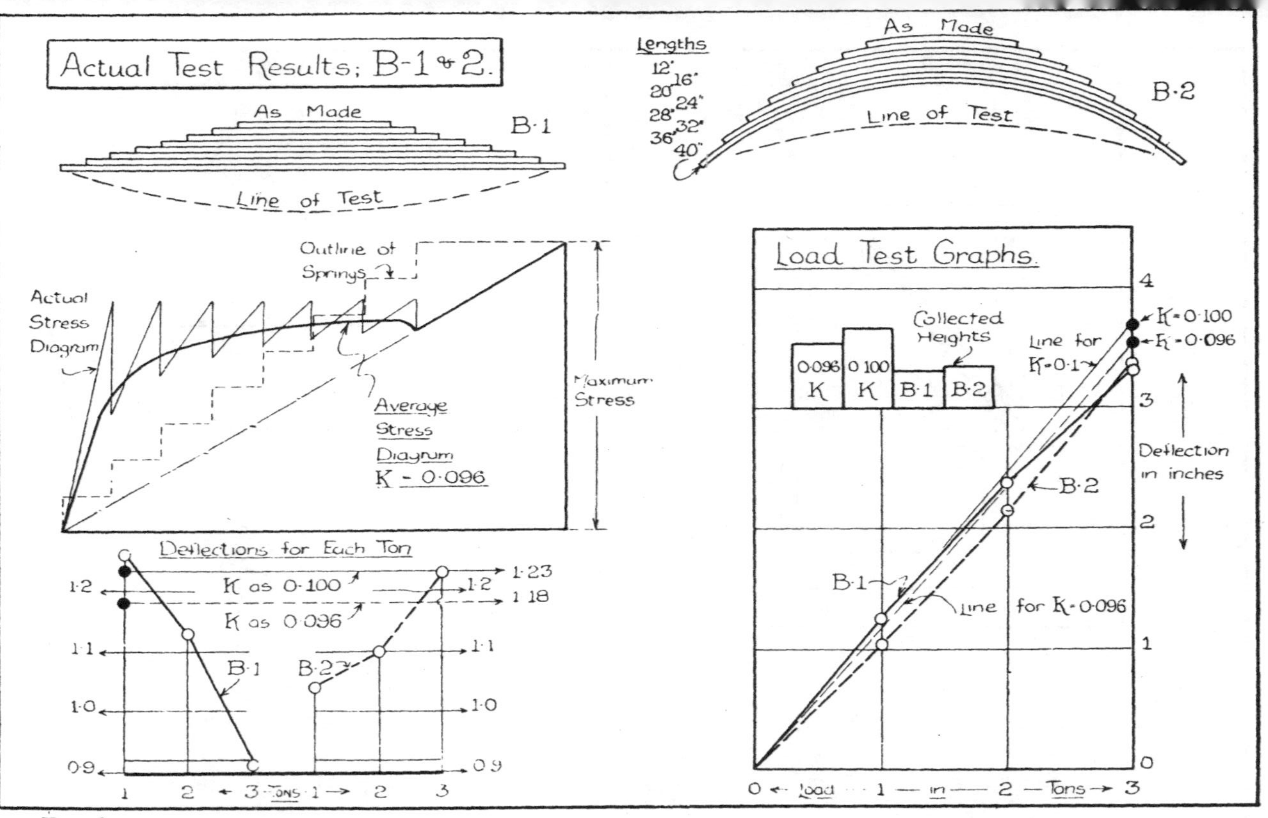

FIG. 89.

Actual Test Results; C-1 & 2.
As Made
C·1
Line of Test
Lengths
5"
10"
15"
20"
25"
30"
35"
40"
As Made
C·2
Line of Test
Maximum Stress
Actual Stress Diagram
Outline of Springs
Average Stress Diagram K = 0·105
Load Test Graphs.
Collected Heights
0·100 K
0·105 K
C 1
C·2
K - 0·105
K - 0·100
Deflection in inches
Line for K as 0·1
C·1
C·2
Line for K - 0·105
0 ← Load — 1 — in — 2 — Tons → 3
Deflections for each Ton
C·1
C·2
K as 0·105
K as 0·100
1·4
1·3
1·29
1·23
1·2
1·1
1 2 ← 3 - Tons - 1 → 2 3

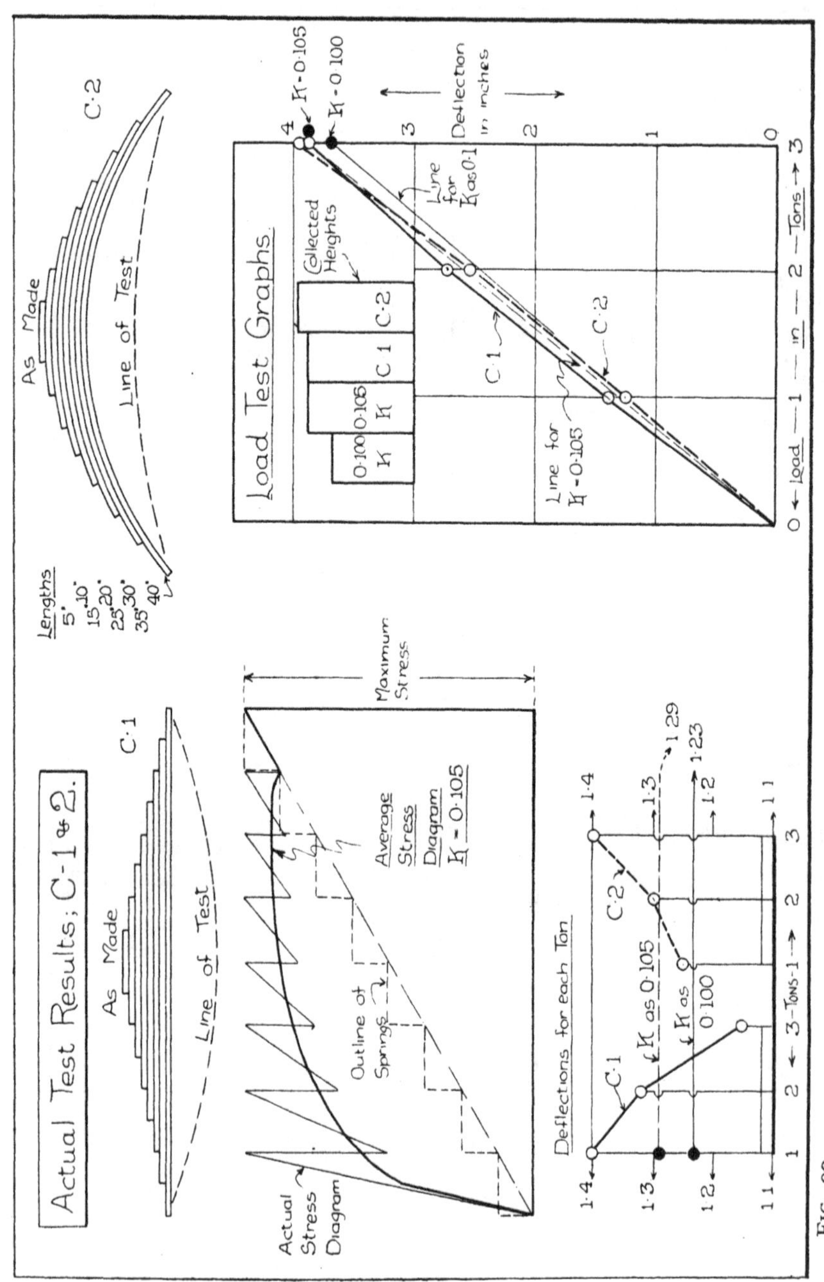

Fig. 90.

Regard the largest variation obtained between average deflection and calculated deflection, which appears as B—1, where, with (K) as 0·100, the difference is 1·23—1·10 or 13 points. The steel is nominally 3 ins. × $\frac{3}{8}$ in., with a (t^3) of 216 accordingly. The percentage difference is 10, and to obtain this it would only be necessary to have the steel thicker by $\frac{1}{100}$ in., or 0·385 in. instead of 0·375 in. The greatest difference between the load lines of the two springs of a series is, as might be expected, with the form A, which is an unnatural spring. The series B and C are as close as can be expected.

The tests just quoted were made on springs fitted with little or no nip, hardened and tempered, and generally treated in the normal way, with steel from stock. They were made, however, specially with a view of studying the deflections obtained, and it might be useful to amplify the confirmations of the deflection formula by reference in the second instance, to divers other types of extreme natures. The springs just detailed were of a length and section size which is used in all classes of work.

Figs. 91 and 92 illustrate a further 8 assorted springs, with the following explanatory notes :—

Fig. 91. No. 1. This shows a spring that was built up for test purposes in connection with general matters, and it is fitted with no nip. In this case, the camber of each separate plate was calculated, and they were made to these cambers —with the result that when put together, all bedded well throughout. The steel was ordinary section, from stock, and the difference between calculated and actual deflections is equally the difference between 0·493 ins. and 0·500 ins., a matter of 7/1000, which is inside practical rolling limits. The deflection constant (K) was taken from a special stress diagram, as the spring was on the light side as compared with the "normal."

Fig. 91, No. 2. This represents a special spring, of one plate, made in connection with a hoop-testing arrangement which will be shown later. Under working conditions, it was intended to transmit a central load of 10 tons. Special oil-hardened steel was employed, and the forged bar was machined to the sizes indicated. The close agreement between actual and calculated results will be noted. Also, the high skin stresses obtained when working with the usual (Z) formula, which is considerably beyond the breaking stress.

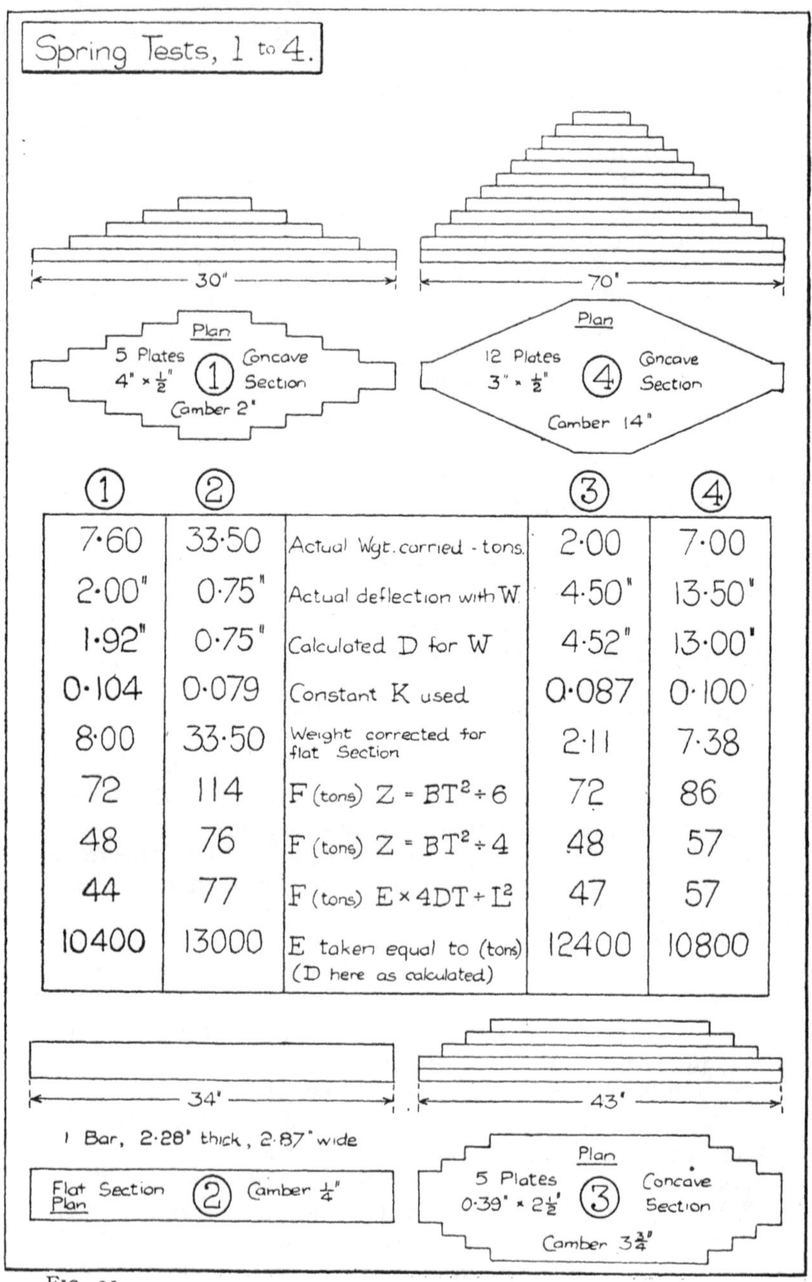

①	②		③	④
7·60	33·50	Actual Wgt. carried - tons.	2·00	7·00
2·00"	0·75"	Actual deflection with W.	4·50"	13·50"
1·92"	0·75"	Calculated D for W	4·52"	13·00"
0·104	0·079	Constant K used	0·087	0·100
8·00	33·50	Weight corrected for flat Section	2·11	7·38
72	114	F (tons) $Z = BT^2 \div 6$	72	86
48	76	F (tons) $Z = BT^2 \div 4$	48	57
44	77	F (tons) $E \times 4DT \div L^2$	47	57
10400	13000	E taken equal to (tons) (D here as calculated)	12400	10800

FIG. 91.

Fig. 91, No. 3. This shows a spring which was designed for a certain type of road vehicle, and included a very long short plate, 34 ins. length, with 43 ins. bearing centres. The back plate had rolled eye ends. The springs were made as desired, but naturally, a large departure had to be made from the true arc to obtain a good shape straight. The constant (K) of 0·087 has been obtained by the means suggested in Chapter VII, of taking proportional weights. In this instance, the " normal " spring would weigh 38 lbs. The present spring weighed 53 lbs. The relative weights are therefore 1·00 : 140. The difference being 40, divide by 3000 (or 3) which is 0·013 to be deducted from 0·100, giving (K) as 0·087. The agreement is quite satisfactory. By consequence (E) is also amended in inverse ratio to (K).

Fig. 91, No. 4. The spring shown here is the standard R.C.H. 12-plate buffing, which is very highly stressed under working conditions. The load given is the test load. In this case all plate ends are speared. Owing to the generally light design, the calculated figure is somewhat less than the nominal actual figure, but in practice they agree, owing to the difficulty of defining the bearing centres, as the ends of back plate of this example are comparatively flat, and bear on shoes. A difference of ½ in. inwards on each end, bringing the effective length to 69 ins. instead of 70 ins. causes the calculated deflection to become 1·78 ins. per ton (14·2 ins. for 7 tons) instead of 1·86 ins. in other words 98·6 per cent. of nominal length causes the unit deflection to become 95·7 per cent. of that calculated on nominal length.

Fig. 92, No. 5. Here is shown an automobile spring of normal design. The back plate was finished with rolled eye ends. The calculated and actual results are within reasonable agreement, the difference being slightly over 2 per cent. The standard formula has been taken for the unit deflection, and a very slight difference in plate thickness or concave would make the discrepancy between the actual and estimated results.

Fig. 92, No. 6. This illustrates the piece of spring steel tested in accordance with the French Railways' specification (Chapter XIX). Here again, the deflection calculated is less than the deflection actual. It is comparative to a difference in thickness between 0·59 ins. and 0·57 ins. It was probably not as great as this, however, but the difference may have been on the " concave."

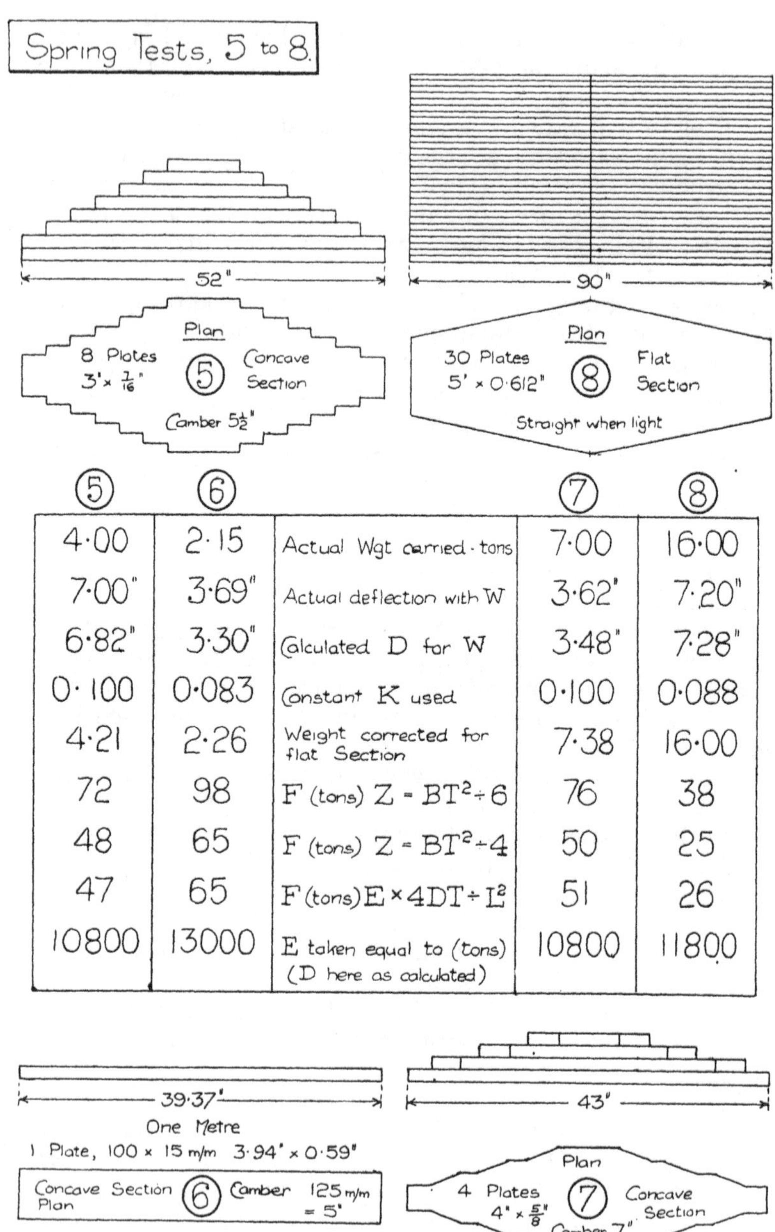

5	6		7	8
4·00	2·15	Actual Wgt carried - tons	7·00	16·00
7·00"	3·69"	Actual deflection with W	3·62"	7·20"
6·82"	3·30"	Calculated D for W	3·48"	7·28"
0·100	0·083	Constant K used	0·100	0·088
4·21	2·26	Weight corrected for flat Section	7·38	16·00
72	98	F (tons) $Z = BT^2 \div 6$	76	38
48	65	F (tons) $Z = BT^2 \div 4$	50	25
47	65	F (tons) $E \times 4DT \div L^2$	51	26
10800	13000	E taken equal to (tons) (D here as calculated)	10800	11800

FIG. 92.

Fig. 92, No. 7. This sketch shows a 4-plate wagon bearing spring, as used here for certain railways, with speared ends It is not easy to dogmatize upon unit deflections in springs such as these, owing to very slight alterations in bearing points affecting the loading results to a comparatively large degree, as indicated in Case 4. The comparison shown, however, is not unreasonably discrepant.

Fig. 92, No. 8. This shows a very special spring as made for a railway dynamometer car, including 30 plates, 5 ins. wide, and 0·612 ins. thick. In this case, each plate was symmetrically tapered in breadth from the middle to the ends, the ends being left 2 ins. wide. Such arrangement is equivalent, therefore, to 40 per cent. of top plates, and the constant (K) in this case is given as 0·093 (Chapter VII, Fig. 28). This correction to 0·093 is, however, for normal concave steel section, whereas these plates were flat rolled, so that 0·093 becomes 0·088 (the ratio of 95 to 100). The agreement between calculated and actual deflection is remarkable, and can only be attributed to the accuracy of manufacture of this spring, which had all the plates most carefully gauged and selected for thickness, and also had each plate load-tested to ensure uniformity of manufacture.

The foregoing examples should be sufficient to show the general accuracy of the unit deflection formula, and also of the correcting constants and methods suggested for abnormal cases. No spring with non-uniform thickness plates is in the series, but the results are always equally accurate, from a practical point of view. It should be carefully remarked that, in the cases of the two springs whose fabrication has been specially accurate, namely, the spring for the hoop-testing device, Fig. 91, No. 2, and the dynamometer car spring, Fig. 92. No. 8 ; the calculation results agree exactly with the actual results, which would seem to show that inaccuracies can always be traced, as already indicated, to rolling variations of thickness or concave, and in doubtful bearing points on plain end springs. The results of the two instances with well defined heads, Nos. 3 and 5, agree respectively, within a negligible percentage fraction for No. 3, and within 2 per cent. for No. 5.

CHAPTER XXIII

REMARKS ON SPRING DESIGNING

On the assumption that the matter in the preceding chapters has been studied with reasonable thoroughness, few doubts should now be present in the mind of the designer as to the most important aspects to be regarded. The deflection per ton (or 1000 kgs.), or pounds per inch of deflection, represent governing features of the desired spring, and this, in the case of railway and tramway springs, particularly for vehicles with high loading capacities relative to their tare weights, has to be based on allowable vertical movement of buffer or coupler centres. In automobile design, " periodicity " is usually the unit of measurement in the designer's mind. This, as before stated, means static deflection, so that knowing accurately the weights to be carried by the springs (not too common a knowledge) periodicity resolves itself into static deflection ÷ weight, giving a unit deflection. Front and rear springs of automobiles are invariably designed with different periodicities to avoid synchronisation, but due regard has to be paid to clearances when the car is loaded. This is particularly a subtle point with private cars, as clearly, if the static deflection under load is to be high, with a correspondingly low periodicity, the purchaser of the car, as he sees it—unloaded—does not appreciate the eccentricity of the mudguard lines relative to the wheels, and probably does not fully comprehend any sound explanation.

Obviously, no rules can be laid down for unit deflections, such are considerations depending on many things. In practice, the following ranges occur :—

Railway Springs.	Locomotive.	0·05 in. to 0·50 ins. per ton.
	Carriage.	0·30 in. to 2·00 ins. ,,
	Wagon.	0·20 in. to 0·60 in. ,,
Tramway Springs.	General.	1·00 in. to 1·50 in. ,,
Automobile Springs.	Truck, Front.	1·00 in. to 2·00 ins. ,,
	Truck, Rear.	1·00 in. to 2·00 ins. ,,
Under full load.	Private Cars, front, total 2 ins. to 4 ins.	
	,, rear ,, 3 ins. to 8 ins.	

The first desideratum for any spring user is to study his necessary designs, and reduce the number of sections employed to the minimum possible. (This naturally turns the designer on to the right road of including only one thickness of plate in any one spring). The dimensional range of sections is not alarming, but the included possibilities are immense. These can be studied with profit, combined with a resolution that designing ingenuity consists of using as few as possible, and misguided ingenuity consists in employing the lot. In practice, railway and tramway sections all fall between $2\frac{1}{2}$ ins. $\times$ $\frac{1}{4}$ ins. and 6 ins. $\times$ $\frac{5}{8}$ ins. Automobile sections go down to $1\frac{1}{2}$ ins. $\times$ $\frac{3}{16}$ in., and up to $4\frac{1}{2}$ ins. $\times$ $\frac{1}{2}$ in. Continental metric sizes fall also within these limits. Nothing seriously has yet been done with standardization of sections in this country, and whilst a movement is on foot, it is one thing for the B.E.S.A. to produce standard section lists, and another thing to prevail on designers to use them. Pending official matter in this direction, it is suggested that spring users attempt something on their own account by reducing their section requirements, particularly in sections of the nature $\frac{9}{32}$ in., $\frac{11}{32}$ in., $\frac{13}{32}$ in., and $\frac{15}{32}$ in. which should be eliminated.

The main item to be determined so far as the spring is concerned, is generally the thickness, and formulæ for essaying this have been given. No doubt should be present as to the type of end finish to use, which should be either a square spear or square cut. The centre fastening arrangement should be, if possible, a downward nib. For railway and tramway work, there is rarely any difficulty in including this. Automobile designers need not include it if they are sure that their centre clamping devices will not slack back. Failing this certainty, they would be well advised to introduce the downward nib, and arrange clips on the short plate, so as to bind together the whole spring.

The design of the suspension ends turns on very many considerations. In railway work, the only dogmatic statement that can be made is " avoid the welded back plate " or " solid-end back." Use always rolled-eyes, jumped ends, or loose washers, to take the suspension arrangements. Automobile makers invariably use the rolled eye type.

If trouble is experienced with failures, it is not altogether fair to blame the steel and heat treatment. This is invariably done, and sometimes correctly done, but there are numerous other aspects which can be regarded. It is not considered good form, even in commercial circles, to return springs for replacement due to fractured hoops, or back plates, when the vehicle of which they once formed a portion, has been wrecked beyond repair, or the locomotive has dropped into the turntable pit. Incidents such as these explain a proportion of " spring failures." Apart from possibilities of these natures, however, the design of any spring giving trouble should be studied before the steel, as such. It might be well, in passing, to sometimes study the manufacturing methods of the plant supplying the spring, as such are not always above reproach, but the manufacturer, on the whole, makes a good job of the springs he has to make ; the unfortunate thing being that he cannot make a good job of the designs. Generally speaking, in this country spring makers are very chary of raising any points after they have received an order, and the drawing to work to. They are also chary of raising any matters before they receive the order, and when they deal with the enquiry, in case they should miss the work owing to venturing to question points which they know are not sound. Manufacturers can claim as a rule a certain knowledge based on lengthy experience of thousands of spring types, whereas the average designer's experience probably only covers a few score. It would therefore, be well at all times, if the designer was prepared to admit that a maker had some little information, which he could, at anyrate, take advantage of assimilating, without necessarily being bound to agree. The foregoing matter is intended to summarize as " in case of trouble study the design before the steel or the manufacturing process." It is not wise to alter specifications of steel quality, whatever variety is being used, without careful investigation—the differences between steels being small compared with the differences between designs which lead to some springs " rusting out," and others failing monthly.

On the Continent spring design is generally as correct as it can be for practical use—as designers appear to have studied all aspects very thoroughly. Also the standardization of sections is assisted by the number of railways which have followed German practice.

In America, spring users do not worry much with designs as regards the detail. They demand a spring of such a length, and such a width, with a certain type of end, and a certain centre fastening, TO CARRY A CERTAIN WEIGHT. Having dictated these points the matter is in the hands of the spring maker, whose designing staff attack the problem with numerous and elaborate data sheets. On occasion, plate thicknesses are specified, but with liberty to the manufacturer to amend, if he first submits his revised ideas for approval.

The extent to which the practice is carried out of leaving matters to the spring maker, is indicated by the number of

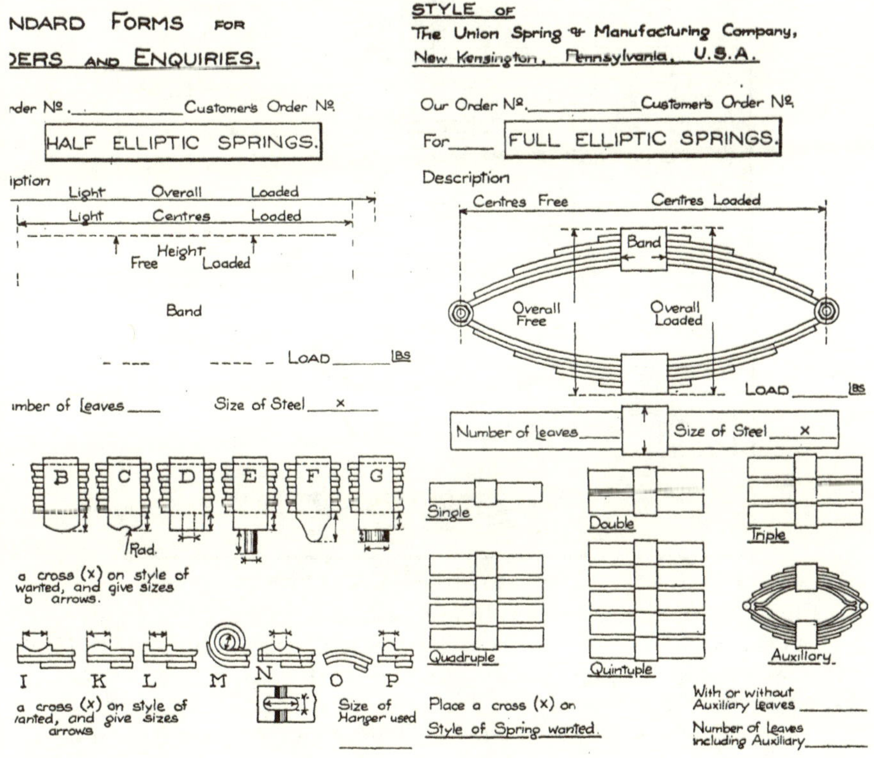

IG. 93.

standard sheets in use. Fig. 93 shows copies of two sheets issued by The Union Spring & Manufacturing Co., one being for engine springs and the other for car bolster springs. With such example, from which it will be seen that the user is not intended to study the lengths of offsets, or the sizes of studs and slits, it should not be a difficult thing to find acceptance for a similar method here, at anyrate, as regards automobile springs. Owing to the large numbers of patterns of vehicles

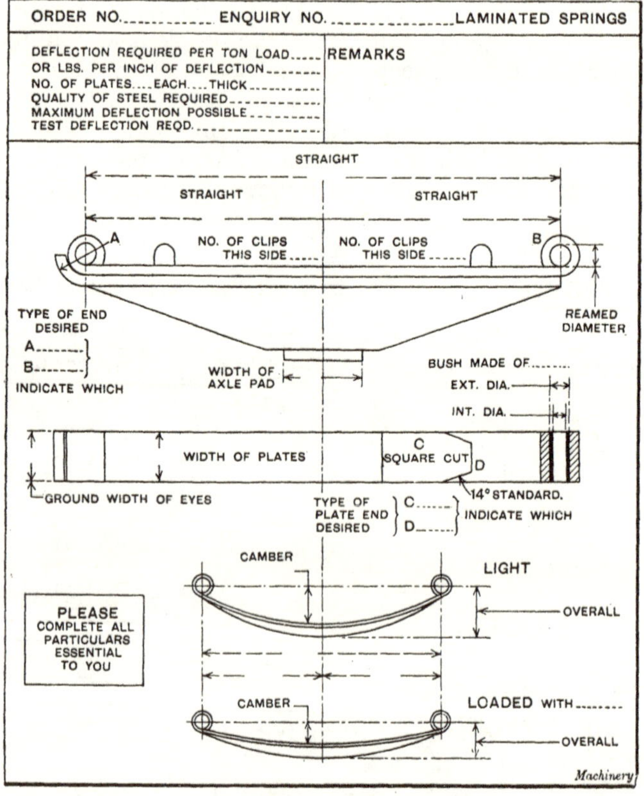
ORDER NO.............. ENQUIRY NO.................. LAMINATED SPRINGS

DEFLECTION REQUIRED PER TON LOAD......
OR LBS. PER INCH OF DEFLECTION........
NO. OF PLATES....EACH....THICK...........
QUALITY OF STEEL REQUIRED..............
MAXIMUM DEFLECTION POSSIBLE...........
TEST DEFLECTION REQD.....................

REMARKS

Fig. 94. Standard Forms—Automobile Springs.

included in the railway rolling stock, it is probably not quite as easy, but the new " grouping system " of the lines should assist some standardization, and ultimately perhaps lead to sheets of this description for railway springs. Automobile manufacturers, however, have a better chance, and it would materially assist everyone concerned if they could see their way to order springs by means of a standard sheet. A suggested one is shown in Fig. 94.

An example of how a spring should not be designed is shown in Fig. 95 Sketch D. Without making any reference to the solid-end back plate, regard the number of plates, the varying thicknesses, the high camber, and the result of all, the consequent weight. Such springs invariably present a distortion of plate section due to the presence of 22 distinct laminæ, from $\frac{5}{8}$ in. down to $\frac{1}{4}$ in. in thickness. This is extremely objectionable, as it assists in the breakage of the

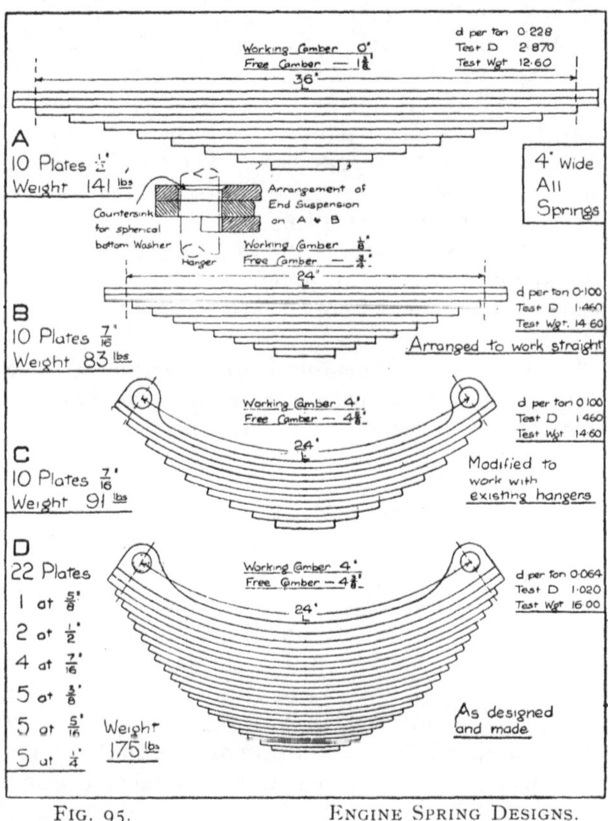

FIG. 95. ENGINE SPRING DESIGNS.

short plate, and causes a very poor job to be made of the hooping, as it is clear that the hoop cannot well be machined to fit a more or less unknown bottom plate curve. This distortion is indicated in Fig. 96.

The long radiused portion of the solid ends stiffen the spring in an unpleasant fashion, and such a design has not a good point in its favour. Needless to say, probably a $\frac{5}{8}$ in. centre hole will be included.

Large numbers of springs of this general type are in use in this country, particularly on works and colliery railways. The justification from which the type has been argued, has doubtless evolved from the fact that the permanent way of such yards is mostly in bad condition, and springs must therefore be very stiff to stand up to the bad road. This argument works circularly—does the bad road break the good spring, or does the bad spring break the good road ? Any permanent way shock should be taken up by the spring, in visible vibration or deflection ; failing which the road has to take the shock. Both permanent way and springs are elastic constructions, and if the spring parallels a metal block, the track has to take its place by performing the double function of delivering and assimilating shock. The result is that the permanent way becomes steadily worse, until the point comes where it re-acts on the springs.

An alternative design to this unhappy spring of the same general construction, and interchangeable (by the use of a longer tail pin under the hoop) is shown in Fig. 85, Sketch C. It will be noted that it is about one-half the weight. Fig. 95—B, shows a still cheaper spring, arranged to work straight, and dispensing with the solid end back. The relative (approximate) costs of these three patterns with D = 100, would be C = 55 and B = 40.

In the top sketch, Fig. 95—A is shown a spring which was designed for works' locomotives, to take the place of such heavy and stiff springs of the type which has just been criticised. When first put into service on a new engine, advices were received from the traffic department that the springs had failed, and on investigation, it was found that this information was based on the fact that the springs had actually been seen to deflect when running over bad joints, etc., which phenomenon had never before been observed. The belief was assisted by the additional fact that the deflection was backwards, which was easily explained as the springs had been designed to work straight under load, which they did, and the B.S. test applied was $1\frac{3}{8}$ ins. "past straight." Many works' locomotives, however, have not sufficient clearance between the tops of the axleboxes and the horn-block crowns to permit a very flexible spring, particularly if the roads are very bad, and care has to be given to this point. In the U.S.A. complaints are sometimes made of the axlebox striking the frame when the normal distance between the two is 4 ins. ! This travel is probably assisted by the fact that springs are officially allowed to run with a certain limited number of broken plates (if these are not top plates).

Regarding the camber of springs, unless special reasons arise in any direction, it is always theoretically best to arrange that the spring is straight, or flat, under full load, as by such design, true deflection and ordinate paths coincide, and the maximum vertical travel is obtained under shock, that is,

Back Plate - Cross Section
Tension Side $\frac{1}{2}''$
Concavity $\frac{1}{32}''$ total = $6\frac{1}{4}\%$ $\frac{1}{2}''$
Concavity $\frac{1}{32}''$ total = $8\frac{1}{3}\%$ $\frac{3}{8}''$
$\frac{3}{8}''$
Concavity $\frac{1}{32}''$ total = 10% $\frac{5}{16}''$
$\frac{5}{16}''$
Concavity $\frac{1}{32}''$ total = $12\frac{1}{2}\%$ $\frac{1}{4}''$
$\frac{1}{4}''$
$\frac{1}{4}''$
$\frac{1}{4}''$
Compression Side

Showing result of cumulative concave on plates of a spring designed with varying thickness plates and relatively high camber.

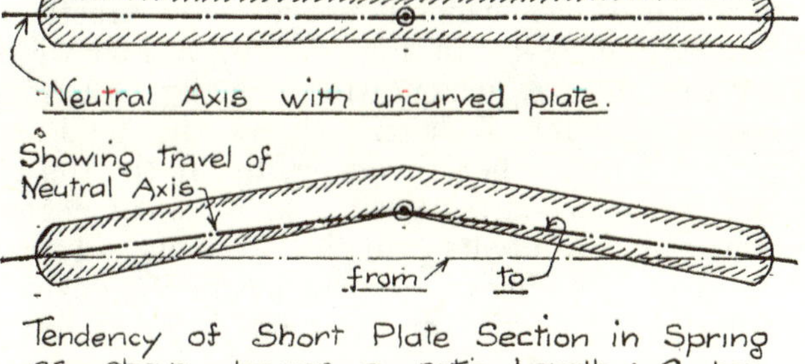

Tendency of Short Plate Section in Spring as above, having a ratio Length ÷ Camber lower than 10.

FIG. 96. EFFECT OF NUMEROUS MIXED PLATES.

the spring is " more sensitive." There are, however, certain objections to this flat loaded spring, and if it is to be employed, special care should be given to hoops or axle attachments. For this reason, it is often recommended that the camber of the spring light should be equal to the test deflection which means that the camber under load will be one-third to one-half the test deflection. The use of a straight loaded spring with ordinary central attachments, causes the spring to become automatically stiffer under a shock, owing to the general shortening which takes place with the backwards spring

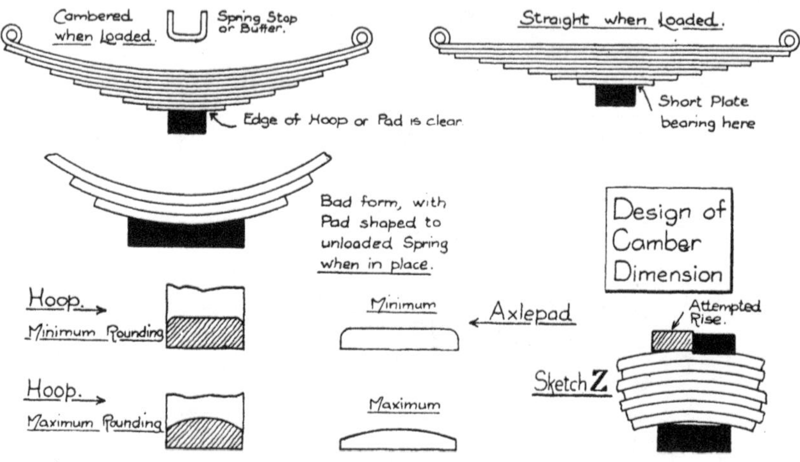

FIG. 97.

movement. The tendency of this is, however, to break the short plates on their bearing planes, and to obtain uniform working, the inside bottom of the hoop or top of the axlepad should be rounded over, which is not always a practical job, particularly with hoops. In Fig. No. 97 are shown some aspects of this matter, which will probably assist in the understanding. Sketch Z herein shows the tendency of an automobile spring, when reversing, to lift the top clamp plates, and this effort is responsible to no small degree for the stretching of clamp bolts (usually too small and too soft) with consequent loosening of the whole clamping device, and the throwing of the maximum stress line into the top plate centre holes, as indicated in Fig. 24. The same effect will take place with a locomotive spring if the hoop is thin relative to the spring capacity. In such cases as these, the short plate will generally break through the centre hole or nib, whereas if the clamp bolts or hoops are on the strong side, a reverse camber under shock will cause failure at the bearing edges of the plate.

Before leaving these general remarks, it will probably be of interest to show detail examples of British, Continental, and American design. The selected springs, illustrated in

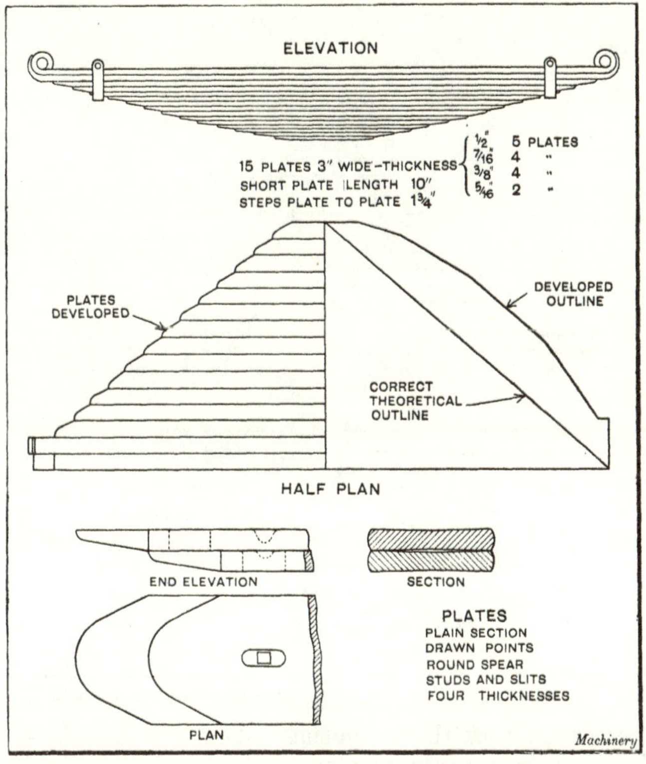

Fig. 98. British Spring.

Figs. 98, 99, and 100, respectively are for a heavy automobile, rear axle, and are all designed to carry the same safe load.

The British spring, Fig. 98, includes 15 plates of four thicknesses. By reference to the developed plan, it will be seen that the spring outline is so far from the theoretical outline that it would not be possible to fit the spring properly with plates all one radius, as numerous radii are required to obtain the necessary nip, decreasing from the top plate to the short plate, which, for machine fitting purposes, involve separate dies or settings for each radius. Such springs, therefore, are not suitable for machine work.

The second plate is wrapped partly round the top or back plate. This is not universal in British practice, but it is far from uncommon. Opinions differ as to its value or otherwise, but when used, as shown, it serves to support the

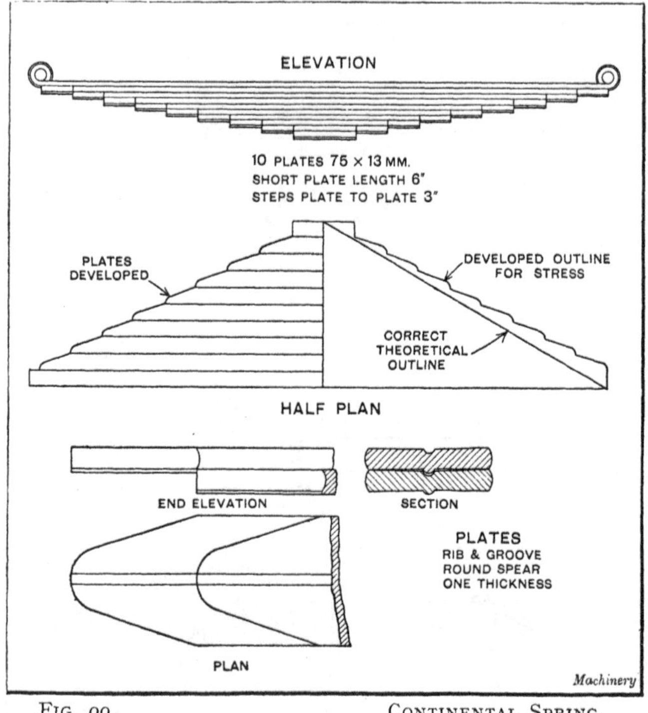

FIG. 99. CONTINENTAL SPRING.

eye, and minimises the seriousness of an eye breakage, and on the whole it is recommended.

The ends of all the plates, except the top and second, have drawn points. These drawn ends are also trimmed, with a round spear of no particular standard. With the exception of the five top plates embraced by the clip, all plates are studded and slitted to check lateral movement —as shown in the view of the plate ends. The centre of the spring will probably be held together by a bolt, passing through a round hole in each plate. The section of the plates is normal.

Continental practice is generally inclined towards the foregoing for pleasure automobiles, but for commercial purposes, diverts along other lines, as indicated by the spring illustrated by Fig. 99. It will be noted that all plates are of one thickness, with a correctly designed short plate, the result being that the amount of material over the true rhombus

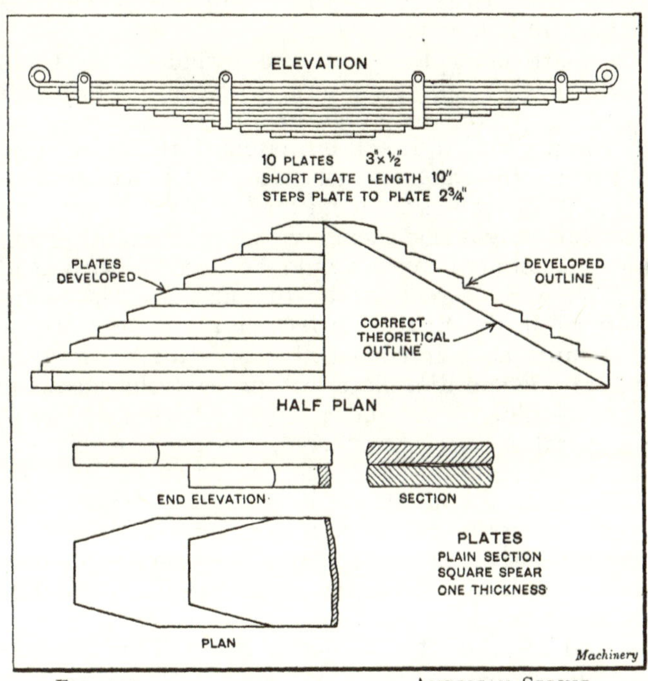

FIG. 100. AMERICAN SPRING.

outline is very little, so that one-radius plates, and machine forming can be employed.

The plate ends are finished with a round spear, and are not tapered in thickness. The steel section used is rib and groove, which obviates the need for studs and slits as side-play checks. A centre bolt is used, and no clips are fitted —Continental practice indulging but little in these details.

American design for all classes of automobiles, pleasure and commercial, is practically identical—only sufficient exceptions occurring to prove the general rule—and Fig. 100 shows a type spring. All plates are the same thickness, the short plate being slightly longer than the Continental pattern,

Q

but not sufficiently long to have any influence on the making of the spring as "one-radius." The plate ends are finished with the standard 14° square spear of the U.S.A., and although the steel section is the normal round-edge and concave, no studs or slits are employed, the four clips fitted being sufficient as checks on the first eight plates—the bottom two being kept from twisting by the centre clamp bolts. An ordinary centre hole is provided.

In the particulars Fig. 101, are included the leading particulars of each spring pattern, with the operations required, and approximate proportionate cost. For the purpose of comparison, it has been assumed that steel prices and labour are the same in each case, and it has also been assumed that the springs are hand-made. As the Continental and American designs lend themselves also to machine making, a further reduction in cost would be accomplished if the manufacture were carried out by this means. The number of operations on the British pattern spring is very obvious, though it must be pointed out that spearing, slitting, and studding, at one end of a plate, will probably be carried out with one handling, but under three different tools. All possible operations are shown cold, though probably some will have to be performed hot—dependent on the quality of steel. The final test of springs is generally the commercial one of price (except in certain high-class automobile practice), and this turns largely on the weight of steel employed. Such final comparison shows the British, Continental, and American designs, to be as 156 : 100 : 106. The similarity between weights, and costs, of the two latter is worthy of remark—the chief difference between them consisting in the round spears and rib and groove steel (Continental) against the ordinary section with square spears (American), and the provision of clips or shackles on the latter.

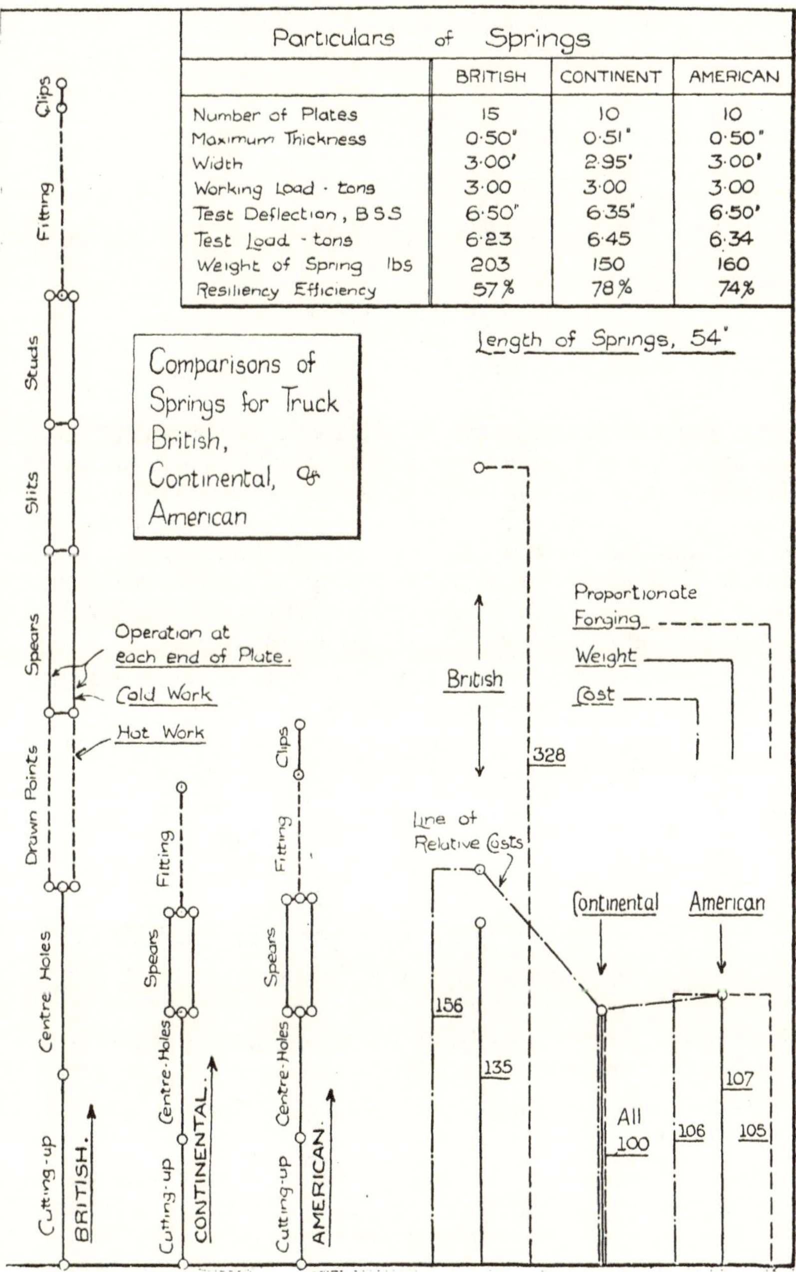

Particulars of Springs	BRITISH	CONTINENT	AMERICAN
Number of Plates	15	10	10
Maximum Thickness	0·50"	0·51"	0·50"
Width	3·00"	2·95"	3·00"
Working Load · tons	3·00	3·00	3·00
Test Deflection, BSS	6·50"	6·35"	6·50"
Test Load · tons	6·23	6·45	6·34
Weight of Spring lbs	203	150	160
Resiliency Efficiency	57%	78%	74%

FIG. 101.

CHAPTER XXIV

AIDS TO SPRING CALCULATIONS

THE designing of springs in detail is really an epitome of the whole subject, and cannot therefore be included in one chapter. Running comments on points of design to be avoided, and other points to be advocated, will be found consistently throughout the section dealing with "Manufacture." It will be of use, however, to reduce spring calculations down to their finest point, and to this end the minimum number of formulæ might be recapitulated and summarised.

The really essential formulæ to have, either in mind, or in a handy adjacent form, are the following :—

A. That giving unit deflection for semi-elliptic springs, as shown in Chapter VII, No. VII.—3.

B. That giving a standard maximum test deflection, as for instance the British Standard, as in Chapter IV, No. IV.—8.

C. That giving the approximate correct weight of a spring for (K) as 0·100, as given in Chapter VII, No. VII.—2a.

D. That giving the approximate actual weight of any spring, as in Chapter XX, No. XX.—4.

E. The essaying formulæ for design, according to thicknesses, are in Chapter XIV, No. XIV.—1 and 2.

F. That giving increment for true (L) according to span and camber, as Chapter V, No. V.—3.

The semi-elliptic is the most common of all types, and it is therefore recommended that only the formulæ relating thereto be kept for instant reference—those relating to other and rarer spring types can be found as desired. Accordingly, the table at the end of this chapter gives certain of these most necessary formula in English, Metric, and "American" units, the latter including lbs., where the English includes tons.

In places where spring calculations are necessary in the regular course of events, such as the works of spring makers, it is convenient to introduce graphs to include certain of the formula aspects. Fig. 102 shows such a graph for the British

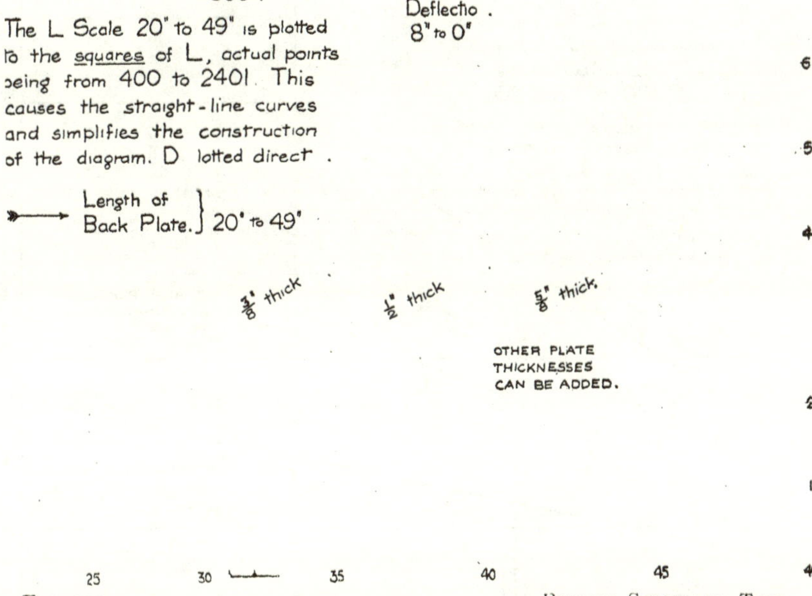

FIG. 102. BRITISH STANDARD TEST.

Standard Test. On the scale of this book, it is not possible to show it of such a size as to be very useful in working, but the construction indication is shown, for 20 ins. to 49 ins., with $\frac{3}{8}$ in., $\frac{1}{2}$ in. and $\frac{5}{8}$ in. plates. An actual working graph for all sizes, would include 20 ins. to 80 ins., for spring length (L) and plates from $\frac{1}{4}$ in. to $\frac{5}{8}$ in. by sixteenths. The method of obtaining the straight line "curves" is stated on the diagram, and by this means, only two points, the first and last—in the case shown 20 ins. and 49 ins.—have to be calculated.

For the unit deflection, a similar construction is employed, as shown in Fig. 103, which gives the deflection per ton for one plate, of the section indicated. The unit deflection of the whole spring is found by dividing this by the number of plates or equivalent plates, if the spring be composed of " varying thicknesses."

The importance of obtaining the correct straight length (L) has already been impressed, and, as many drawings include only span and camber, a separate calculation is needed

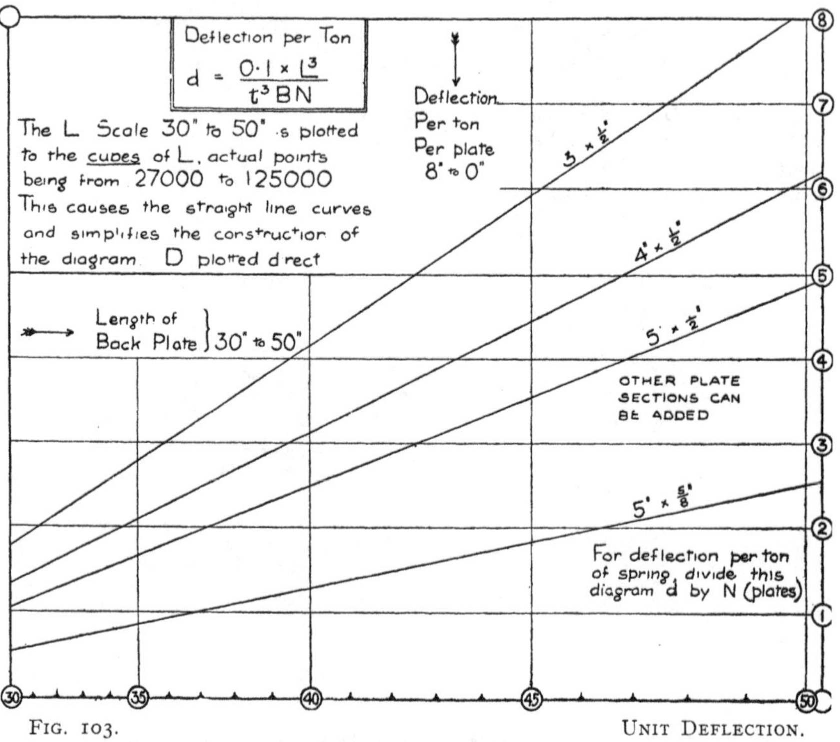

FIG. 103. UNIT DEFLECTION.

for (L) before the unit deflection or test can be obtained. The graph for this is relatively complicated, and is shown in Fig. 104, which has been based on the particulars of arc increments in Molesworth. The " master curve " is plotted direct from the table (page 698, 25th edition), and the increments over the length range (top datum line) of 25 ins. to 45 ins. are worked along each main ordinate. To find the increment on any length of span between 25 ins. and 45 ins., with ratio span ÷ camber (or S/C as on graph) between 6 and 16, follow the procedure needed for the example shown, namely, a span of 36 ins. and camber of 4 ins. The ratio is 36/4 = 9, which is referred to on the bottom datum line of S/C. Follow this ordinate till it reaches the master curve, and from thence divert horizontally until the 36 ins. down-coming ordinate is reached. The point thus obtained falls between the 1 in. and $1\frac{1}{4}$ ins. secondary curves, and by estimate, one would call this $1\frac{3}{16}$ ins., which is the amount of the increment necessary. In other words, the "straight"

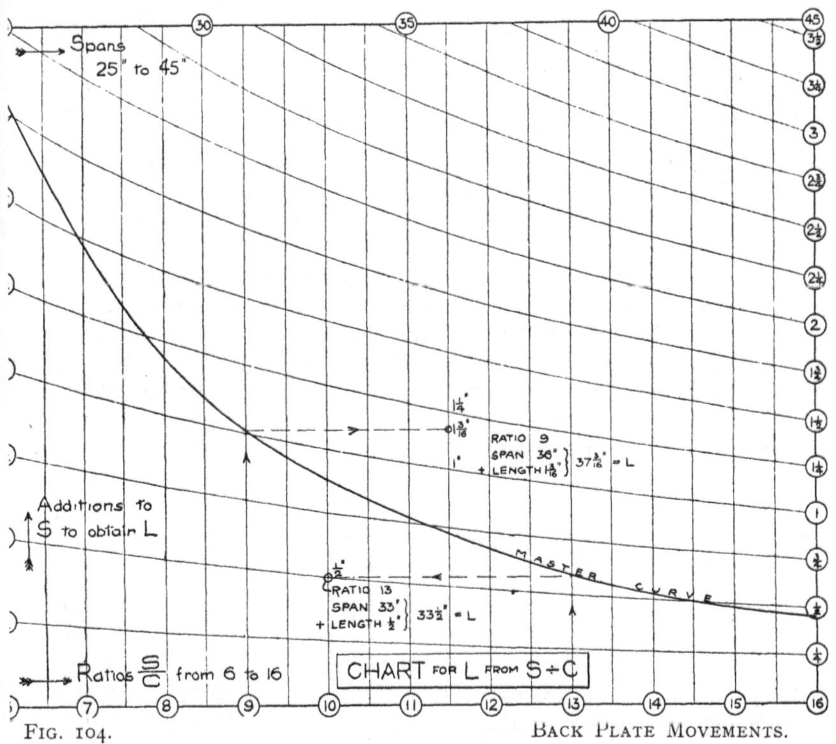

FIG. 104. BACK PLATE MOVEMENTS.

length of an arc of 36 ins. span (or chord) and 4 ins. camber (or versin) is 36 ins. plus $1\frac{3}{16}$ ins. In Chapter V were indicated corrections necessary for "headed" backs, with solid-ends or rolled eyes, and these must be taken account of. Recapitulating, for rolled eyes, add the thickness (T) of the top plate, so that the ratio for the bottom datum line becomes span ÷ (camber plus thickness of back) ; and for solid-ends, add $\frac{1}{2}$T, making the ratio span ÷ (camber plus half-thickness of back).

Alignment charts can be used instead of the graphs suggested, but on the whole, it will be found that the latter are the easier to work from. Other valuable accessory particulars for spring calculations are a list of figures corresponding with (t^3) on the lines of the following :—

T							
T =	$\frac{3}{8}''$ =	$\frac{12}{32}''$ =	$\frac{24}{64}''$			= 0·375".	t^3 = 216
T =					= 0·38"		t^3 = 225
T =			= $\frac{25}{64}''$		= 0·39"	= 0·390"	t^3 = 243
T =				= 10 mm.		= 0·394"	t^3 = 250
T =					= 0·40"		t^3 = 262
T =		= $\frac{13}{32}''$ =	$\frac{26}{64}''$			= 0·406"	t^3 = 274
T =					= 0·41"		t^3 = 282
T =					= 0·42"		t^3 = 305
T =			= $\frac{27}{64}''$			= 0·421"	t^3 = 308
T =					= 0·43"		t^3 = 325
T =				= 11 mm.		= 0·433"	t^3 = 333
T =	$\frac{7}{16}''$ =	$\frac{14}{32}''$ =	$\frac{28}{64}''$			= 0·437"	t^3 = 343

A complete list drawn up on the above lines, is a considerable timesaver, as it includes, apart from normal sizes, $\frac{1}{32}$ in., $\frac{1}{64}$ in., and $\frac{1}{100}$ in., with the millimetres running throughout. In the event of steel being rolled " thick " or " thin " to specified size, the effect on the unit deflection can be readily determined from the gauged section without it being necessary to calculate (t^3) in sixteenths (say) 0·42 ins. which may have been supplied for $\frac{7}{16}$ in. The figures above show that 0.42 is equivalent to 305, as against 343 for $\frac{7}{16}$ in., which indicates that the spring will be weaker to the extent of the 305 : 343 proportion.

A further diagram which is of value relates to the weight of springs for estimating purposes. It is based on the Formula XX.—4, and illustrated in Fig. 105. An alignment chart is shown in this case, as a matter of interest. The first ordinate necessary is the scale ($L + L^s$) in this case made ranging from 30 ins. to 70 ins., and plotted as the logarithmic values of 30 to 70. An ordinate representing (b) or breadth is next set up, at a convenient distance, in the diagram = 2X, This distance is bisected, and an intermediate ordinate erected, which gives the product of ($L + L^s$) and (b). At a further convenient horizontal distance from this " product " ordinate, two further are erected, at equal distances (Z) apart, the first of which will be the final weight figure required, and the second, the total thickness of the spring (TT). All values on the scales are logarithmic. The " weight " ordinate really represents the product of ($L + L^s$), (b), and (TT), but as (K) is a constant of 6 for nearly all springs, the division by (K) can be made on this weight ordinate, so that instead of the high extreme being $350 \times 10 = 3500$, it is shown as $350 \times 10 = 3500 \div 6$, or 580 (approx.). Scales, and intermediate graduations, can be made as desired. To work from the diagram a spring with, say, ($L + L^s$) as 40 ins.,

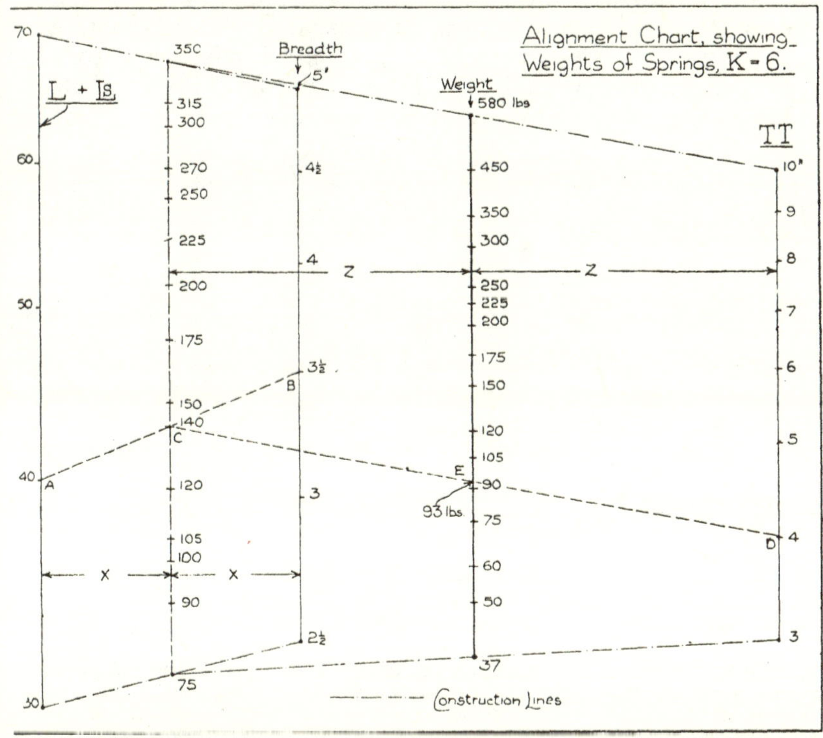

FIG. 105.

(b) as $3\frac{1}{2}$ ins., and (TT) as 4 ins. (or 8 plates @ $\frac{1}{2}$ in.), place a straightedge from 40 (1st scale) to $3\frac{1}{2}$ (3rd scale), and the product of 140 is on the 2nd scale. (This need not be read, and can be in arbitrary units—in this diagram, however, the correct graduation facilitates the construction of the " weight" ordinate.) From this point on the 2nd scale, line to 4 on the 5th scale, and the answer of 93 lbs. is found on the 4th scale. This example is illustrated.

In the design of new springs, the most important unknown factor is invariably the thickness. Lengths and widths are usually more or less settled features, and the number of plates is always readily adjusted—but on the thickness mainly turns the solution of the design. These calculations can be simplified to a remarkable extent by the standardization of sections as far as possible, and formulæ can then be derived for essaying (bn) in terms of a standard thickness. For instance, suppose American lines were followed in the design of a set of springs as shown for a usual British locomotive pattern—Fig. 106. In the U.S.A. ½ in. thick for such work is used invariably, and taking this thickness, the formulæ for finding (bn) is XIV.—2, recapitulated :—

$$bn = 0{\cdot}175\text{WL}.$$

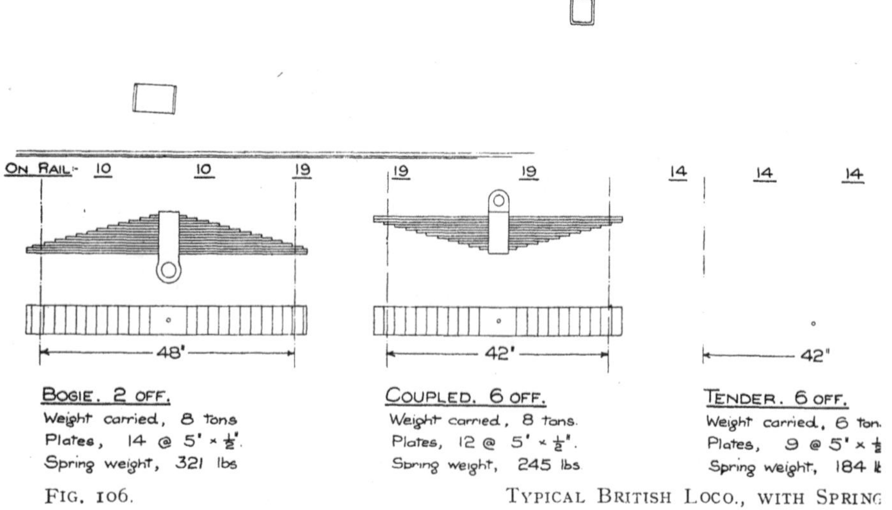

FIG. 106. TYPICAL BRITISH LOCO., WITH SPRING

Applying this to the three patterns of springs, the following results are obtained :—

Bogie Springs.	(bn) = 67·4,	which may be	14 × 5, 15 × 4½, or 17 × 4
Coupled ,,	(bn) = 58·9,	,,	12 × 5, 13 × 4½, or 15 × 4
Tender ,,	(bn) = 44·2,	,,	9 × 5, 10 × 4½, or 11 × 4

As 5 ins. is possible for nearly all engine work, this width is clearly the best to use, and if the springs are calculated throughout, the following figures appear (Plates all 5 ins. × ½ in.) :—

Bogie Springs :—14 plates, test load = 2·08 times working load.
Coupled ,, :—12 ,, ,, = 2·03 ,, ,,
Tender ,, :— 9 ,, ,, = 2·03 ,, ,,

The essaying formula given is based on a factor of 2·00 for test load ÷ working load, and if a higher factor is required, the figure 0·175 must be amended to suit.

By the use of the graphs, the design of each type can be accomplished equally quickly. From the B.S.S. Test graph, a ½ in. plate, 42 ins. long, can deflect 3·9 ins. From the unit deflection graph, 5 ins. × ½ in. × 42 ins. deflects 2·9 ins. per ton. The working load for these (coupled) springs being 8 tons, the test load should be 16 tons, which should correspond to a deflection of 3·9 ins. or 0·245 ins. per ton. The division of 2·9 ins. (deflection of ONE plate, 5 ins. × ½ in.) by 0·245 in. gives the required number of plates, exact answer being 11·8.

A simple list of equivalent plates in the event of "varying thickness" plates being employed in a spring, is given below, and this can be amplified with advantage by the inclusion of millimetre, and oddment sizes. However, the given list is generally sufficient for most heavy shops.

EQUIVALENT PLATE THICKNESSES

One Plate of Thickness given below =	¼ in.	5/16 in.	⅜ in.	7/16 in.	½ in.	9/16 in.	⅝ in.
¼ in.	1·00	—	—	—	—	—	—
5/16 in.	1·95	1·00	—	—	—	—	—
⅜ in.	3·40	1·70	1·00	—	—	—	—
7/16 in.	5·40	2·75	1·60	1·00	—	—	—
½ in.	8·00	4·10	2·35	1·50	1·00	—	—
9/16 in.	11·40	5·80	3·35	2·20	1·40	1·00	—
⅝ in.	15·65	8·00	4·60	2·90	1·95	1·40	1·00

In designing laminated springs for normal purposes, the minimum number of plates employed should be four. For certain purposes, it might be desirable to design a spring with a single plate (as Fig. 91/2) and this is occasionally done. One plate is, of course, the irreducible minimum, but there is no obvious maximum, and frequently springs are made of thinner plates than needful, with resultant increased weight and cost. A well-designed spring—with all factors regarded—of $\frac{5}{8}$ in. thick plates, will give as much service as one twice the weight of $\frac{1}{2}$ in. thick plates. One important feature of a laminated spring must not be overlooked in this connection, namely, the safeguard provided by its being made up of a number of independent units. It is in this respect that it is superior to a coil spring for bearing purposes, as, unless a group of these are arranged—in which case they parallel the independent units of the plate spring—failure of the spring results in either the whole becoming solid under load (in the case of a soft spring) or the load being carried unsatisfactorily on a fractured spring, which causes immediate drop in height of framework, and greater deflection, until the two halves become useless. The breakage of any one plate in a laminated plate, apart from the back plate, is a matter of relatively small moment on (say) a 10-plate spring, and such breakage is frequently not discovered until the hoop is removed, as on a uniform thickness spring, it will only increase the deflection by 11 per cent. Generally, however, if broken plates are known, the spring should be replaced, although, in the U.S.A. definite permission is accorded to run engine springs with up to three and four broken plates. Such practice, however, invites further breakages, as additional stress is thrown on to the sound plates. The 4-plate spring suggested as the minimum is quite satisfactory in working, and is used in this country for railway wagons. On the Continent 5-plate and 6-plate springs for heavy automobiles are not uncommon. Clearly, however, the fewer the plates, the less perfect can be the spring shape, unless very long spears are employed—which are not as a rule, practicable. The following is suggested as the minimum number of plates for various classes of work :—

Railway :—	Locomotive	8 plates.
	Coaching Stock	6 ,,
	Freight Stock..	5 ,,
	Buffing Springs (generally 70 ins. or over)	10 ,,
Tramways :—	Passenger Cars	6 ,,
	Freight Cars	5 ,,
Automobile :—	Front Springs	6 ,,
	Rear Springs	8 ,,

No spring should be designed with plates thicker than $\frac{5}{8}$ in. unless some very exceptional requirements have to be met. Plates of $\frac{3}{4}$ in. and $\frac{7}{8}$ in. are occasionally used, but it must always be remembered that the thicker the plate, the less susceptible it is to uniform heat treatment.

The following concentrated summary of important points concludes this chapter on " aids to design." The reason for some of these has been already emphasised, and others are treated at length in the Manufacturing Section.

a. Springs of plates of uniform thickness are the only perfect type, both theoretically and practically. They reduce the number of sections to be stocked for repair purposes —which is an important practical item in favour of such design.

b. The ends of spring plates should either be left square cut, or speared (tapered in width). For springs with less than 3 ins. offset, square cut ends are recommended.

c. The offsets of spring plates should be uniform throughout and all plate lengths should be multiples of the short plate.

d. Two, or at the most, three, full length plates, are practically sufficient for all types.

e. Rib and groove steel is a good section, for plates over $\frac{3}{8}$ in. thick. Studs and slits are unnecessary for railway work on springs under 36 ins. long. Back plates should never be studded, and if not checked laterally by suspension arrangements, clips should be fitted to the 3rd or 4th plates. Automobile springs should always be clipped as alternative to studding and slitting. Clips for these springs also take rebounds.

f. All qualities and treatments of steels give the same unit deflection, as the modulus of elasticity (E) remains constant, on uniform section, at 13000 tons. (20,000 kg. per sq. mm.)

g. If the spring is correctly designed, ordinary water-hardening steels will be found satisfactory for most work.

h. " Upward nibs " should never be employed as centre fastening means. The most generally satisfactory forms are the cotter hole or downward nib.

i. Every endeavour should be made to avoid the use of welded back plates. Plain ends—with loose washers—or small " jumped " ends are recommended for locomotive work, and rolled-eyes for all other classes of work.

j. In designing " band " hoops, if at all possible, maintain uniform thickness on all sides.

k. Special hoops, of lugged and other patterns, as required for locomotive work, should be machined from the solid steel block.

l. Specify always the testing of the completed spring, and adhere to the requirements of the British Standard Specification. This has been adopted as a rule of testing by all leading spring makers and railway engineers of the " British School," and has proved itself most satisfactory from both points of view.

m. For the testing of spring steel, analysis results and camber tests on B.S.S. lines, can be considered sufficient.

n. The purity of steel, as exemplified by analyses with less than 0·035 per cent. Sulphur or Phosphorus, has to be paid for. These elements can be allowed up to 0·07 per cent. without deleterious effects.

	British	American	Metric
d=	$\frac{0\cdot1\,L^3}{nbt^3}$	Symbols have the same significance as throughout the text. Metric 'tonne' = 1000 KG.	$\frac{0\cdot0155\,L^3}{nbT^3}$
D=	$\frac{L^2}{900\,T}$	←→ Increment for L. Common to all.	$= \frac{2\cdot65\,D^2}{S}$
TEST W=	$\frac{0\cdot177\,nbt^2}{L}$	$\frac{10^5\,nbT^2}{L}$	$\frac{0\cdot071\,nbT^2}{L}$
w=	$\frac{[L + L^{S}] \times b \times TT}{6}$		$\frac{\leftarrow \text{SAME}}{220000}$

W here is Working Load with Test Load as twice this Working Load.

For Essaying Thicknesses

$$bn = KWL$$

INCHES	$\frac{5}{8}$	$\frac{9}{16}$	$\frac{1}{2}$	$\frac{7}{16}$	$\frac{3}{8}$	$\frac{5}{16}$
K=	·112	·138	·175	·228	·311	·450
M/M	15	13	12	10	9	8
K=	·125	·167	·196	·281	·348	·440

ESSENTIAL FORMULÆ.

British in Inches & Tons, American inches and lbs., Continental m/m & tonnes.

LAMINATED SPRINGS

PART B :—MANUFACTURE

CHAPTER XXV

SPRING STEEL—MANUFACTURE AND ANALYSIS

In the previous Section it has been necessary, from time to time, to touch upon the rolling and qualities of spring steel, and these important points will now be gone into at greater length.

Spring steel qualities—for trade purposes—divide under three heads, "Water-hardening," "Oil-hardening," and "Alloy." By the first two distinctions, "straight" carbon steels are always implied, that is, steels which contain no purposely added alloying metal, such as chromium or vanadium. Before dealing in detail with each of these grades, it might be of interest to recapitulate briefly the processes of steel manufacture which will have to be referred to, five in number, namely—

(1) Crucible.
(2) Bessemer Acid.
(3) Bessemer Basic (Thomas Steel).
(4) Open Hearth Acid (Siemens-Martin Steel).
(5) Open Hearth Basic (Martin Steel).

(1) Crucible Steel.

In this method, the charge of specially selected high-grade material is melted up in small crucibles, and then "teemed" into ingots. Such ingots can be made of any weight by pouring from a sufficient number of crucibles (Krupp's by a high organization, have poured 2,000 crucibles into one ingot, weight about 50 tons) but modern British practice has practically discarded crucible work except for tool steel, where one crucible, of 40 to 50 lb. weight, makes one ingot. Before the rise of what might be termed the bulk processes, such as the Bessemer and Open Hearth, all steel was made by the crucible method, and to comparatively recent dates, many specifications retained this clause in connection with spring steel manufacture. A certain amount of steel made in the crucible is still used in America.

(2) BESSEMER ACID STEEL.

In this process, non-phosphoric pig-iron, molten from a separate cupola or taken direct from the blast furnace, is poured into a " converter " usually between five and ten tons capacity. Air is then blown through the liquid iron until all the carbon is removed. The molten metal is then re-carburized to the necessary temper by the addition of spiegeleisen or ferro-manganese (pig irons high in manganese) and poured into ingot moulds of any required capacity, usually ten to fifteen cwt. The bulk of spring steel in this country was made by this method up to about fifteen years ago, and there continues to be a fair amount produced, although the process is being gradually displaced by the higher capacity Open Hearth furnaces, with a lower working cost per ton. By the Bessemer process, pig iron is converted into steel in about twenty minutes.

(3) BESSEMER BASIC STEEL.

Known on the Continent as " Thomas Steel." Practically the same as No. 2, the difference being in the pig iron used, which is of lower grade and cheaper quality due to its being high in phosphorus. This deleterious element is removed by the basic process, but not by the acid. (The terms " acid " and " basic " refer to the lining of the converters, the former being ganister and the latter dolomite). The procedure follows No. 2 except that an " after-blow " is necessary before re-carburization, during which the amount of phorphorus present in the pig iron is reduced to about 0·05 per cent. A considerably quantity of lime has to be added to the molten pig to assist the lining in the elimination of phosphorus. This steel is rarely employed in this country for spring purposes, but a quantity is made on the Continent for such use.

(4) OPEN-HEARTH ACID STEEL.

Known also as Siemens-Martin. Non-phosphoric pig iron with a proportion of selected scrap is charged into the " bath " of a fixed gas furnace, capable of attaining an extremely high temperature. The charge is then melted, and owing to the

Nº	Steel Process	Hardening Medium	Analyses. Per Cent, % Carbon, C		Manganese, Mn		Silicon, Si		Sulphur, S		Phosphorus, P		Chromium, Cr		Vanadium, Va		Molybdenum, Mo		Spring Temper Tensile	Elastic Limit	Brinell Ball dia	As Rolled Tensile
1	Bessemer	Water	0·45	0·50	0·80	1·00	—	0·20	—	0·08	—	0·08	—	—	—	—	—	—	65	55	3·5	43
2	"	"	0·40	0·45	"	"	—	"	—	"	—	"	—	—	—	—	—	—	60	50	3·6	39
3	Open Hearth	"	0·70	0·75	0·40	0·60	—	"	—	0·06	—	0·06	—	—	—	—	—	—	70	60	3·4	50
4	"	"	0·55	0·60	0·60	0·80	—	"	—	"	—	"	—	—	—	—	—	—	65	55	3·5	45
5	"	"	0·50	0·55	0·40	0·60	—	"	—	"	—	"	—	—	—	—	—	—	65	55	3·5	43
6	"	Oil	0·75	0·80	0·60	0·80	—	"	—	"	—	"	—	—	—	—	—	—	65	55	3·5	52
7	"	"	0·90	1·10	—	0·50	—	—	—	0·05	—	0·05	—	—	—	—	—	—	82	60	3·1	60
8	"	Water	0·35	0·40	0·40	0·60	—	0·20	—	0·04	—	0·04	0·70	1·10	0·15	0·25	—	—	90	75	3·0	50
9	"	Oil	0·45	0·50	"	"	—	"	—	"	—	"	"	"	"	"	—	—	90	75	3·0	52
10	"	"	"	"	0·90	1·10	1·80	2·20	—	"	—	"	—	—	—	—	—	—	85	70	3·1	50
11	"	"	"	"	0·40	0·60	0·80	1·00	—	"	—	"	0·70	0·90	—	—	—	—	90	75	3·0	50
12	Openhearth	British	0·50	0·80	—	—	—	—	—	0·05	—	0·05	—	—	—	—	—	—				
13	"	Standard	0·45	0·70	—	—	—	—	—	0·05	—	0·05	—	—	—	—	—	—				
14	Openhearth	Railway	0·50	0·80	—	—	—	—	—	—	—	—	—	—	—	—	—	—				
15	or Bessemer	All Water	0·45	0·70	—	—	—	—	—	—	—	—	—	—	—	—	—	—				
16	Open Hearth	Water	0·45	0·50	0·70	0·90	1·50	2·00	—	0·10	—	0·10	—	—	—	—	—	—	80	65	3·2	44
17	"	Oil	0·90	1·05	0·30	0·50	—	0·20	—	0·05	—	0·045	—	—	—	—	—	—	82	60	3·1	60
18	"	"	0·70	0·80	"	"	1·75	2·00	—	0·045	—	"	—	—	—	—	—	—	89	78	3·0	55
19	"	"	0·42	0·52	0·75	1·00	—	0·20	—	0·04	—	0·04	1·00	1·20	0·12	0·18	—	—	89	82	3·0	50
20	"	"	"	"	"	"	—	"	—	"	—	"	"	"	—	—	0·25	0·40	89	82	3·0	50
21	"	"	0·50	0·60	0·70	0·80	0·45	0·55	—	0·045	—	0·045	0·70	0·80	—	—	—	—	85	75	3·1	50
22	"	"	0·55	0·65	0·50	0·80	—	0·50	—	0·06	—	0·05	0·45	0·70	—	—	—	—	85	70	3·1	50
23	"	"	0·45	0·55	"	"		"		"	—	"	1·00	1·40	—	—	—	—	90	75	3·0	48
24																						
25																						
26																						
27																						

relative slowness of the process compared with the Bessemer, samples can be more frequently taken, and the molten metal better controlled, until the necessary quality is obtained, when it is poured or teemed into the ingot moulds. This process is at present chiefly used in this country for the production of spring steels, but is very rare on the Continent and America.

(5) Open-Hearth Basic Steel.

Known on the Continent as "Martin Steel." The process is practically the same as No. 4, except for the employment of the relatively cheap, low grade, phosphoric pig iron, and the consequent necessary addition of large quantities of lime. In modern high production plants, the molten pig is charged direct into the bath from the blast furnace. An additional attraction of this process is the large amount of miscellaneous scrap material which can be used in the charge. Not largely employed so far in this country for the production of spring steel qualities, but universal for this purpose on the Continent and in America. Open Hearth furnaces are made from 10 to 80 tons capacity (electric, 3 to 10) and ingots up to 3 tons weight (for re-rolling purposes) are poured.

Occasionally steel is specified as "electric" which indicates that it must be melted in electric furnaces—the metallurgical processes being either Nos. 4 or 5. Light selected scrap is employed to a great extent for this method of melting.

In modern practice, "water- hardening" spring steel is made by either the Bessemer or Open-Hearth processes, and "oil-hardening" and "alloy" steels exclusively by the Open-Hearth.

The principal so-called "alloy" steels employed for laminated spring work are known as "Chrome-Silicon," "Chrome-Vanadium," and "Silico-Manganese." The advantage claimed for them as against the straight carbon steels is the high elastic limits which can be obtained with proper treatment. Representative analyses of various carbon and alloy steels are given in the accompanying table, and the following remarks apply.

No 1. A Bessemer steel, suitable for all classes of springs, and when all matters are taken into consideration, namely, ease of manufacture, cost, and service satisfaction, probably the best steel that has been produced for laminated springs. British quality.

No. 2. A similar steel, used for machine made springs (in this country) and welded back plates. The lower carbon content renders it more suitable than No. 1 for the latter purpose, as it burns less readily. A similar reason applies also to its employment for press-made springs. British quality.

No. 3. A Siemens steel for all springs for water hardening. It has been used extensively, but is not recommended, as the following No. 4 is generally better.

No. 4. A standard British open-hearth steel for all classes of springs. This is lower in carbon than No. 3, and accordingly better for manufacture, as it is less liable to heat damage, and more easy to "forge," (nib, stud, slit, etc.) necessary operations being done cold.

No. 5. A similar steel to No. 4, but for use in welded backs, and therefore of lower carbon content.

No. 6. An oil-hardening steel, employed in this country when oil-hardened carbon springs are specified for railway work—which is not often.

No. 7. The American standard specification for carbon steel for railway springs. It will be noted that the carbon is high in comparison with No. 6. With this quality, all forging must be done hot—which partly accounts for the fact that U.S.A. practice has reduced such operations to a minimum.

No. 8. A "chrome-vanadium" steel, suitable for water-hardening, as has been made here. It is not, however, frequently used. A steel of this nature is very difficult to "fit" after it has been water-immersed, the hard quality rendering it necessary to obtain a high tempering heat in order to remove warp, etc.

No. 9. A usual "chrome-vanadium" alloy as made in this country, chiefly for automobile springs.

No. 10. A "silico-manganese" steel, oil-hardening, which is favoured by some spring makers, and gives very good results.

No. 11. A "chrome-silicon" steel, favoured by other makers here, which also gives good results. Both this and No. 11 are generally restricted to automobile springs.

No. 12. British Standard Specification, No. 6, for all plates except welded back plates. Railway springs.

No. 13. British Standard Specification, No. 6, for welded back plates.

No. 14. British Standard Specification "without analysis," No. 6a, for all plates except welded back plates. Railway springs.

No. 15. British Standard Specification "without analysis," No. 6a, for welded back plates.

No. 16. A very usual Continental quality of "silico-manganese," water hardening. Gives excellent results for all classes of springs.

No. 17. American automobile spring quality, "carbon."

No. 18. American automobile spring quality, "silico-manganese." From the European point of view, this is very low in manganese.

No. 19. American automobile spring quality, "chrome-vanadium."

No. 20. American automobile spring quality, "chrome-molybdenum." This is a comparatively new steel, and has not yet found a footing in this country or the Continent.

No. 21. American automobile spring quality, "chrome-silico-manganese." In this country, it would be called a "low chrome."

No. 22. British equivalent of No. 21, "low chrome." Used for automobile springs to a certain extent.

No. 23. A British quality, "high-chrome," for automobile springs.

The tensile results given on the table are approximately those which can be obtained before and after treatment, but such are not as a rule specified, and it is recommended that they should not be: a general analysis specification is quite sufficient for the steel-maker.

Analyses Nos. 1, 2, 3, 4, and 5, all fall in accordance with the British Standard Specifications for Springs and spring steel, except as regards the sulphur and phosphorus contents. Both these elements were permitted, during the late war (1914—1918), to range considerably higher than the old standard of 0·035 per cent., engine springs being allowed to include 0·06 per cent., and carriage and wagon springs 0·07 per cent. The post-war specification has been amended to allow 0·05 per cent, which stands at present. On the whole, it would have been a considerable advantage from the steel makers' point of view, if this war-time limit of S. and P. had been tacitly permitted to remain, as the restriction to the lower figure simply means that a very excellent steel, as regards chemical composition, has to be arranged for. It has not yet been shown that any better spring is produced by this low S. and P. content, than with the higher figure of 0·06, for Engine work, and 0·07, for C. and W. work—and certainly many hundreds of thousands of springs have been made, worked well, and finally rusted out with nearer 0·1 per cent. of S. & P., than 0·07. There is of course, a British Standard Specification without analysis, and thousands of springs are made to this.

One of the leading authorities on steel metallurgy, Mr. Harry Brearley, contends that the limitation of sulphur to the figures generally given, is quite needless. He traces the objection to the pure iron-carbon steels, in which the sulphur present existed as iron sulphide, with a melting point lower than that of steel. This sulphide formed non-metallic envelopes around the crystals of the ingots, which envelopes fused when the steel was raised to a forging or welding heat, and caused the metal to break away along these paths of fusion. It was therefore concluded, and with reason, that sulphur caused steel to become " hot-short." These early steels were, however, very free from manganese, and at a later date, when the bulk steel processes came into existence, manufacturing commercial steels relatively high in manganese (0·40 to 1·00 per cent.) it was found that the sulphur content existed no longer as a straight iron combination, but as a manganese combination, namely, sulphide of manganese or double sulphide of manganese and iron. These combinations have a higher melting point than steel, and it appears that, instead of forming envelopes round the solidified ingot crystals in the manner of the iron sulphide, they float in the still molten steel as spherical or ellipsoidal particles, which are finally caught and held in the freezing ingot. In working such steels at forging and welding heats, there is no tendency for the sulphides to melt out, as their melting point is higher than steel, and in hammering or rolling the steel, these sulphide particles elongate in much the same way as blowholes or slag inclusions. As referred to later in this chapter, slag streaks are by no means a disadvantage in the case of steel bars to be used for laminated springs, and, in view of the analogy which is claimed between these and the sulphide streaks, it would not appear that the latter can be harmful. Most commercial steels of to-day contain less than 0·10 per cent. of sulphur, and if a specification is desired for this element, probably 0·08 per cent would suit everyone concerned.

Generally speaking, however, the British Standard Specification is rational, and above all, it is a standard, so that the older ideas of every engineer having his own analysis specification are fading away, as they should, and within the reasonable limits of the specification, manufacturers have a free hand. Actual manufacturing conditions have to work within finer limits for Carbon than 0·50 to 0·80, as such would give far too much variation for the spring fitters to manipulate satisfactorily on a given order, and five points of Carbon is generally regarded as a practical range (such as 0·55 to 0·60) which can be readily made by the steel melter, and handled by the fitters.

Tensile tests are often specified to be taken from the steel bar, but this is not very useful, a reliable analysis being quite sufficient to indicate whether the steel is of suitable spring quality. A rough formula for obtaining the unhardened tensile strength of Carbon steel (in finished form) is as follows: Ultimate tenacity, lbs. per sq. inch = 42,000 + 1000 lbs. for every 0·01 per cent. of Carbon (the quantity 0·01 per cent. is often known as a "point"). For example, 0·45 Carbon, low Manganese, tenacity would be 42,000 + 45,000 = 87,000 lb. or 39 tons per sq. inch, which is not far off the actual results. Two other elements, Phosphorus and Manganese, enter into this formula, but the former is necessarily kept so low that the effect on the tenacity can be eliminated in normal steels, and the resultant of manganese is complicated by its bearing a relationship to the carbon content. Up to about 0·50 per cent. manganese the effect is small, with 0·45 carbon and 0·80 manganese it is about $2\frac{1}{2}$ tons, and, crudely, above this point it is two tons for every total one per cent. of carbon plus manganese.

In ordinary carbon steels, manganese is the chief element which causes a good spring steel. Beyond one per cent. however, it may be dangerous, owing to such high content tending to produce water cracks in the spring plate.

Water hardened springs are still most in favour in European practice for railway work, as such steels are cheap, and workable within fairly large temperature ranges. To get the best results out of the much higher priced alloy steels, a very strict temperature control is necessary, as such steels have a very restricted temperature range within which these "best results" obtain. There is no doubt regarding the wonderfully high elastic limits and deflection tests accordingly possible, with alloy steels, when great care is taken in their manufacture, but under normal economic conditions, this is not possible, and furthermore, it is almost consistent practice to oil-harden these steels, which involves a more costly manufacturing process, unless very special plant is employed.

Steel with 1·00 to 1·10 per cent. Manganese and a high silicon content (silico-manganese steel) is not liable to develop cracks when water-treated, is somewhat better when so treated than ordinary carbon qualities. It is, however, frequently oil-hardened so as to obtain superior results, and the carbon raised accordingly.

Hard water is always more liable to cause water cracks than soft water, and doubtless a tributary cause to the general excellence of Sheffield made springs is the fact that the local water supply is of the latter variety. It has always been a Sheffield belief that the superlative excellence of the cutlery products of the town has been (and still is) due to the peculiar properties of the water used for hardening, and there is unquestionably some foundation for this.

For automobile springs, European manufacture includes both oil and water hardening carbon and alloy steels. Silico-manganese, water hardened, is largely employed on the Continent, whilst chrome-silicon and chrome-vanadium, both oil hardening, are popular in this country. The large use of water-hardening steels, for both automobile and railway springs, is due to the more highly skilled labour available in Europe. Oil-hardening steels are comparatively "fool-proof," it being much more difficult to crack these in the quenching process, and such quality is therefore more suitable for use with the high-production methods of the U.S.A., which are manipulated by unskilled or semi-skilled labour.

Oil hardened carbon or alloy steels are, it will be noted, almost universally used in the U.S.A. for all classes of springs. Such treatment with carbon steels, obviates water cracks, but nevertheless the author's opinion is that a standard British Bessemer Acid Steel, water hardened, by good spring fitters, is the best quality of steel for use in railway springs, and also for all other springs. The inherent unsoundness of Bessemer steel, owing to its inclusions of slag streaks, without doubt assists its longevity in an article subject to nothing but direct bending stresses, such as a laminated spring. Such defects are fatal in a coiled spring, but except as regards retarding plate breakages, are negligible in laminated work. The reason would appear to be much the same as that which presents itself with puddled iron, which is a compound of iron fibre with slag and iron oxide as the divisional media, and which is known (when of high quality) to be more reliable than steel for certain important purposes. All classes of steel can include defects due to "pipe", blowholes, etc., as such defects are due to the method employed in the casting of the ingots. In rolling into bars, however, such "pipes" etc., become elongated, and are then equivalent in effect to slag and sulphide streaks. Special care is, however, usually employed in the casting of expensive alloy steels, and such are therefore, as internally sound as it is possible for them to be produced. It must be mentioned that unsound steel as regards pipe and slag inclusions can be obtained by the open-hearth process equally with the Bessemer—to a certain extent it depends upon the amount of "crop" taken off the ingot. It has been common practice, however, to remove little or no "crop" for ordinary bar rolling, and in such circumstance, steel produced by the Bessemer process gives a higher percentage of "unsound" material than that produced by the open-hearth, owing to the former process casting smaller ingots than the latter, with accordingly a greater number of "crop ends" per 100 tons.

The explanation of this phenomenon seems to be as follows. Steel is only theoretically homogeneous, and in practice, there must be a weakest part, where the co-efficient of molecular coherence is least. Under repeated bending stress, it follows that at this point molecular coherence first ceases, and it is the starting point for a potential break. With the more imperfect steel, the discoherence has not proceeded far before a slag streak or "pipe" is reached, which at once checks the spread of the incipient fracture. Consequently, the next weakest molecular structure has to fail, and this will not necessarily be adjacent to the original crack or even in the same plate. This second fracture will be pulled up in much the same way as the first, the result being, that when one plate finally fails completely, others will be found with odd cracks, but still doing their work. With the absolutely sound steel, the fracture having once started proceeds through the plate, with resulting complete rupture.

The elastic limits given on the analysis table are those suitable for spring work. It must be understood that varying heat treatments on all the steels shown, particularly the alloy steels, will give widely different results, but to go higher with elastic limits than those noted is not advisable for springs.

As was shown on the diagram, Fig. 8, the highest possible elastic limit obtainable on a dead hard sample, possessing no ductility, coincides with the breaking stress of a steel piece, and represents a result on a hardening quality when "dead hard," and consequently at a maximum of brittleness. Taking an ordinary water-hardening, 0·55 per cent. carbon steel, the breaking stress—and coincident elastic limit—when dead hard, is in the neighbourhood of 150 tons per square inch; whereas in the soft "as-rolled" condition, these are 45 and 25 tons respectively. That is to say, the value of Breaking Stress/Elastic Limit (or yield point) can range from 1·00 to 1·80. Put another way, the percentage value, Elastic Limit : Breaking Stress, varies from 56 per cent. when soft to 100 per cent. when hard, and all possible variations between these limits can be obtained by varying heat treatments. In practice, the treatment accorded to such a quality of steel as above noted, would give 70 tons tensile, with 60 tons elastic limit, or equal to 86 per cent. To obtain a higher percentage than this, whilst theoretically increasing the possible shock range owing to a higher elastic limit, really causes the steel to be more susceptible to shock, with increased risk of early failure through fracture, owing to the decreased toughness.

The point of camber testing on springs, with regard to specification requirements, has been touched upon already, and the camber test is generally regarded as the best shop method for proving the suitability or otherwise, of any steel that may come through for spring making. It is generally dealt with on different lines to the specification test, in the sense that a piece of standard length is made to a standard camber for the thickness, pressed straight, and the loss noted. A greater or lesser loss, or it may be, complete fracture, or complete flattening, gives the necessary information as to its suitability. Such camber tests have generally grown up with the shop, and each shop has its own particular ideas as to dimensions, which is by no means objectionable, as the results obtained are comparative, based upon long known results which have been recognized as producing the most satisfactory springs. Fig. No. 107 gives a diagram indicating

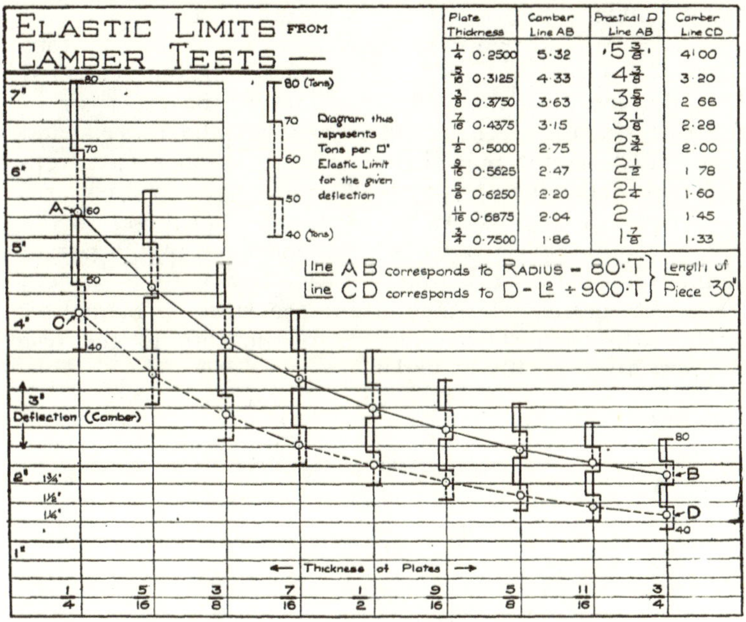
ELASTIC LIMITS FROM CAMBER TESTS —
Plate Thickness | Camber Line AB | Practical D Line AB | Camber Line CD
1/4 0·2500 | 5·32 | 5 3/8 | 4·00
5/16 0·3125 | 4·33 | 4 3/8 | 3·20
3/8 0·3750 | 3·63 | 3 5/8 | 2·66
7/16 0·4375 | 3·15 | 3 1/8 | 2·28
1/2 0·5000 | 2·75 | 2 3/4 | 2·00
9/16 0·5625 | 2·47 | 2 1/2 | 1·78
5/8 0·6250 | 2·20 | 2 1/4 | 1·60
11/16 0·6875 | 2·04 | 2 | 1·45
3/4 0·7500 | 1·86 | 1 7/8 | 1·33
Line AB corresponds to RADIUS = 80·T
Line CD corresponds to D = L² ÷ 900·T
Length of Piece 30"
80 (Tons)
70
60
50
40 (Tons)
Diagram thus represents Tons per □" Elastic Limit for the given deflection
7"
6"
5"
4"
3"
2"
1"
1¾"
1½"
1¼"
Deflection (Camber)
A
B
C
D
Thickness of Plates
1/4
5/16
3/8
7/16
1/2
9/16
5/8
11/16
3/4

FIG. 107. ELASTIC LIMITS FROM CAMBER TESTS.

roughly elastic limits corresponding to what one might term "stable" deflections on a 30 in. test piece, in other words, the deflection at which the piece can be pressed flat without any perceptible practical loss, measured in a practical manner. The elastic limits noted are based upon the ordinary methods of making camber tests which have been before alluded to, namely, pressing the test pieces flat between two rigid beams, which is approximately equivalent to the actual spring working.

From actual testing, the author has confidence in the general accuracy of the elastic limits per deflection per thickness given in the diagram, particularly as it is a known fact that the obtaining of elastic limits on alloy steels from tensile tests is not very reliable. For example, the following were actual results obtained on certain qualities :—

Elastic Limit (extreme fibre) off spring made of ordinary Bessemer Steel (ordinary calculation) 76 tons
Ditto. (author's calculation) .. .. .. 51 tons
Elastic Limit (extreme fibre) off spring made of alloy Steel (ordinary calculations) .. .. 114 tons
Ditto (author's calculation) .. .. .. 75 tons
Elastic Limit obtained off the same alloy steel (hardened and tempered tensile) .. .. 35 tons

Steels of spring qualities, in the hardened and tempered state, do not give the definite point on the stress-strain diagrams which is given by unhardened or milder steels, and there is little doubt that this characteristic calculated from increment deflections by carefully noting the point of set (or yield point, equivalent for all practical purposes to the elastic limit) is more reliable than endeavouring to obtain the limiting elastic stress from a tensile test piece.

The determining elastic limit of the steel is, of course, on the extreme surfaces, and its relationship to the test requirements generally as under—

No.	Test Required.	Actual Test.	Elastic Limits Ordinary Calculations.	Author's Calculations.
1	British Standard, Springs $L^2/900T$	—	68	46
2	Suggested for Alloy Springs $L^2/700T$	—	87	58
3	British Standard-Carbon Spring Steel. Radius 80T	—	93	62
4	Elastic Limit as shown by 15 per cent. drop on No. 3	—	79	53
	French Specn. Spring Steel			
5	"Allongement Elastique," 6 m/m.	—	76	51
6	"Allongement Elastique," 6½ m/m.	—	83	55
7	"Allongement Elastique," 7 m/m.	—	89	59
8	"Allongement Elastique," 8 m/m.	—	101	68
9	Tensile tests—Carbon Steels	60–75	—	—
10	Tensile tests—Alloy Steels	70–90	—	—
11	Camber tests—Carbon Steels	—	75–90	50–60
12	Camber tests—Alloy Steels	—	105–120	70–80

Nos. 1 and 2 are test requirements on made Springs, and fall short of stressing up to the elastic limit (revised calculations) as they should do. Test No. 3 stresses up to and generally over, the elastic limit, which is shown by the final stable deflection obtained after the first loss, as indicated in No. 4. This loss may be anything between 10 and 20 per cent. according to the tempering and quality.

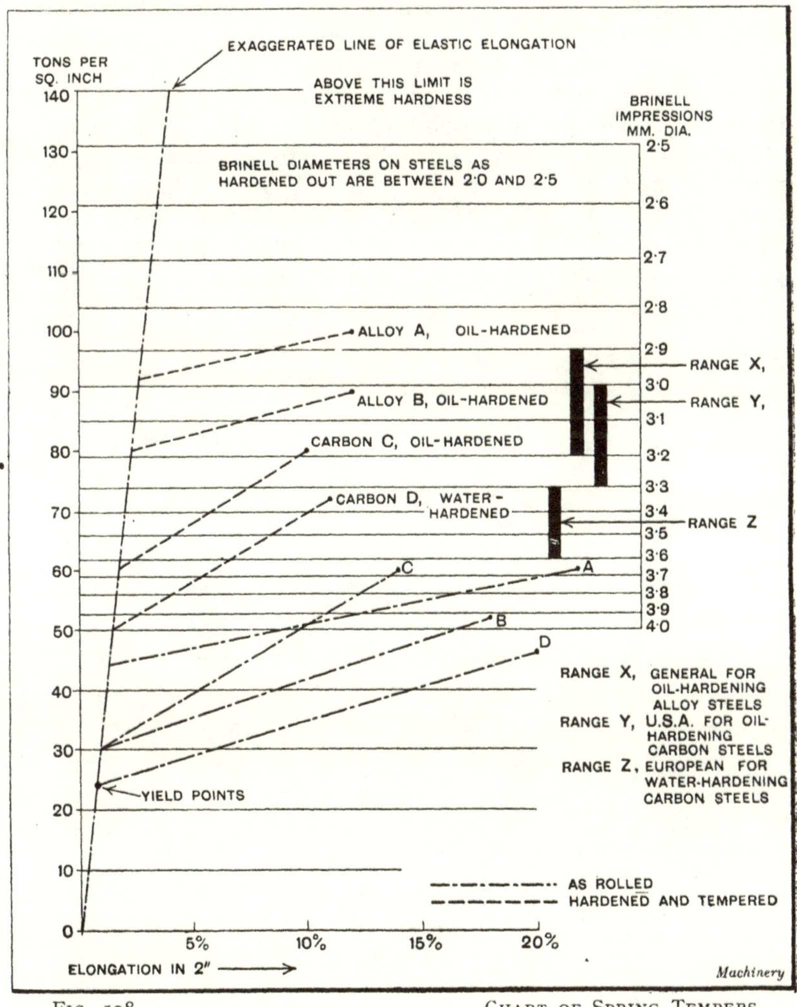

FIG. 108. CHART OF SPRING TEMPERS.

It will be seen that on the British Standard Spring Test (Carbon) if the factor X (test load/working load) be taken as two, the working load equals 23 tons skin stress, which is about 40 per cent. of the absolute elastic limit of 55 tons, assuming every plate is perfectly treated to this point.

A chart showing the usual range of spring tempers worked to with the steels usual in spring practice, is given in Fig. 108. It will be noted therefrom that the alloy and carbon "oil-hardening" varieties are worked with Brinell ball diameter limits of 3·3 mm. to 2·9 mm. (hardness numbers 340 to 444). Not infrequently alloy steels are found (on broken plates) to be still harder, and including 2·8 and 2·7 mm. dia. The author's point of view, which is shared by other authorities, is that the ideal figure to be aimed at is 3·2, range given as, say, 3·1 to 3·3 mm. When the hardness is greater than this, the steel is by no means as suitable for withstanding the shocks (or impacts) which are received, particularly in the case of road vehicle springs.

CHAPTER XXVI

SPRING STEEL—SECTIONS AND ROLLING

THE range of sections in use for the manufacture of all classes of springs, is from 1 in × $\frac{1}{8}$ in., to 6 ins. × $\frac{5}{8}$ in. For certain small work, sections less than 1 in. × $\frac{1}{8}$ in. are employed, and on rare occasions, as bumper springs, sections wider than 6 ins., and up to 8 ins. and 9 ins. There are several types of spring steel, which are included in the various section drawings, Fig. 199, and described as follows :—

No. 1. These show respectively the small section and large
No. 2. section mentioned above, as being the general limit sizes. They are of what might be termed the "standard section" having round edges, and being usually slightly concave or "hollow" on the width—this does not, however, seriously obtain on the smaller sections.

No. 3. This is a full round edge, rolled more often in conjunction with rib and groove steel, and used for top and bottom plates which are embraced by hoops with large radius corners. The rolling of the full round on the section obviates the necessity for grinding the plates to suit the hoop.

No. 4. A "standard" section of normal size.

No. 5. These four sketches show standard U.S.A. edge radii always taken as being equal to the thickness of the plate. There is no particular standard here, or on the Continent.

No. 6. Shows two plates, 100 mm. × 10 mm. with rib and groove of the French style.

Spring Steel Sections.
1
2
1" × 1/8"
6" × 5/8"
3
1/4"
3/8"
1/2"
5/8"
4
5
6
100
10
10
10
7
3½
Edge
10
Mm. Sizes
3½
3½
3
90
20
13
13
9
5
13
Millimetre Dimensions
4
4
3
8
75
8½
7½
2¼
22
22
22
22
10
11
Millimetre Dimensions
8
7
2
12
13

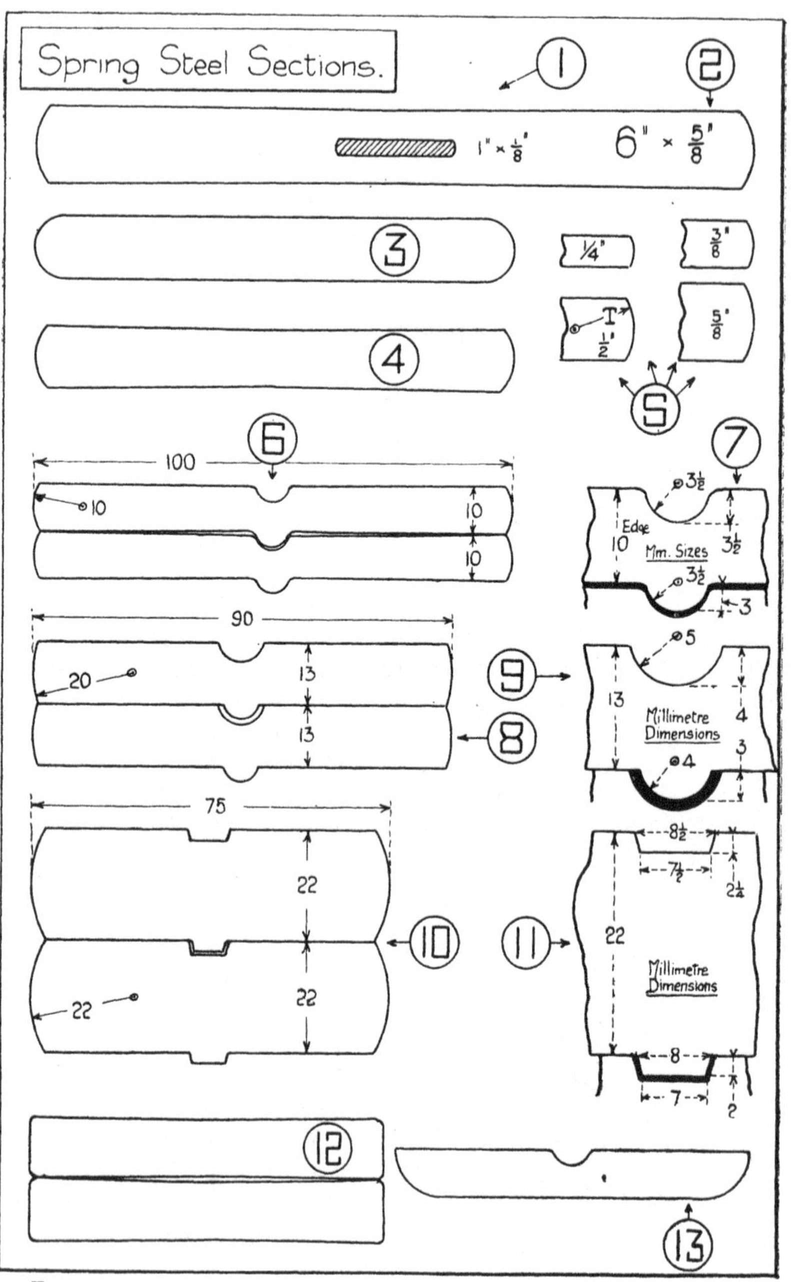

FIG. 109.

No. 7. Gives detail sizes of No. 6 rib and groove, and shows also (in heavy black) the clearance between plate and plate.

No. 8. Illustrates the German Railways Standard section, 90 mm. × 13 mm. of which all springs have to be made, unless special approval is obtained for variation.

No. 9. Shows the detail sizes, and clearance, of No. 8.

No. 10. Is a special (French Railway) section for heavy buffing springs. The rib and groove herein is of the flat form.

No. 11. Gives detail dimensions, and clearance, of No. 10.

No. 12. Two plates of "square-edge" section, rolled either as shown, or of rib and groove form, more usually of the latter.

No. 13. Is a special groove form, for short plates or packing plates of springs with a large corner radius in the hoops. Continental style.

The rib and groove sections are chiefly employed on the Continent, and are not specially favoured here, or in the U.S.A. The great point of their use is that the side check imposed throughout the plate lengths obviates the necessity for studs and slits, and consequently reduces very substantially the number of potential breaking planes in the spring. To give satisfaction in fitting, however, it is clear that the rolling must be of great accuracy, in order to ensure the centralization of the distinguishing feature. Rib and groove steel should not be used on plates less than $\frac{3}{8}$ in. thick, owing to the possibilities of over-stressing the rib due to its distance from the neutral axis of the section if any normal profile is employed, although it is obvious that if the rib and groove are proportionately reduced, no objection can be raised to them on the smallest sizes, although they might not then be of great value. It can be taken that the employment of this section makes no perceptible practical difference to the unit deflection of the springs. The German Railway section is the most desirable of the patterns of rib and groove shown, as the clearance is greater than with the French sections, which makes it a more pleasant matter for both the rolling mills and the spring fitter. One great difficulty of rib and groove steel lies with the fitting, and it necessitates great care in setting up rolling mill guides, etc., to ensure that no trouble will be experienced, and that the line of rib and groove will be straight, and not wavy, as is sometimes the case. Trouble is often experienced in fulfilling orders for steel bars of the rib and groove type owing to the supply of inaccurate or incomplete particulars of the existing springs for which the stock material is required. In the absence of definite standards, it is a necessity to provide a sample piece of existing spring steel if satisfaction is to be obtained.

All the sections are generally rolled concave to facilitate fitting. If they were rolled flat, and had to be fitted to a large camber, the bent plate would not be flat, but slightly buckled, and consequently it would not be possible to make a good close fit of the plate edges. The concave is therefore introduced to compensate for the possibility of this slight buckling due to bending, and to ensure that the plate edges will be in close contact. The maximum allowable concave in general practice is $\frac{1}{64}$ in. per side or $\frac{1}{32}$ in. in total thickness. As previously mentioned, engine springs are made from steel which is rolled with the minimum possible of concave, for normal designs being about $\frac{1}{100}$ in. per side. Such steel is generally rolled when the concave rolls have worn down, in British practice these rolls being turned new to the standard of $\frac{1}{64}$ in. per side on whatever width, say 4 in., of which the greatest quantity is then required.

Springs can be fitted from absolutely flat steel, but if they have much camber the rounding of the plate does not improve their appearance and entails a great deal of additional hard work to the spring fitter.

An unfair advantage is taken of the fact that concave is a practical necessity in spring bars, by buyers in this country of certain types of railway springs, stipulating that the springs must fall within a certain weight. Generally the stipulated weight is such that the concavity has to be over the correct maximum to obtain it, and any slight variation in plate lengths or spear shape will bring the spring over weight, which is a loss to the manufacturer, as only the specified weight is paid for. Such springs have to be rolled as "extra concave" to keep down the weight, and consequently, difficulties occur in their carrying the correct load. The evil is specially pronounced with the R.C.H. standard "private owners' wagon" springs (bearing and buffing) and it should certainly be prohibited for springs of this class to be ordered within a stipulated weight. They have to be made to the R.C.H. drawing, so the manufacturer has no opportunity of putting in excess weight in the way of cutting longer plates, etc. A certain check could be put to this practice by specifying the limits of weight within which each type of private owners' spring, should fall.

The rolling of spring steel accurately to size is a very exacting business, and when accomplished, highly creditable to the producers. It is often overlooked that spring steel bars are a hot rolled product, and drafters of specifications have a tendency to include " tolerances on section " rather on the basis of what they would like, than what they can hope to get. At times, suggested tolerances are such that whilst one end of a bar would be right, the other end would be wrong, owing to the difference in heat between the first and last ends as they go through the rolls. In this country no standard margin is fixed, but it is generally assumed that the variation in thickness should not exceed $\frac{1}{128}$ ins. (0·0078 in.). Even this small variation will make a difference of 0·078 in., or $\frac{1}{12}$ in. on a ten-plate spring. In practice, railway spring steels are not objectionable up to $\frac{1}{100}$ in. (0·01 in.) variation. Light springs for pleasure automobiles should be nearer to the $\frac{1}{128}$ in., owing to the effect thickness variations have on carrying capacity. In width, a variation of $\frac{1}{40}$ in. (0·025 in.) should be the limit of tolerance, not because of any load carrying feature in this case, but because of the irregular appearance of springs with plates of varying widths. and also the difficulty of making a workmanlike job of such springs if they are hooped. The above tolerances are of course " plus " and " minus ".

American railway practice dictates as a general thing, a thickness limit of 0·01 in., and a width limit of 0·03 in., both plus and minus, which is sound practice.

The Society of Automobile Engineers (New York) puts out the following specification :—

" The finished bars shall be of double concave section, with round edges. The radii of the arcs of the two concave surfaces shall be of equal length. Rolls to produce the round edge shall be turned to a radius equal to two-thirds the thickness of the bar."

This specification also includes particulars of maximum and minimum concaves (on each side) varying from 0·004 in.

minimum and 0·009 in. maximum (nominal 0·007 in.) for 1½ in. wide bars, to 0·023 in. minimum and 0·031 in. maximum (nominal 0·029 in.) for 5 ins. wide.

The section tolerances are of interest, as below :—

0″ to 2¼″ wide.	Width, plus $\frac{1}{32}$″, minus 0.	Thickness, plus or minus 0·005″
2¼″ to 3″ ,,	,, plus $\frac{3}{64}$″, minus 0.	,, plus or minus 0·006″
3″ to 5″ ,,	,, plus $\frac{1}{16}$″, minus 0.	,, plus or minus 0·007″

The British Engineering Standards Committee are suggesting a specification dealing with automobile spring steels, in which the rolling tolerances are as follows :—

0″ to 2″ wide.	Width plus or minus 0.010.
Over 2″ wide.	,, plus or minus 0.015.
Up to and including $\frac{5}{16}$″.	Thickness plus or minus 0.005″
Over $\frac{5}{16}$″,	,, plus or minus 0.008″

The standard of concavity laid down provides that the surfaces of the spring plate shall have such radius that on a plate 2½ ins. wide, it is 0.015 thinner in the middle than at the edge.

Usual Continental specifications generally allow 0·5 mm. (0·02 in.) on width, and plus or minus 0·02 mm. (0·008 in.) in thickness. Total concavity of two faces is to be 0·30 mm. (0·012 in.) and the rib and groove axis must not deviate more than (plus or minus) 0·30 mm. from the bar axis.

From the ingot the spring steel billets are rolled, varying in size and weight according to the finished section required. They are then re-heated and put through the spring mill, a diagrammatic representation of the rolls of which is given in Fig. 44. Billets are rolled from 3 in. to 6 in. wide, and 1½ in. to 2 in. thick, and sheared off in from one to two cwt. pieces, They are then re-heated and passed through the roughing thickness and edging rolls, being worked continually to gauge till the correct width and thickness for finishing is obtained. The final passes are through the finishing edging and concave rolls, which should give the dimensions desired. As previously mentioned, great care is required to obtain the exact sections, to the limits which should be imposed for the purpose of getting the best springs. The roller's gauge is made, of course, for the hot bar, and is based upon the known contraction from the usual rolling temperature, from which it will be perceived that the uniformity of the reheating billet furnace, and the withdrawal of the billets therefrom at a constant temperature, play a very important part in the accuracy of the finished bar.

The type of rolling mill shown is in general use in this country, where a very high finish is demanded and obtained on spring steel bars. The more ordinary type of mill, known as "the collar and groove" mill, does not include the edging process, which is of value, apart from the edge rounding, inasmuch as it squeezes the scale off the bar, instead of

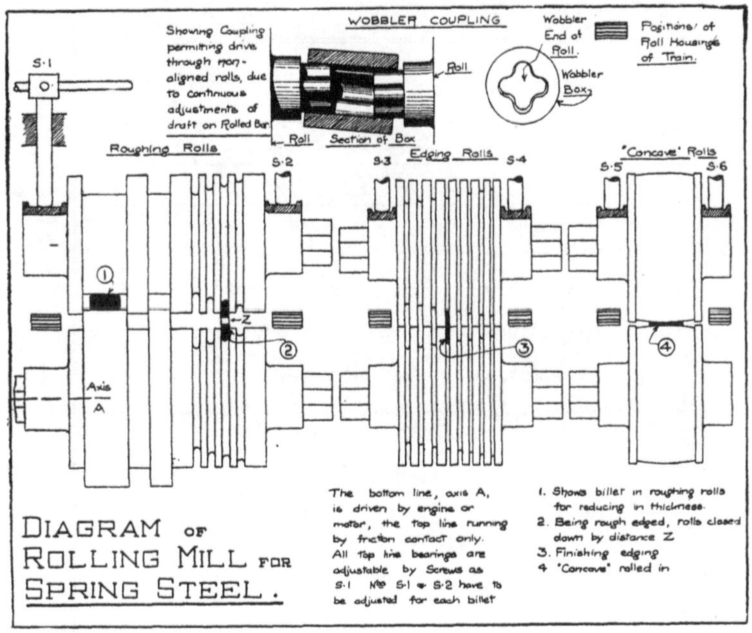

FIG. 110.

tending to squeeze it into the bar. This latter mill, if very well set up, has advantages, however, in the rolling of rib and groove steel, but in such case the section will be square edged, unless a form of universal mill is used, which employs vertical axis rolls to form the edges. American spring mills are of the high speed pattern, and frequently of the continuous type.

In this country, the time is certainly not yet ripe for anything in the manner of high production spring mills. The number of sections in use, and the relative smallness of most of the orders, negatives any advantages to be obtained from the high capital expenditure. In the U.S.A. a minimum basis of 100 tons per section is generally required before that section will be put in hand. The railway trade there generally orders steel 4 ins., 5 ins., and 6 ins. wide, and all widths ½ in. thick—which totals six very simple sizes. The automobile trade cannot contain itself within these limits, but the greater number of sections required is balanced by the large

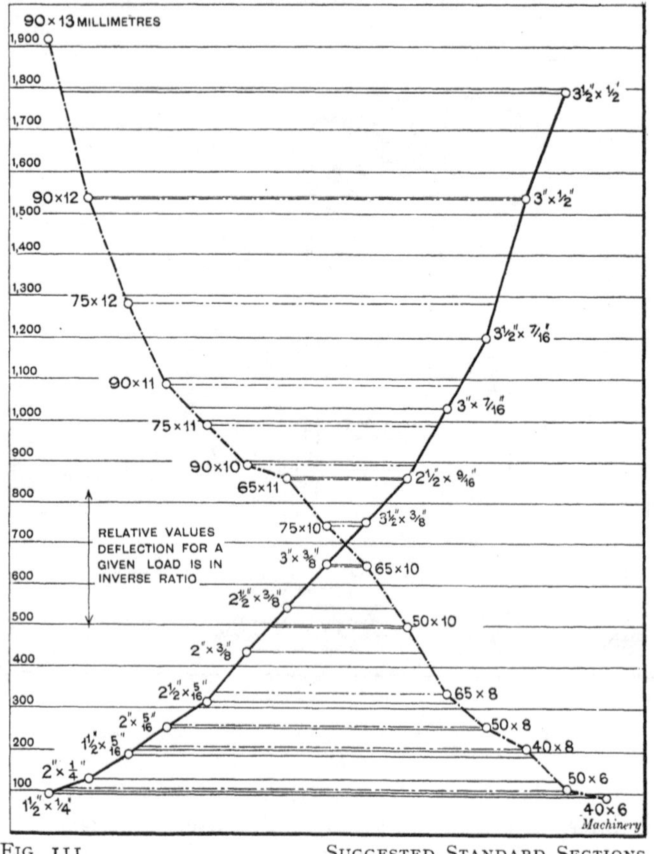

FIG. 111. SUGGESTED STANDARD SECTIONS.

size of the orders. In this country, it is not unusual to place an order with the mill for 10 cwts. which will include 5 sizes, and probably more material is rolled to waste in the trials necessary to obtain the correct section than is ultimately supplied for the order. The standardization of sections would provide a very useful work for the British Engineering Standards Committee, for instance, and as such became adhered to, manufacturing costs would fall. In Fig. 155 is given a graph showing suggested sections for the automobile trade, with their relative values as regards unit deflection. For the normal railway trade of this country, the following sizes would be found sufficient :—

- 2½ ins. wide, × 5/16 in. and 3/8 in. thick. (Narrow gauge stock.)
- 3 ins. wide. × 3/8 in. 7/16 in, and ½ in thick.
- 3½ ins. wide, × 7/16 in. and ½ in. thick.
- 4 ins. wide, × ½ in., 9/16 in., and 5/8 in. thick.
- 5 ins. wide, ½ in. and 5/8 in. thick.
- 6 ins. wide, ½ in. and 5/8 in. thick.

The above totals 14 sections, and probably, by agreement, could be reduced to 10 sections. The use of a minimum number of standard sections simplifies design, manufacture, and maintenance, being particularly obvious in the latter instance as it reduces the stocks required for repair.

A similar type of road vehicle, made by three different British leading makers, includes no less than 18 different sections for front and rear springs on the three trucks, made up as follows :—

Vehicle		Spring		Width		Thicknesses		Sections
Vehicle	A.	Front	Spring,	2¼″	wide	—2	thicknesses.	4 sections.
,,		Rear	,,	3½″	,,	—2	,,	
,,	B.	Front	,,	2⅜″	,,	—3	,,	7 sections.
,,		Rear	,,	2 ″	,,	—4	,,	
,,	C.	Front	,,	2½″	,,	—3	,,	7 sections.
,,		Rear	,,	3″	,,	—4	,,	

All the above could have been designed equally efficiently on two widths, 2½ ins. for the front, and 3 ins. for the rear—and strictly logically, there is no reason why 3 ins. should not be used for both front and rear. The employment of every known sectional size spread over high carbon, low carbon, silico-manganese water-hardening, silico-manganese oil-hardening, high chrome, low chrome, chrome silicon, and chrome vanadium, steel qualities; all these being probably further complicated with restrictions of sulphur and phosphorus contents, makes a variety which is more interesting than practical, and the difficulty of carrying such stocks is readily apparent. To carry out quick deliveries efficiently, the spring maker should stock at least 200 different steels, with resulting additions to the cost of each spring produced.

Let it be finally noted that the production of any standardized steel sections, when over 2 ins. wide, should not include the thicknesses $\frac{9}{32}$ in., $\frac{11}{32}$ in., $\frac{13}{32}$ ins., etc. These are generally used in designs which include three or four thicknesses in the same spring, and they are always liable to be mistaken for the adjacent sizes in sixteenths, with the result that the designer's ideas are considerably spoiled by the plates being mixed instead of being in the perfectly graduated order intended.

The usual length of bars as rolled and supplied in Europe, is 14 ft. 0 ins. to 18 ft. 0 ins., representing one billet. In the States, the rolled bar is from 100 ft. 0 ins. to 300 ft. 0 ins. long, cutting into handling lengths of about 16 ft. 0 ins. The more ends which are present for a given weight, the more waste is entailed, and, a point of not less importance, the more laterally bent stock is present (known here as "dog-legged" bars). Great care must be taken with the hot straightening if satisfactory stock is to be supplied to the spring shop, and provision is required on the hot banks of good straightening plates permitting of uniform cooling. Bars bent laterally give considerable trouble to the hand fitter—who is expected to rectify them—but substantially more trouble to any machine fitting process, as they will probably not be detected until they are in the form of the curved and finished plate. With the supply of bars cut from long lengths, as in the U.S.A., this trouble is reduced to a minimum.

CHAPTER XXVII

THE FORGING SHOP—CUTTING-UP

THE manufacture of the spring steel bar having been dealt with, the next stage is to follow it through to the spring fitter. From the spring mill, the bars are passed to the forging shop, which carries out any, or all, of the following operations—

(*A*) Cutting into the lengths required the long bars from the mill. Such cut ends will be " square."

(*B*) Tapering, or drawing, the plate ends on the thickness.

(*C*) Punching centre holes, nibs, or making other forms of central registering points and hoop attachments.

(*D*) Spearing into various forms the ends of the plates.

(*E*) Punching studs and slits towards plate ends to prevent lateral movement, sinking ribs and grooves, or providing other means for this purpose.

(*F*) Grinding as required.

The use of separate or combination machines for operations *A*, *C*, *D* and *E*, turns upon production required, the lay-out of the shop, and the ideas of the management. One arrangement favours a double machine for cutting up and centring, that is, two rams in the one frame, with a triple machine to follow, consisting of three heads, doing successively the spearing, studding and slitting. Such an arrangement is very suitable for a railway repair shop, as all requirements are combined in the two machines. Another arrangement includes all the essential work on one complete unit—so that one machine, with one drive, will cut up, centre hole, roll the plate ends, spear, and stud and slit. This type is also of use to a repair shop, or small manufacturing plant, and the advantages of a multiplication of such machines for high productive work are many, as the number of drives is reduced, and the handling of plates is at a minimum. The one-machine one-operation method has also its advocates, as the breakdown of a machine, or the removal of worn or broken punches and beds stops only one operation, instead of two or three. More driving arrangements are required,

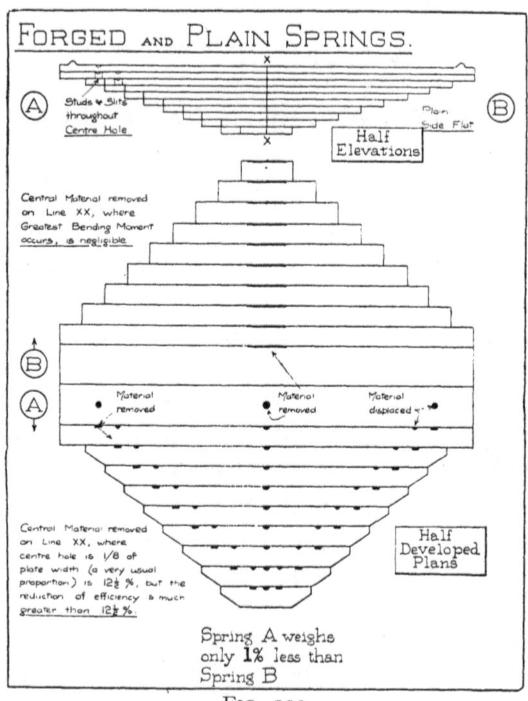

FIG. 112.

but the system is somewhat more elastic, as frequently studded and slitted plates are not speared, or speared plates are not studded and slitted, so that three heads may be running for one operation. Spring plate forging machines of British make are not, as a rule, fitted with stop motions on the rams, the great art being to train the operator, and speed the machines, so that every stroke can be caught.

The numbers of operations that can be performed have already been indicated on the typical road vehicle spring dwelt on in Chapter XXIII, and other possibilities of forging and its effects (excluding drawn points) are shown in Fig.

112. Spring A is the usual design given for engine springs. Sometimes the centre hole is substituted by a nib, which will increase the efficiency of the design if downwards, and substantially reduce it if upwards. It will be observed that with centre holes, studs, and slits, forty-six reduced or distorted sections are obtained, each of which is a potential breaking point ; well ahead, from a time point of view, of any steel breaking point in the spring. On the spring regarded as a solid beam, there are nineteen planes reduced from the correct section, giving stress increases accordingly. Spring B (flatted one side only) has only ten positions of reduced section throughout the ten plates, and as this reduction involves only the removal of the rounded edge of the spring plate it is negligible. Taken on the solid beam, there is (practically) only one line of reduced section. The spearing has no effect, one way or the other, but as an attempt at weight reduction it is of no value—with short plates of the correct length—as the short plate must be maintained longer than the full width of the hoop, which retards the commencement of the spearing, causing, in this particular case, the first five bottom plates to be longer than the corresponding square cut plates.

The one operation that is essential to all plates is that of cutting to length. After this, numerous combinations are possible, as shown diagrammatically in Fig. 113. This indicates the amount of variety which is possible in forging work, from a mechanical point of view, and additional complication is introduced by the fact that certain specifications stipulate certain processes must be done hot ; also certain sizes of holes and plates necessitate hot forging. No. 1 routing is obviously the process to be aimed at, merely (cold) cutting, and (cold) centring, and for relatively short engine springs, or rib and groove steel, any other operations are refinements only. Nos. 9 and 10 routings are very common, however, and the cost of manufacture runs up accordingly, with no practical equivalent value.

Hot forging, as regards holes, nibs, slits, and studs, whilst frequently specified, is of little value, as it makes no appreciable difference to the efficiency of the spring plate. Any structural deformation, caused by cold forging in the material adjacent to holes or nibs, is heat treated back to normal when the plate is made hot for curving and hardening. Hot forging is sometimes favoured by the manufacturer on certain qualities of steel, or for other reasons, not because of any peculiar affection for the gentle treatment of the spring steel concerned, but in order to make the work easier on the forging machines and tools. Generally, however, all work that can be forged cold is forged cold, unless, of course, the specification stipulates " hot forging." From the economic point of view there is not a great deal to choose between hot and cold

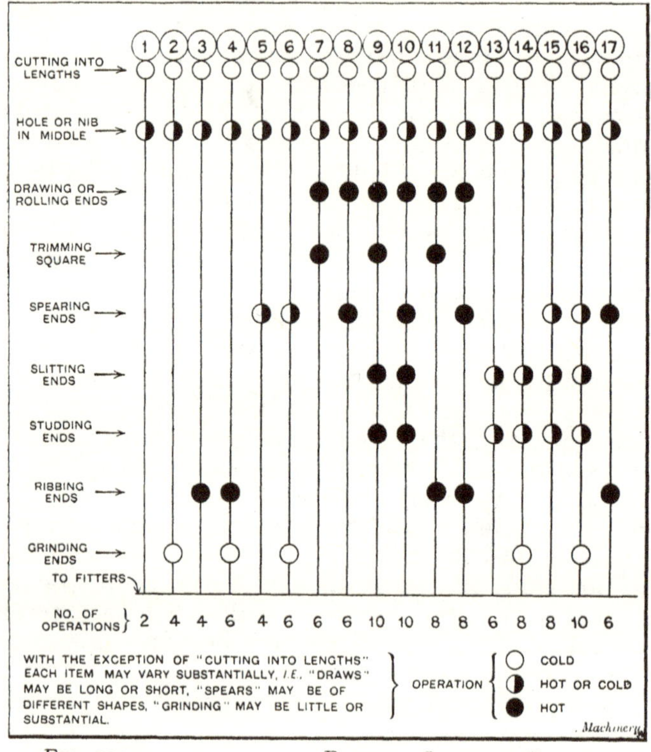

FIG. 113. ROUTING CHART OF PLATES.

forging—the former requires fuel, but reduces the wear and tear on the press tools and presses, and furthermore, permits the use of lighter machines, which are cheaper to purchase, and require less power to run. The spearing, cold, of a 0·60 per cent. carbon steel plate, 3 ins. × $\frac{1}{2}$ in., requires a pressure of about 120 tons (this of course will depend on the extent of the spear), whereas when heated, less than one-tenth of this pressure is needed.

In this country, steel is supplied from the mills " as rolled." and as certain alloy qualities are air-hardening to a greater or lesser degree, it becomes impossible to punch or spear the plates cold—in fact, they are often difficult to shear cold. Necessarily, therefore, alloy or 1·00 per cent. carbon steels have to be forged hot. In the U.S.A. where oil-hardening steels are used exclusively, the bars are usually annealed, and consequently (with the exception of the carbon steel) can be cold forged ; as the actual hardness of well-annealed alloy steels is not seriously different to straight carbon steels of 0·50 per cent. to 0·60 per cent. carbon content.

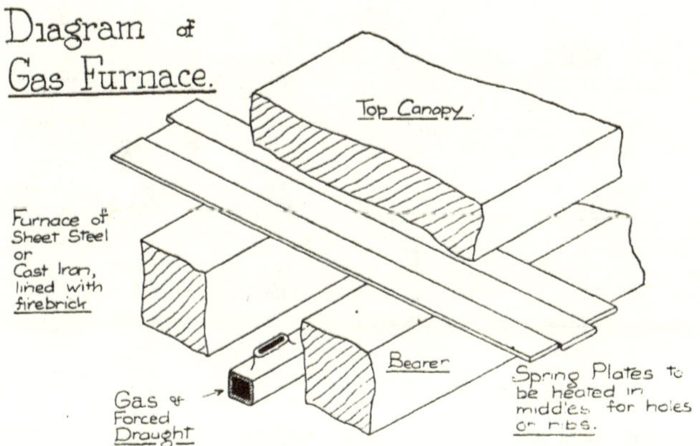

FIG. 114.

The furnaces used for heating plates for forging are generally of a simple nature. In this country and the Continent, coal is the usual fuel, with gas and oil as occasional heating mediums. In America, of course, oil, with its consequent advantages, is used almost entirely. A small oil furnace can be arranged on a travelling bogie, and is then of an elastic nature, as it can be taken from point to point as required, and only needs coupling on to the shop air supply. Drawn points, studs and slits, etc., only require the plate ends heating, and a furnace 12 ins. deep is quite sufficient for this purpose. Centre holes and nibs require a special arrangement to ensure only the minimum amount of plate middle being heated, and Fig. 114 illustrates the usual method. Every care should be taken to avoid overheating, or scaling the plates by too much air supply—and the heats employed should be the lowest at which it is possible to do the work.

Furnaces fired by oil or gas under blast are frequently not too pleasant for the worker, as no doors can be provided, and the blast drives the heat through the openings necessary for the plates—which are put through at a rapid rate. A very good protection common in the States, is the provision of a continuous row of small chain lengths, hanging vertically, down which water is continually trickling. This does not interfere with the easy insertion or withdrawal of plates,

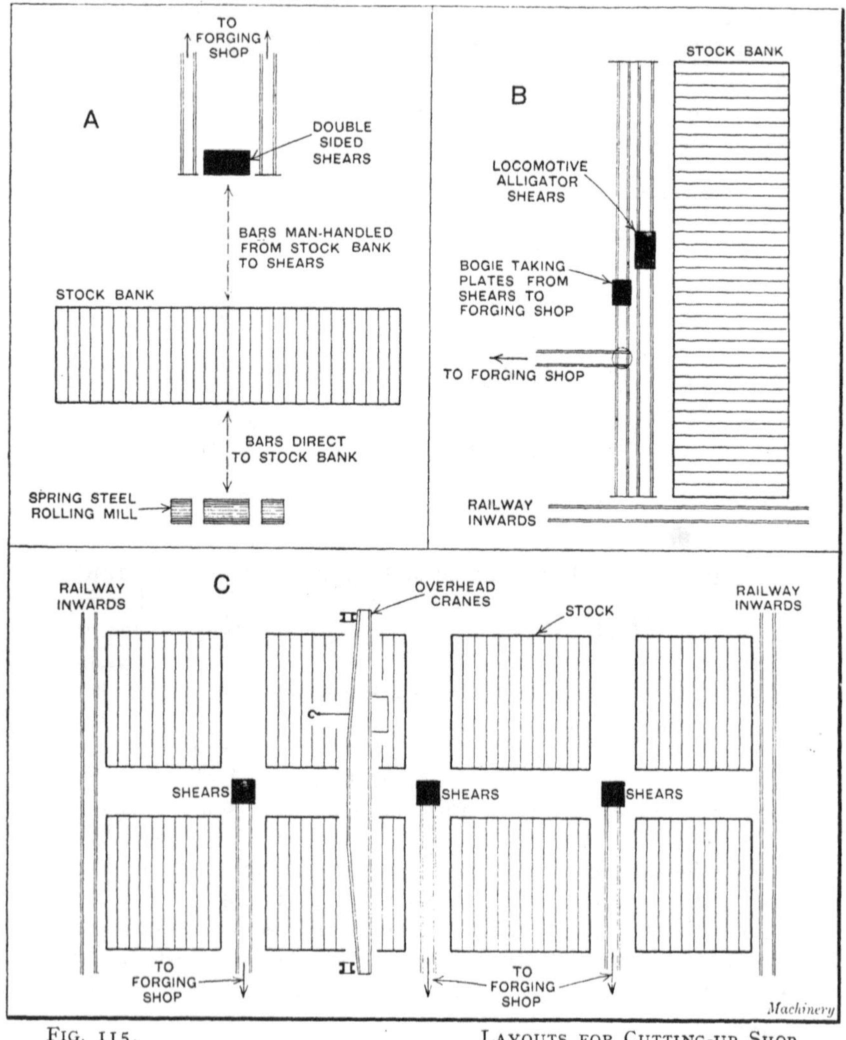

FIG. 115. LAYOUTS FOR CUTTING-UP SHOP.

and considerably improves the working conditions—which fact is reflected by the high production obtained. Many U.S.A. furnaces, are, in addition, provided with water-cooled doors—the casting for the latter being hollow, and having a continuous water stream circulating to waste.

The lay-out of the cutting-up shop is a matter of the first importance, particularly with firms who are doing a large production. In this country, and the Continent, where most of the leading spring manufacturers roll their own spring steel, the stocks carried are not very large, spring shops with a production of 100 tons per week frequently running on a stock of 200 to 300 tons. Manufacturers who buy their steel from outside sources limit their stocks as much as possible, owing to the large variety of sections, and the obliging character of the steel rollers, who can, owing to the relatively small mills, and large number of them, deliver orders with remarkable speed when it suits them, and consequently the spring makers purchasing rarely stock more than 200 to 300 tons. The high-production shops of the U.S.A. are chiefly confined to the automobile trade, as in the railway trade, apart from locomotive springs, a proportion of tramcar springs, and passenger car bolster springs, all stock runs on coils. These automobile spring shops have not infrequently capacities of 200 to 500 tons of finished springs per week, nearly all buy their steel from mills generally far distant from the spring works, and these mills will not roll less than 100 tons per section. As the result, the stocks held are enormous, varying from 10,000 to 30,000 tons. More attention has therefore been given to stock layouts in the U.S.A. than has been given in Europe.

Some typical arrangements are shown in Fig. 115. Sketch A herein illustrates an ideal known to the writer as existing only at one European firm, and this in Sheffield, the rolling mill being a prolongation of the spring shop. The rolled bars are laid on the floor in piles of quality and section as soon as they are cold from the mill. As required, they are cut up at the fixed double-sided shears, which is the first machine of the forging shop, the remaining machines being all adjacent. The amount of handling is reduced to a minimum, and the spring department, being always on the spot, can, and does, exercise no little control over the rolling.

Sketch B shows a lay-out with a specially made steel rack (instead of the usual iron stanchions) and a travelling shears running immediately in front. By this means, the shears are transported to the bars, instead of the bars to the shears—a method of considerable labour-saving importance. The arrangement of the rack (illustrated in Fig. 116) provides for the bars being the same height as the shear blades. The particular arrangement shown is English, but similar schemes are in operation in the U.S.A.

Sketch C is a typical American layout. The steel is brought in on the railway at each end of the stock warehouse, and handled from there by overhead cranes to its respective

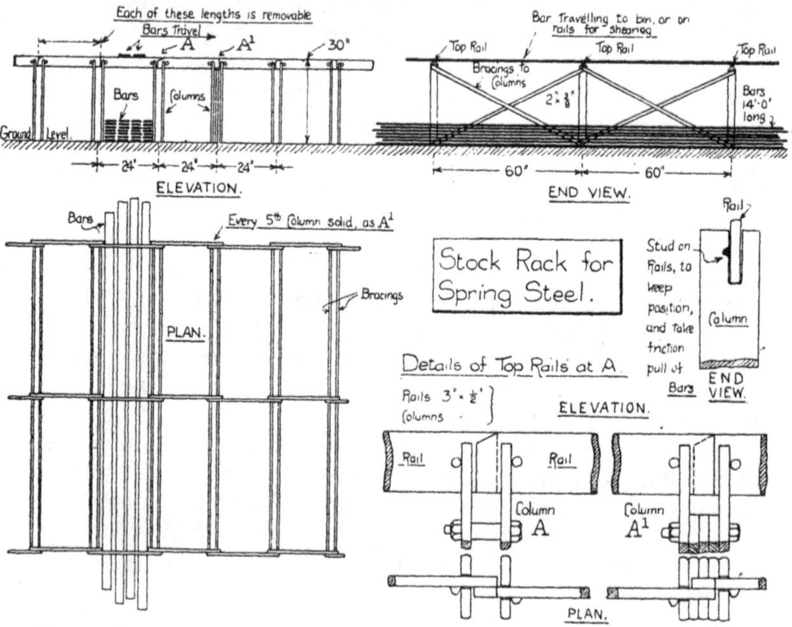

FIG. 116.

position. When required, the cranes carry the bars to the shears fixed as shown. At times, these shears are carried by the cranes to the bar positions.

The steel rack referred to in connection with Fig. 115—B is shown in Fig. 116. It is of spring steel construction throughout with top rails removable as required, permitting thereby vertical access to any particular bin. The bars could therefore be brought in on a flat wagon, and hauled across the rails until the special bin for the particular lot is reached, when they drop therein.

A shears of the double-sided pattern, English design, is illustrated by Fig. 117. These are of the typically substantial character of leading British machine-tool design, and include nothing likely to get out of order. In position, gantries are attached on each side, with the necessary stops for cutting

FIG. 117. DOUBLE-SIDED SHEARS. J. BUCKTON & CO., LTD., LEEDS.

a series of bars to fixed length. Shears of this character will cut up material at a far greater speed than can be handled by the following machinery. The only disadvantage to be urged for the pattern is that in the event of the failure of any part of the electrical equipment, both sides are stopped—or alternatively, both sides are moving when only one side is cutting. However, such probability has to be balanced against the extra capital cost of another motor ; or of providing stop motions to cut out the non-operating blades—

Fig. 118. Shearing Machine. Usines de Braine-le-Comte, Belgium.

both of which entail additional complication. Particularly so far as the actual machine parts are concerned, great simplicity and great solidity should be the ideals aimed at for spring shop work—and the shears illustrated certainly fulfil both conditions.

A Continental make of spring shop shears is shown in Fig. 118. This is a very compact type, and has a stop motion fitted. A tunnel guide is provided for the bars, which ensures a square cut vertically and horizontally. This arrangement does not find much favour in this country—the square cutting being a measure of the skill of the workman. In the period, now disappearing, when drawn points were the rule, the final end trimming was done after the hot process, so that the absolute accuracy of the sheared work was not a great matter. General practice here is to reverse the bar after each cut, so as to obtain both cuts the same way, as Fig. 119, and this could not easily be done if a guide was in position.

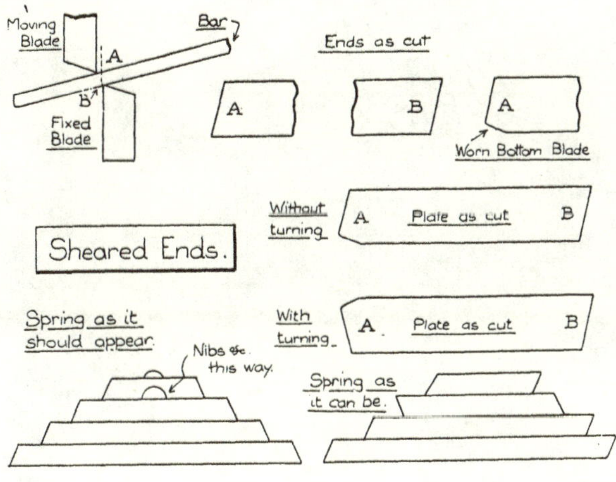

Fig. 119.

The two patterns of shears just illustrated are known as "guillotine" shears, and are almost universally employed in this country and on the Continent. In America, however, cutting-up is divided between this type, and the "alligator," a stationary form of which is illustrated in Fig. 120. It is generally considered here that the "guillotine" form gives the better cut, but the quality of this is determined more by the condition of the shear blades than by the type of shear, and there would appear to be little matter of argument on the latter ground. The machine shown in the picture is of U.S.A. manufacture, and requires practically no foundation. American machines for "black work" of the type of spring work, are characterised by solid design and simplicity, and, having been designed and manufactured by very practical American designers, are also characterised by what in this country would be called "lack of finish." This negative feature does not, however, cause them to work less efficiently—although British buyers do not always appreciate this point. The motion of the shears is reminiscent of the old "helve-hammers," it being in the form of a cam on the driving shaft, which gives a positive cutting action, and a

FIG. 120. ALLIGATOR SHEARS. JOHN EVANS SONS, PHILADELPHIA.

negative return, due to the weight of the tail end of the rocking arm, assisted by the plate spring. A special reinforcing piece is provided adjacent to the shearing head, which prevents any tendency to spring when cutting.

An English form of alligator shears, of the locomotive pattern, is shown in Fig. 121. This is heavier in general design than the foregoing, as it is intended to cut 6 ins. × $\frac{5}{8}$ in. spring steel without any "kick," and weight is required to hold it to the rails. The motor takes its current from overhead wires, the controller being fixed in some convenient position at one end of the track. A cut-out switch for the current is provided on the shears, so that the motor can be stopped instantly should occasion require. This provides a more satisfactory method of obtaining power for locomotive shears than the alternative sometimes used of flexible cables, plugged in at points adjacent to the position of the machine.

FIG. 121. LOCOMOTIVE ALLIGATOR SHEARS. F. BREARLEY, BINGLEY, YORKS.

The drive is similar to the shears in Fig. 120, except that there is no need for an assistant return spring. The tail end of the rocking arm works between substantial guides, to maintain the blade alignment, but in this respect, No. 120 is somewhat better, as the reinforcement is close to the blade. The travelling wheels of small diameter are loose on the axles—these being a driving fit in the frame, and non-rotating accordingly. The wheels are bushed with roller bearings, and double flanged. Movement of the shears from point to point can be done by two men.

An American pattern of travelling alligator shear is illustrated in Fig. 122, intended to cut up to 4 ins. × ½ in. steel, In this machine, the rocker bearing is made of great width, which obviates the necessity for any reinforcing head. The stop gauge is shown in position, and the locomotive movement is arranged to be worked through the geared hand wheel.

FIG. 122. TRAVELLING ALLIGATOR SHEARS. COULTER & McKENZIE, BRIDGEPORT, U.S.A.

In motor driven shearing machines, it is always of value to include a variable speed motor—as by this means, the cutting speed can be regulated according to the lengths of the bars. Short plates, for instance, perhaps 8 ins. long, can be cut at a much higher rate than plates of 25 ins. to 60 ins. long. A speed limit of from 20 down to 10 strokes per minute will be found suitable, if the machine is powerful enough to handle every stroke on a given size. If a fixed speed is provided, it should not exceed 18 strokes per minute.

Before the cutting-up of steel for any particular line of springs is commenced, the forging shop is provided with a "measure," generally a piece of flat about 1 in. × $\frac{1}{4}$ in., having marked on it the varying half-plate lengths. These are worked out from the straight length of the back plate in conjunction with any leading plate length dimensions which may appear on the drawing. The stop is then set on the length gauge attached to the shears, and the width gauge (if any) having been fixed in position, one lot of plates will be cut off according to the quantity required. The length rod stop is then adjusted to the next length, and so on, until the whole of the varying lengths have been cut up. According to the system of the shop, these cut lengths are then placed on railed bogies, electric trollies, or wheelbarrows, and taken to the next machine operation, which is generally centre holing or nibbing.

CHAPTER XXVIII

CENTRE FASTENINGS

THE type of centre-fastening means employed is of the greatest importance, and is one of the chief design factors which goes to prolong or curtail the working life of a spring. Necessarily, with railway springs, the hoop is attached by the centre fastening to the spring plates in general, and a detailed description of various fastenings is reserved to be dealt with in connection with the spring hoops. Automobile springs are invariably holed through the centre, the bolt acting to fasten together the plates, and its head registering in the axle or axle-pad.

The ideal fastening would, of course, leave the plate full strength in the middle, without the removal or displacement of any material. There are certain designs effecting this, but they do not come immediately within the present remarks. The general means employed are variations of centre holes or centre nibs. From the spring fitter's point of view, the centre hole is the best fastening, as plates can be collected on a pin for fitting purposes, and with round holes, plates can easily be revolved to assist even heating. From an efficiency point of view, whilst opinion is divided as to the relative values of the round hole (punched or drilled), cotter hole or downward nib (hot or cold), it is unanimously agreed that the upward nib is the worst possible form. In order to obtain definite information on these matters, the author some time ago carried out a lengthy series of tests on hardened and tempered spring steel, and very valuable information was obtained therefrom. The steel was all from one cast, and rolled into 4 in. by ½ in. normal section. The nibs, holes, etc., were all of standard sizes for the plate widths. One series of tests was carried out on pieces hardened and tempered in the usual way, and the other series was carried out on pieces carefully heated in a gas muffle to a definite temperature of 900° C., hardened at 820° C., in water, and tempered as usual by the "sparkle" method. The pieces were supported on bearings 10 in. centres, and then deflected until breakage occurred, the breaking weight and deflection at rupture

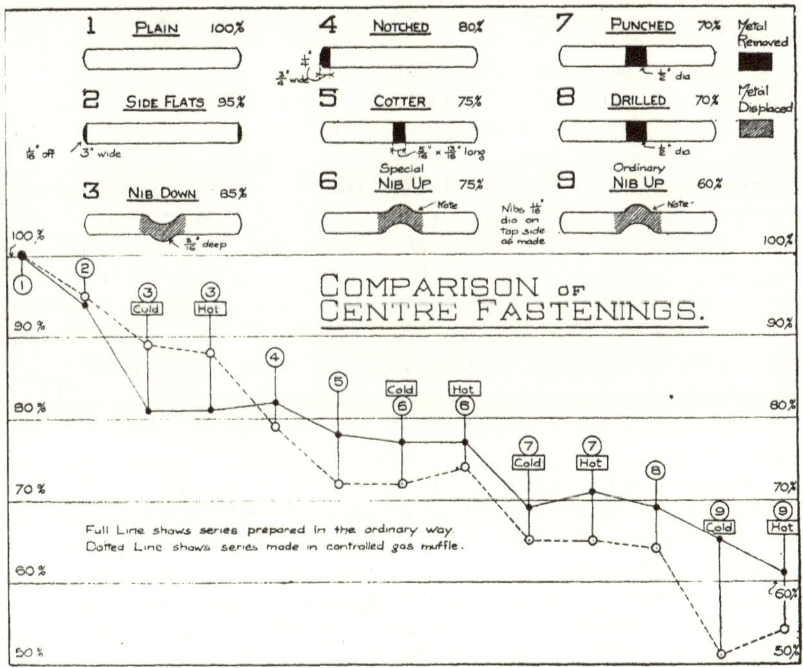

FIG. 123.

being carefully noted. From these figures the efficiencies were gauged over the two series (several pieces were broken in each series, without any serious divergencies in results for the same type of fastening). Fig. 123 gives details of the fastenings tested with their percentage values compared with the solid plate. Generally, it will be noted that the pieces prepared in the ordinary way gave the better results, but a further series might reverse, as very slight varieties of treatment will cause a fair difference in the test results.

The order of value may be re-iterated here :

(1) Solid Plate, 100 per cent.
(2) Milled each side (Side Flat), 95 per cent.
(3) Downward nib (normal type, hot or cold) 85 per cent.
(4) Side-notched (hot or cold), 80 per cent.
(5) Cotterhole (hot), 75 per cent.
(6) Upward nib (special type, hot or cold), 75 per cent.
(7) Punched hole (hot or cold), 70 per cent.
(8) Drilled hole, 70 per cent.
(9) Upward nib (usual type, hot or cold), 60 per cent.

It will be seen that the punched or drilled (round) hole has only 70 per cent. the efficiency of the solid plate—this with a hole removing only 12½ per cent. of the plate sectional area. The explanation of this was as given in Chapter XVII. Larger or smaller holes will of course, reduce or increase the efficiency. Therefore it is obvious that all plate holes should be kept as small as ever possible. To a certain extent, the size of such holes, for a definite diameter of rivet, is contingent on the accuracy of rolled width, as if relatively large variations are permitted, and the hole is very near the rivet size in diameter, the lateral movement of the uneven width plates will choke the rivet passage. Rivet holes should not be allowed to exceed by more than $\frac{1}{16}$ in. the diameter of the rivet.

The great advantage of the rivet or cotter form of fastening, apart from its practical fitting value, is the simplicity of the hoop fixture, as the rivet is merely driven through plates and hoop and headed up. The downward nib possesses a high efficiency value, but is not generally regarded as quite as simple for attachment to the hoop, the mere fact of a downward nib being employed driving certain designers into packing plate and set screw schemes, which are expensive and not every useful, as the hoop fastening can be made almost as simple as the rivet type. One disadvantage of the nib in all its forms is that, unless care is taken in the forging, the nibs will ride on the plate adjacent, which tends to hasten breakage.

Features of interest present themselves with the side flat and notched type, though these are not very popular, as the hoop fastening is generally more complicated than with the other patterns. Neither of these particular patterns are very suitable for automobile work as usually produced—although they can be made to give great satisfaction in service if care is taken with the manufacture. The side flat should be $\frac{1}{8}$ in. in depth, owing to the tendency with road vehicles, particularly those with the Hotchkiss drives, to pull their springs through the centre fastenings. No better arrangement can be found than the side flat, but it necessitates either a U-hoop or a box hoop—the first of which is the more popular with automobile designers, although there is no reason why the box hoop should not be more freely used. The low cost of ordinary springs, compared with the total cost of the vehicle, justifies a small extra amount of expenditure on a really reliable fastening—as spring troubles are amongst the

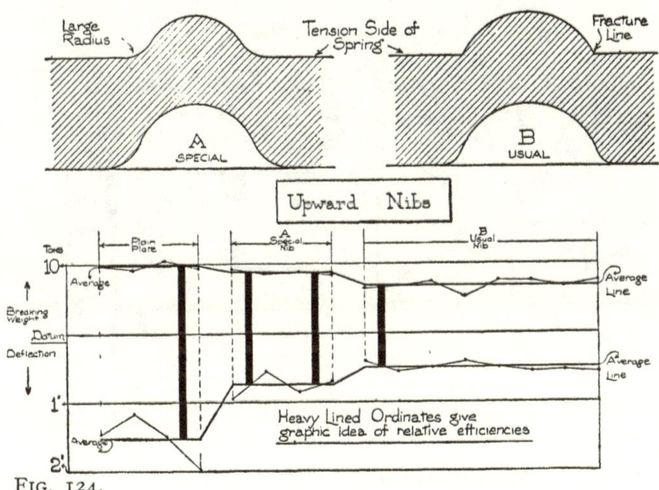

FIG. 124.

most prominent which afflict the road vehicle user, and probably the first designer who will take the responsibility of introducing a novel and sound centre attachment will find himself amply repaid by the enhanced reputation of his general production.

The upward nib, as usually made, is, as before remarked, easily the worst possible type. Unfortunately, it is very commonly used for locomotive springs of the underhung type, as set-screws do not appear easy (to the usual designer) to insert in the top of the hoop, underneath the axlebox T-hanger, whereas they are easy to get in the bottom. The upward nib is made subservient to the easily fitted set screw, and breakages are then blamed on to the manufacturer, or the steel, or the bad weight distribution, or anything else, except the inherently thoroughly bad design. Upward nibs of an unusual pattern, as No. 6, are fairly efficient, but nibs are generally regarded as small details which no one cares to include in specification requirements. These requirements will possibly lay great stress on hot forging, or tapering the plate ends, or grinding a neat radius thereon, none of which are in any way material points, but the shape of the upward nib for heavy engine springs will not be stipulated. Such a nib form as the special one sketched requires care in the making of the punch bed, as it is most important that no

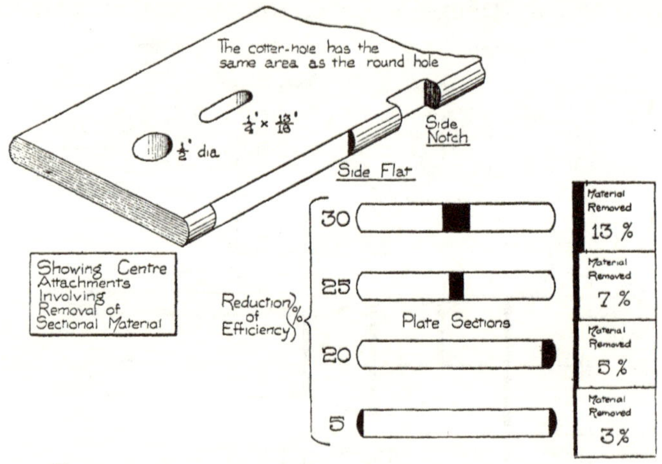

FIG. 125.

sharp line occurs anywhere. Fig. 124 gives an enlarged detail of these upward nib types, with certain results obtained therefrom, which should be sufficiently convincing to damn entirely the usual type.

With the fastening forms requiring the removal of material, such as Nos. 2, 4, 5, 7 and 8, the reason for the relatively high efficiencies of Nos. 2, 4, and 5, is due to the fact that Nos. 2 and 4 remove small areas of lateral metal, and No. 5 weakens the spring a minimum in the centre section by placing the major axis of the cotter longitudinally. Centre round holes, as Nos. 7 and 8, remove metal at the most valuable part of the section, which point is illustrated in Fig. 125.

A general review of all considerations results in allotting the first place for general simplicity, security, efficiency, and economy, to the downward nib form, and for railway springs this is very usual—it being employed at least as much as the rivet or cotter. American practice includes it almost invariably for locomotive springs, and British and Continental engineers favour it to a high degree for all classes of railway work. It has, however, two disadvantages for automobile work, one being the liability of the spring backs to move relative to the axle, owing to the cumulative effect of numerous plates and excessive nib clearances, and the other being in the fact that it does not hold the plates together, as does a bolt, so that when removed from the vehicle, all or certain of the plates are loose. The first point is illustrated in Fig. 126, which makes clear the movement referred to. This can be obviated with care in the forging of the centre nib, as whilst the material in the depression will always be somewhat larger than in the embossment, it can be kept down by skilful arrangement so as to make a perfect fit. The latter, however, does not make things too happy for the spring fitter, as the fit is sometimes too perfect to allow the nib to properly bed in the depression. A total end movement of $\frac{1}{32}$ in. can be allowed as a " fit," and this will be found a great improvement upon a similar movement of $\frac{1}{8}$ in. This point is not of serious importance in railway springs, so long as the clearance is within $\frac{1}{8}$ in. The second disadvantage referred to can be obviated by clipping or shackling the short plate, so that the spring can be bodily removed, with the clips holding the plates. A great advantage of the downward nib form for automobile springs is its resistance to shear, which is very much higher than that of a rivet of safe size for the plate width. An attempt to combine the two forms is largely used in the U.S.A., and known as the " cup " or " dowel," which consists of a large diameter shallow depression with a small bolt hole, and is certainly a good shear resisting medium, whilst permitting the spring plate to be held by the through bolt without the need of short plate clips.

In Fig. 126 are shown dimensioned a number of standard centre depressions, which are described as follows :—

A. shows a " cup " centre, flat bottomed. Theoretically, the top depression equals the bottom embossment, but actually a small clearance is of course present, which does not exceed as a rule $\frac{1}{64}$ in.

B. is a similar " cup " but of curved shape. There would not appear much advantage in this over (A). Neither of these forms are used to any extent outside America.

A
B
C
Bolt Hole for 3/8"
A
PLAN
Centre Nibs & Dowels.
D
E
H
G
F
Metric Sizes
J
Possible Effect of Clearance
Top Plate (Back)
Relative Movement

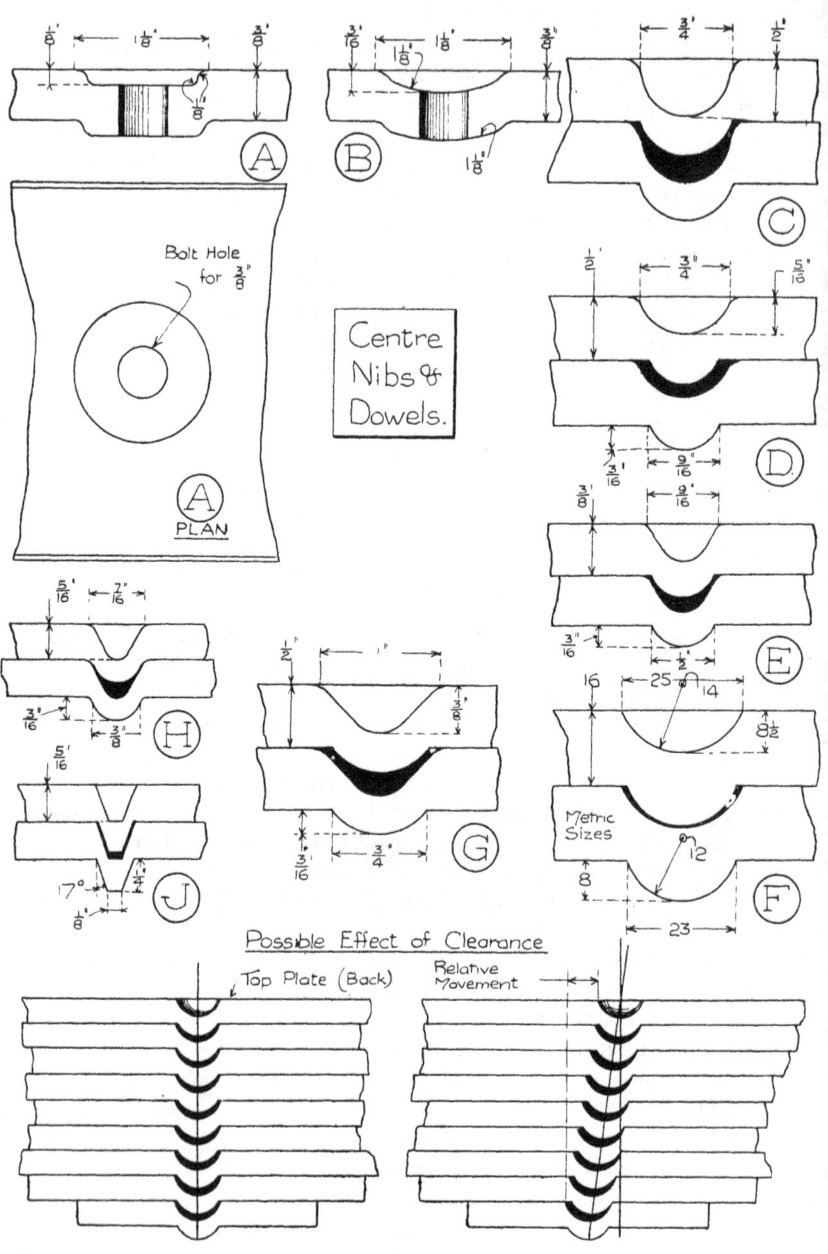

FIG. 126.

C. is a centre nib for railway springs—forged hot, of dimensions usual in the U.S.A.

D. is a form of nib which can be forged cold—and is used in European work.

E. another European form, with little clearance—which is a very good pattern for all classes of springs.

F. the standard form of the German Railways for locomotive springs.

G. shows a cold forged nib, suitable for railway springs.

H. is a standard American automobile form, with no clearance.

J. is a similar form to (H) but square bottomed. Both (H) and (J) are hot forged.

It will be observed that some of the forms shown are very little better than a hole, as the depression is nearly or quite equal to the plate thickness, notably C—E—G—H—J. Such nibs as these will not give the 85 per cent. efficiency recorded for the best practical form, which approximates to D—F. Their advantage over the hole lies in their presenting spring steel as a shear resistance, instead of a probably soft steel rivet or bolt.

CHAPTER XXIX

SIDEPLAY CHECKS

STUDS and slits are provided in plates to check lateral movement. Some check of this sort is necessary with springs over 36 ins. long, if no clips are provided such as should be the case with automobile springs, but it is becoming increasingly general in railway and tramway work to dispense with these details in relatively short springs. The alternative side-play check, involving the use of rib and groove steel, has been touched upon, but such section is not very popular in this country, although there is at present a tendency to employ it for new designs. A bastard type of this is sometimes made, which is the subject of a very old patent. The ends only of the plates are pressed into rib and groove form with this method, but apart from the U.S.A. where a few instances still present themselves in heavy commercial automobiles, the arrangement has fallen into disuse.

One very important point deserves to be specially noted, namely, back plates should never be studded. Designs showing this are frequent, and can only be the result of thoughtlessness or misguided consistency, inasmuch as in most cases, the back plate, and two or three following plates, are held from lateral movement by the suspension gear or spring shoes. (See Fig. 127.) Generally, there is no need to commence studding until the third plate is reached.

Studs in a back plate are commonly starting points of fracture, particularly if they are in solid-end (welded) backs, as they are generally located at the point where the steel may have been overheated in the back making. The efficiency of any plate is—with ordinary sizes—reduced 20 per cent. at the point of location, so, if absolutely imperative, care should be taken to place studs or slits at positions in the plate where there is a surplus of material over and above the theoretical minimum required.

The possible different arrangements of slitting are three, "open slit," "ordinary slit," and "secret slit." The first is very usual in engine spring work, the second is common over the whole of spring work, and the third is very frequently used, from the point of view that the spring is likely to suffer less from weathering influences if the studs and slits are entirely concealed, which has certainly some logic. These

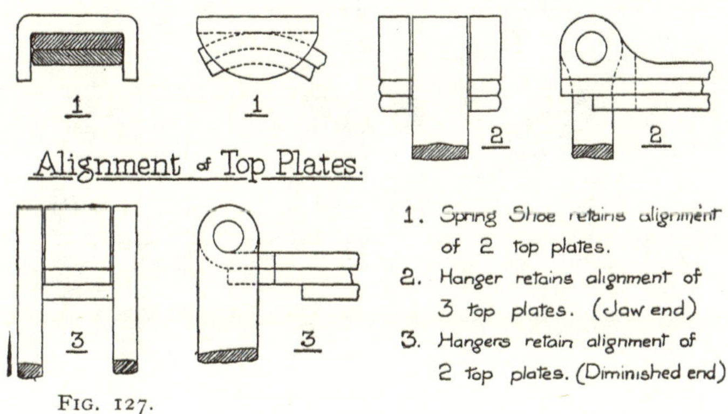

FIG. 127.

three types are shown in Fig. 128, with various forms of studs and slits. The "open slit" is the safest from the stress point of view, as there is no danger (if the spring is reasonably designed) of its reducing any high stress carrying area. The second and third require careful placing, as before indicated, and reference to Fig. 129, showing a typical arrangement of plates with their bending moments, sufficiently indicates the reason therefor.

The best type of stud and slit is indicated on Fig. 128, with a round-ended slit, and small well-defined stud. The worst type is also indicated, with a square slit, and deep stud. Such as the latter can frequently be knocked off with a hand hammer, and the square slit cracks from its corners.

It is a point often overlooked that the displacement of material on the bottom side of a plate under studding is never equal to the displacement on the top side. To obtain a deep stud therefore, the punch must go nearly through the plate. This same point .obtains, as has been already remarked, in centre nibbing, and of necessity there is, on account thereof, a certain amount of play between the depression and the impression—the object aimed at being to so design the shape of the punch and bed that this play is as little as possible, consistent with absolute freedom of the nib.

Makers of road vehicle springs have appreciated the weakness of the stud and slit as lateral checks, particularly when no clips have been present on the spring. The continual vibration gradually wears the stud until it is of little value,

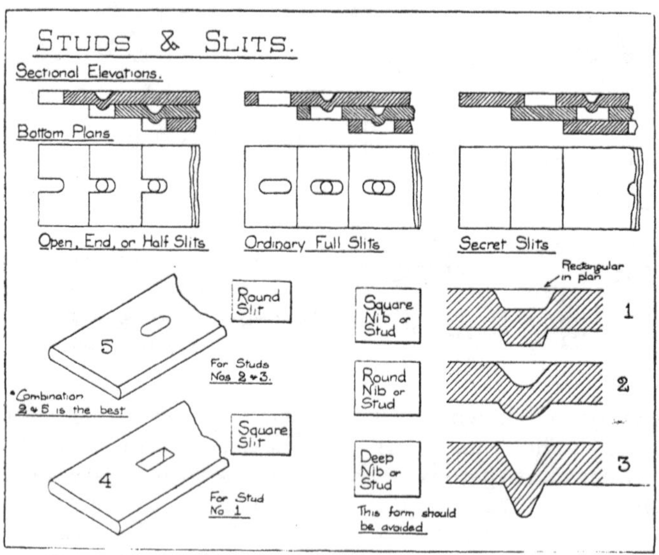

FIG. 128.

and the plates shift accordingly. Two alternative side-play checks are accordingly in use, one of the rib and groove form (previously referred to) and the other of a lugged form, made by thinning, spreading, stamping, and bending, the plate ends. Both these forms are in use to a limited extent, but neither are much liked, owing to difficulty of replacing plates —as it is a difficult job to make the rib or the lugs away from the specialized machinery, whereas any repair shop can introduce a stud or slit. These alternatives are illustrated in Fig. 130.

Generally speaking, British practice springs present the maximum of studs and slits, Continental but a moderate quantity, and American a minimum. In automobile work, where the forces tending to create side movement are greater than with railway work, studs and slits should be entirely dispensed with in favour of good clips, which with ordinary designs serve the dual purpose of taking rebound, and preventing side-play ; and with centre nib designs can include the additional function of holding the plates together.

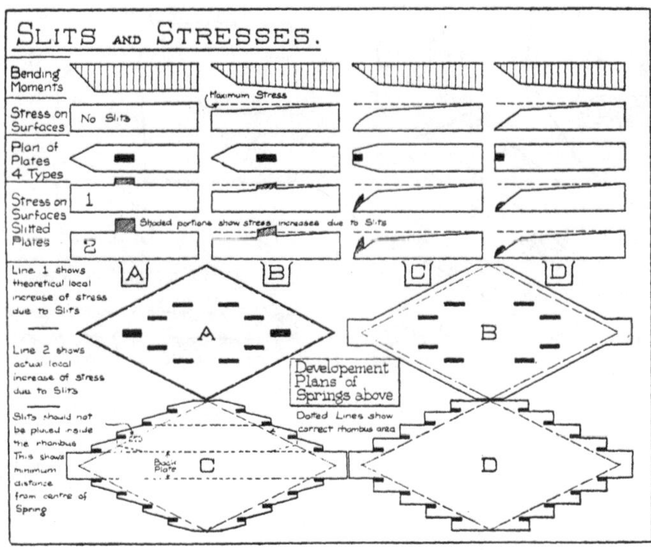

FIG. 129.

If it is essential to employ studs and slits, a certain amount of intelligent attention should be given to the quantity included, as in most cases, there is no need to put these in short plates. Every stud and every slit is a potential breaking point, particularly with certain types of springs, and an examination of those taken out of service will reveal a high percentage of fractures through the slit.

A few standards of studs and slits are shown in Fig. 130, as follows :—

A. shows three views of a usual British pattern, cold forged, and suitable for railway springs.

B. shows a European automobile spring type, hot forged—with the result that the stud is a very neat fit to the slit.

C. is a modified stud and slit, known as the "saw and bead," and this is still used to a certain extent for road vehicle work. The lateral fit of the "bead" to the saw cut is practically exact, but the device retains all

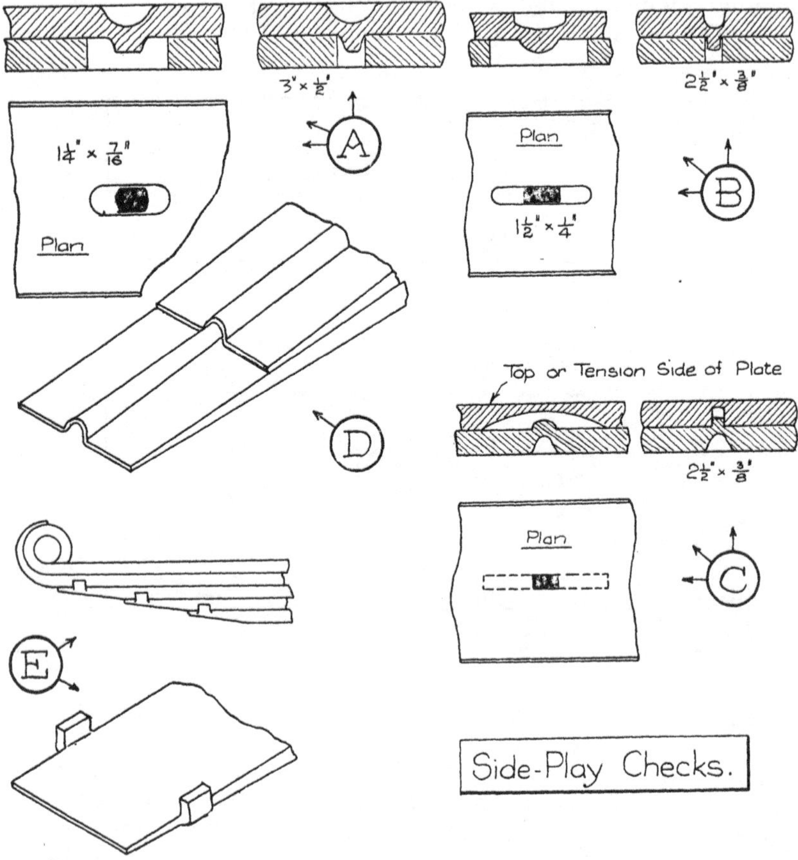

FIG. 130.

the disadvantages of the usual form, inasmuch as it both removes and distorts potentially active material. The slit is made by the periphery of a thin circular saw, working on a cold plate.

D. is the rib and groove form previously referred to—this

can be stamped or rolled, and extends about 6 ins. from each end of each plate.

E. shows the lugged form of plate.

Great care is needed in the placing of studs and slits to ensure free movement as the plates flatten out under load, otherwise certain plates may be checked in their free movement, and stresses diverted, until some failure occurs. Indifferent positioning of studs also makes extra work for the fitters, and it is not an uncommon thing to have to either partially reduce the stud size, or grind it entirely away, in order that plates may bed themselves properly.

CHAPTER XXX

SPEARS AND DRAWN POINTS

THE spearing of plates has been previously touched upon in Chapter XVI, and various types were shown in Fig. 58. Spearing—of correct pattern—causes the plate-to-plate distribution of the load to operate with maximum of freedom, by equalizing the stress throughout the plate length, due to the removal of superfluous material. Except for long carriage or automobile springs, however, spearing is not practically essential, as railway springs are in general too stiff for any difference in effect to be appreciated between speared and square cut plates. Spearing is more often regarded as an end finish, and much ingenuity is displayed in giving every possible dimension of the spear on the working drawing and varying the shape each time for the leading bogie, coupled wheel, trailing truck and tender springs. Theoretically, the length of the spear should be varied according to every design, practically such variation is of little value and quite intolerable. So long as the developed plan of a spring falls on, or a reasonable distrance outside the rhombus plan previously explained, the difference in deflection per ton is undetectable with commercial measurements between the various spear shapes employed, and plates being square cut.

Cold spearing is the heaviest job that falls to the lot of spring forging machinery, which is one point in favour of hot work. It must always be borne in mind that unfortunate incidents will occur in the best regulated plants, and on occasion, steel qualities which are not spring steel qualities, are rolled down into spring steel sections. When such changeling material is introduced to the forging machinery as cold work, and the substitute is perhaps, 1·2 per cent. carbon file steel, something has to happen, particularly if the section being worked is ½ in. or ⅝ in. thick. In a spearing machine when such occurs, the tools are invariably spoiled, and on occasion, the machine frame is broken.

Spearing tools are relatively costly items, which is one point in favour of standardization of spears—a small enough matter as it affects the spring, but a matter of considerable moment to the manufacturer who retains in stock a considerable quantity. Additionally to the stocking, the time involved in changing from one spear to another is not inconsiderable, and adds its quota to general shop charges. The

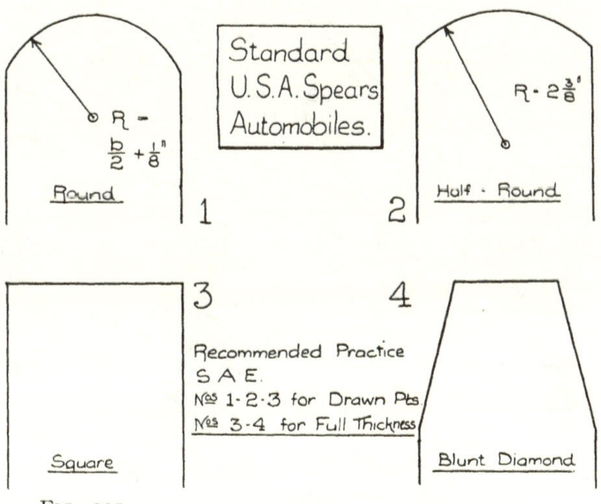

FIG. 131.

U.S.A. have certain standard spear forms for the automobile trade, which are illustrated in Fig. 131. These are the recommended practice of the Society of Automobile Engineers (New York), and are all unexceptionable forms. The most used is the "blunt diamond," No. 4, or, as called in this country, the "square spear." Over here, where appearance is made to count with a value altogether disproportionate to its importance, it is frequently stated that the square spear does not present a good finish. Fig. 132 shows an inverted view of a spring so finished, so that this "point" can be judged for itself. In connection with the half-round standard spear, No. 2, Fig. 131, the radius shown for this will take everything up to 4½ ins. wide, which is a section not often required for road vehicle work.

In Fig. 133 are shown some practical aspects of spearing, which should go far to substantiate the value of the square spear standard of the U.S.A. Sketch A herein shows how

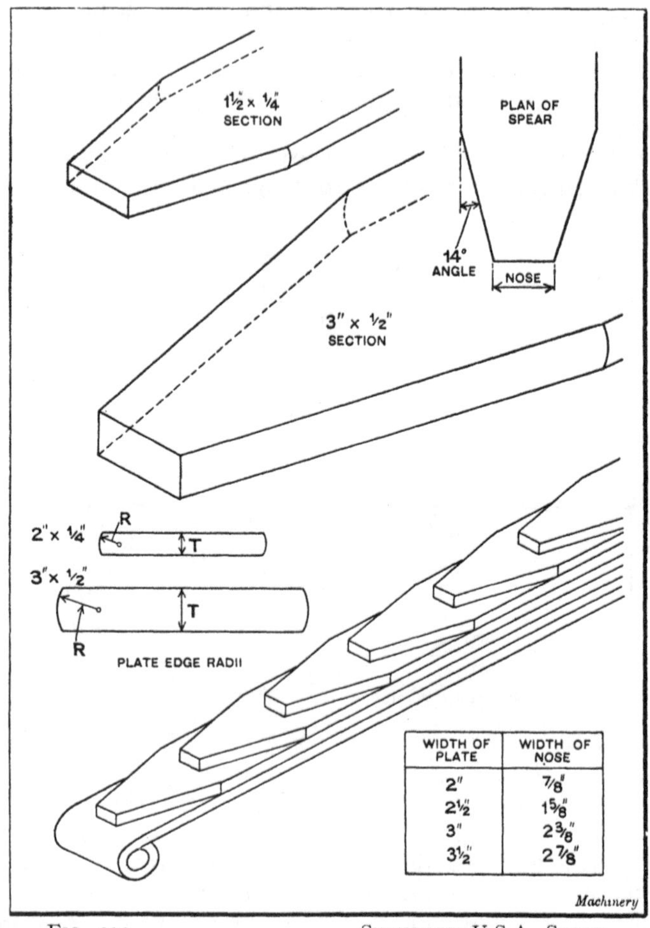

WIDTH OF PLATE	WIDTH OF NOSE
2"	7/8"
2½"	1 5/8"
3"	2 3/8"
3½"	2 7/8"

FIG. 132. STANDARD U.S.A. SPEAR.

bars are cut up in some works there. A double spear and punch is employed, and the parting and spearing of two ends are performed at one stroke. This procedure involves a small amount of waste between bar and bar, but it saves a great deal of handling, and also gives very square ends, as the cut, instead of being a shear cut, is a punch cut. Needless to say, this operation has to be done cold, and the importance of oil-hardening steels being well annealed before cutting up is obvious. Sketch B shows a round spear tool set, consisting of the bed and blade, the latter being the moving element. The disadvantage of this spear form can be seen by the comparison of it with the square spear tools of Sketch C. These latter can be in the form of two straight shear blades for the moving head, and they will work over a large series of spear dimensions. The tool design involves far less machining, and the tools are also easier to trim-up when this is needed. The round nose, Sketch B, is the portion which

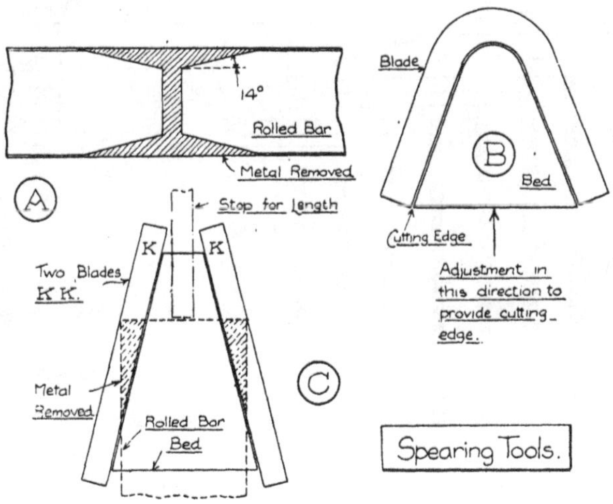

FIG. 133.

cuts the middle width of the bar, which presents the maximum of segregation (frequently carbon) and therefore tends to damage the nose, which is the most difficult part to return to cutting perfection. No nose is needed for the square spear, and the variety of plate end dimensions which can be dealt with under the one head, and with the one tool setting, is indicated in Fig. 134. The only amendments needed to the general arrangement are the adjustment of length and width gauges, which are very quickly performed.

Designs of springs which include very long overlaps should always be speared, but generally speaking, the usual length of existing spearing tools and arrangements will not permit of this being done usefully. Plate ends of this nature have therefore to be machined down to size, which can be carried out with fair speed on efficient machinery, although not at the same speed or cost as machine spearing. However, springs of this nature are exceptional, and can generally support the enhanced price.

The tapering of plates in thickness at the ends, otherwise the "drawing of points" is a subject which has called forth much argument. With the developed spring plan enclosing the theoretical rhombus, it is a matter of no importance (except as regards cost) whether the plate ends be speared,

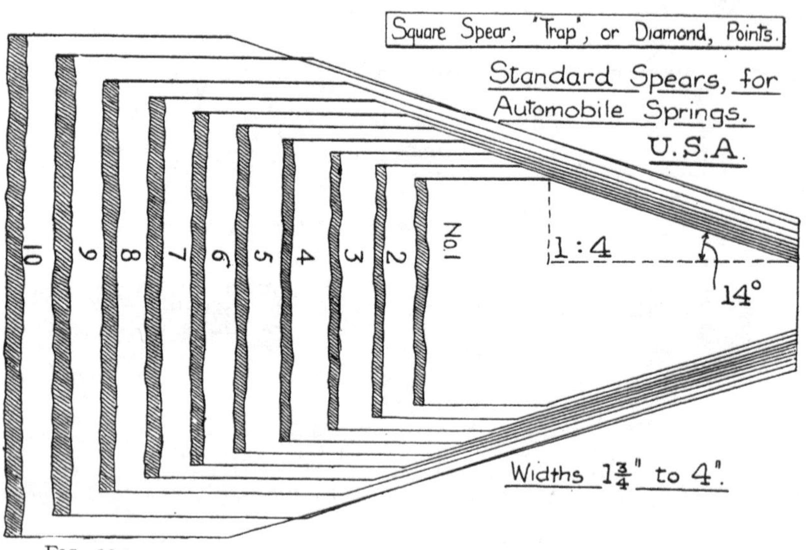

FIG. 134.

drawn or square. With drawn points falling inside the rhombus, however, the spring becomes weaker. Long and complicated calculations could be made on this subject, but, as from the author's point of view, such a spring is wrongly designed, it does not appear very useful to go deeply into the matter. When drawn point overlaps drawn point, the subject becomes still more involved, both practically and theoretically. To make a decent job of a heavy spring with drawn points overlapping is extremely difficult, and when such spring is under load, most of the points, fitted "on" with such difficulty, promptly come "off." The amount of drawing usually shown on engine spring designs is negligible, owing to the generally relatively small steps from plate to plate, and it is sheer waste of money carrying it out. The buyer, of course, pays, but it is certainly worth while to study as to whether this requirement cannot be generally dispensed with, resulting in a cheaper spring, and speedier production.

As before emphasized, the draw should be a perfect parabola, on thickness only, and any spearing of width spoils

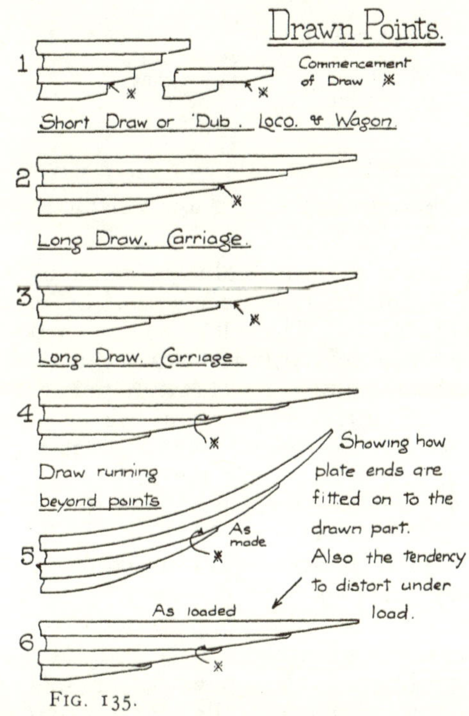

FIG. 135.

the whole of the idea. Actually, the draw is a straight taper, and the nearest point in practice which is obtained to the parabola, is got by grinding the ends. Fig. 135 shows various aspects of drawn points, in the hope that illustrations of this nature will at no distant date take their place amongst historical descriptions of obsolete processes.

The drawn point is gradually disappearing in U.S.A. automobile spring practice, as it is not possible to handle high production work through plate forming machines if the plates have "rolled ends." More adjustment to the machinery is necessary, and considerably more hand-fitting is required to make a passable spring from the machine-formed plates. Attempts have been made here to manufacture automobile springs of English design with machines of American pattern, but have had to end in failure, chiefly owing to this difficulty. A careful study of springs for road vehicles was made at the Motor Show (Olympia, London) of 1920, and the following figures were obtained :—

British.	Commercial Vehicles.	90 per cent. had " drawn points."
Continental.	,, ,,	25 per cent. ,, ,,
American.	,, ,,	24 per cent. ,, ,,
British.	Pleasure Cars.	93 per cent. ,, ,,
Continental.	,, ,,	91 per cent. ,, ,,
American.	,, ,,	23 per cent. ,, ,,

All the above percentage figures show a reduction during the last two years.

The drawn point is nearly always made by the use of tapering rolls, the alternative process being by swaging under a quick-acting hammer. The rolls, however, make the better job. The plate has necessarily to be raised to a heat in excess of the forging heat required for spearing, holing, or nibbing ; and consequently, there is always a risk of damaging the steel. Furthermore, in the treatment process involved in spring making, the thinned points are the first parts of the plates to arrive at a hardening heat, and they have then to soak, or probably get very much hotter, while the middle portion of the plate is arriving at its correct temperature. This additional time tends to decarburise the thin points. After quenching, the tempering is performed, and here again, the thin points become hotter than the body of the plate and are softer accordingly. The above remarks do not apply with the same force when controlled furnaces are used, as in American practice, but in this case, comparatively few drawn points are made. In European practice, where furnaces are generally uncontrolled, and the heating of plates depends entirely upon the spring fitter, drawn points are predominant in the automobile trade, and have to suffer as above. Consequently, more " nip " is generally introduced, and the short plates break more frequently. From a random lot of 50 springs, 8 plates, 3 ins. × $\frac{5}{16}$ in., which were taken out of service, the following figures were obtained :—

Twenty-three springs, drawn points and centre nib. Oil-hardening steel.

Fifteen broken plates were found, Nos. 7 and 8 (short plate), or 33 per cent.

Twenty-seven springs, square cut, full thickness, centre-hole. Water-hardening steel.

Five broken plates were found, Nos. 7 and 8 (short plate), or 9 per cent.

These springs had been in the same service, and the difference is worth noting. Most of the water-hardened springs had been running twice as long as the other type, so that the discrepancy in their favour is even more marked than would appear from the figures given.

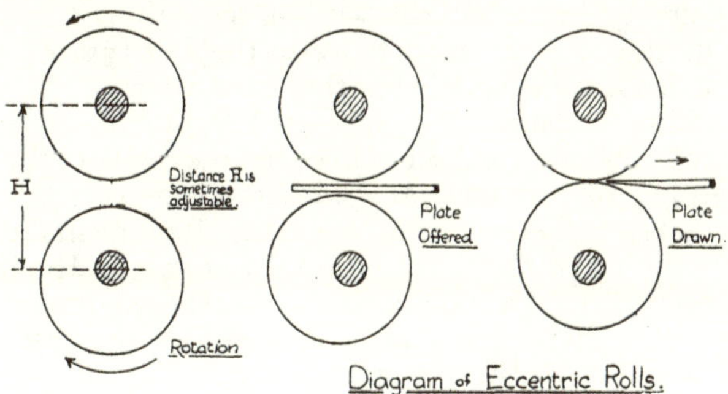

FIG. 136.

There is not the need for the additional amount of " nip " that a fitter usually introduces into a spring with drawn points, but he finds it easier to include it with drawn-point plates than with full-thickness short plates, and as a certain school consider plenty of " nip " makes a lively spring, the maximum possible is put in—with results as indicated.

A diagram of the rolling device in use for the manufacture of the drawn point is shown in Fig. 136. End stops are generally fitted to regulate the amount of draw, and very long tapers can be obtained by a series of successive rollings.

CHAPTER XXXI

SPRING CLIPS AND REBOUND PLATES

THE small spring fitment known as a "clip" in the automobile industry, and as a "shackle" in the railway trade, is as much a matter of great importance to the first type of spring as it is a matter negligible for the second type. An automobile spring is consistently subjected to severe shock, and consequent rebound, and in the latter movement, the axle is attempting to part company with the chassis. The only holding devices checking this are the attachments of the spring to the chassis and the axle respectively. The spring itself, however, is not a solid beam, but composed of numerous plates, so that the axle is really only held (in the absence of clipping devices) by the pin which secures the rolled eye to the chassis shackles, or dumb irons. The general effect of a rebound is therefore to lift the back plate off the remainder of the spring, with sometimes serious results. To effectively combat this effect, every automobile spring should be substantially clipped, in accordance with one or other of the clip forms shown hereafter.

Clips in themselves are of little value unless they are properly placed and fitted. They always perform the very useful function of replacing the need for studs and slits, with the weak planes throughout the spring which occur through the use of this latter arrangement. To act properly as "rebound clips" they must, however, hold the spring solidly against the reverse of the shock, so that the back plate is not acting by itself to take this effect. Very little attention seems to be paid by automobile spring makers to this item, and whilst it is frequently claimed that "so many" rebound clips are fitted, they are more often than not useless for such purpose, and even dangerous. Fig. 137 shows various forms and aspects of clips of the "U" pattern which are most ordinarily fitted. In these sketches :—

No. 1. shows the clip provided with a small hole, and the spring plate end nibbed to fit thereto. This is a very good form, and easy of attachment.

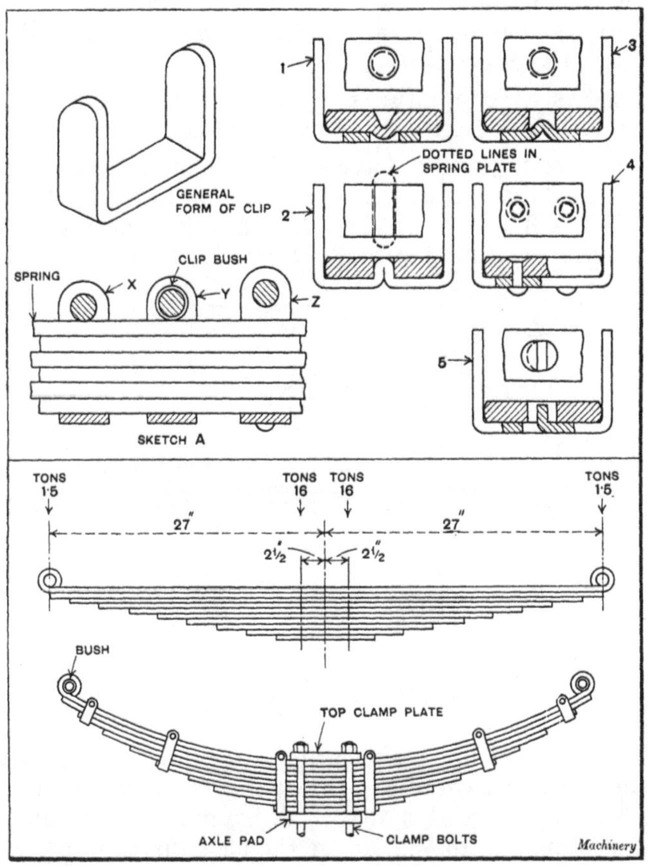

FIG. 137. CLIPS FOR AUTOMOBILE SPRINGS.

No. 2. shows a form with a slit in the spring plate, and the clip doubled over, full width, and inserted therein. This is a very good form, but the clip requires more manufacture than No. 1.

No. 3. shows a similar arrangement to No. 1, but with the positions of the hole and nib reversed. No. 1 is the better form, as the hole in the clip makes a convenient registering point for the bending of the latter.

No. 4 shows the more usual form, namely, rivets inserted into the clip bottom, and countersunk into the plate. One rivet was at one time standard practice, but owing to breakages of the same permitting the clip to wander, nearly all springs over 2 ins. wide have two rivets per clip.

No. 5 shows a good form, the clip having a punched lug which inserts into a hole in the plate.

Types Nos. 1, 2, 3 and 5, have special advantages, inasmuch as that unless the bolt at the top is close on to the back plate these clips will not be held. It is therefore necessary for the punching or drilling of the bolt holes in the clip legs to be

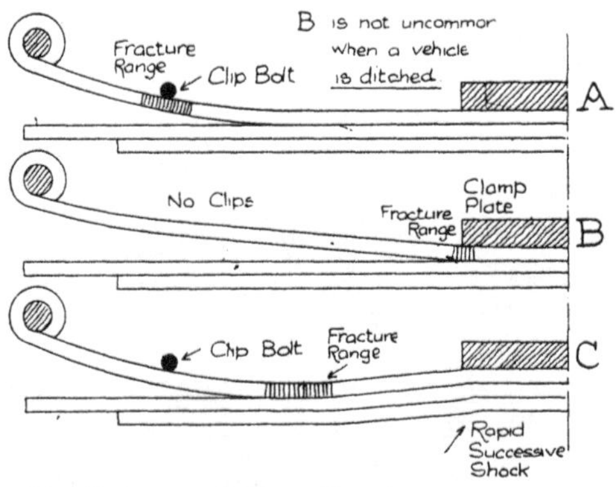

Fig. 138.

exact, so that the slip on the spring is as shown at X, Sketch A. The rivetted form of clip is responsible for many failures through being fitted as shown at Z, Sketch A, with the bolt anything from $\frac{1}{8}$ in. to $\frac{1}{2}$ in. above the back plate. The result is that when a rebound arrives, the back levers itself against this bolt, which is, of course, rarely less than 6 ins. to 10 ins. from the rolled eye, and, if the steel be on the hard side, the end snaps off. In springs made without clips, this effect is sometimes achieved by the back plate levering itself off against the sharp edge of the central clamp. Investigations show that a large number of back plate failures is due to bad clip fitting, or no clips, and it is suggested that well-fitted clips should always be provided on automobile springs at the rate of one per 10 ins. of length, which would give in general, 6 per rear spring, and 4 per front spring. American makers are ahead in the provision of quantities of clips (without reference to their fitting), British practice is becoming alive to the advantages, and strangely enough, Continental practice makes little use of them. To a certain extent—particularly in heavy Continental springs—their need as sideplay checks is discounted by the use of rib and groove steel.

The bottom drawing in Fig. 137 shows how clips should be fitted to a heavy automobile spring. By the use of those on the short plate, the spring can be made with a centre nib fastening instead of a bolt hole, and yet remain a complete entity, easy to handle.

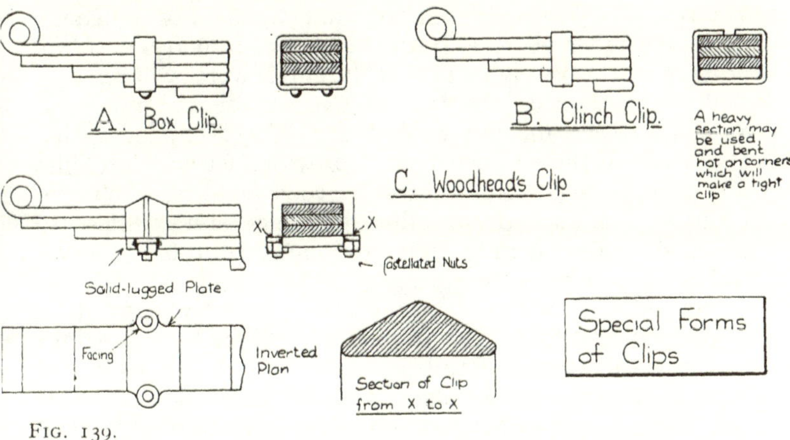

FIG. 139.

In Fig. 138 are illustrated the breakages of plates through the rebound leverage, and will suggest methods of eliminating this. It also shows a frequent cause of trouble due to vehicles running into ditches, etc., and allowing thereby the weight of the axle to be suspended from the back plate, which, with unclipped springs, causes the back to either bend or break. It must be stated that the chief reason for the absolute breakage of back plates through rebound causes is due to their being too hard—if 3·2 dia. Brinell or softer, they will bend and set, but the vehicle can at anyrate get home, whereas if harder than 3·0 dia. Brinell, they will invariably break off short. An investigation of quantities of broken backs has shown that nearly 50 per cent. failed owing to one or other of the leverage causes shown, combined with a hard temper.

Additionally to the U-clip forms which have been illustrated, types of box clips are also in use. These are shown in Fig. 139, wherein Sketch A illustrates a true box clip, welded up, put over the spring, and rivetted. The disadvantage of this form is the weakness of the rivet attachment, and the somewhat greater expense of making the exact size of the clip, as if not a tight fit, it is of no value. Sketch B shows a "clinch clip" rivetted at the bottom, and the top ends bent over. This is of no great value, as the leverage of the back will raise these ends unless a very heavy section is provided, which is not usual. The best form which has been so far evolved is probably that shown at Sketch C (patented and registered by Jonas Woodhead and Sons, Ltd., Leeds). It is somewhat expensive, but nevertheless provides absolute solidity, and the strong section of the clip is worthy of remark. It will be noted that the bolt lugs are made as an integral part of the plate and not welded on. A well-clipped spring is a necessity if the motor vehicle is provided with a Hotchkiss drive (that is, no radius or torque rods are fitted), but to efficiently take such drive, clips should be started close to the rolled eye—generally they are mounted too far down towards the centre of the spring.

Long railway carriage springs are frequently clipped or shackled on the ends. These shackles originally came into use in the early days of railways owing to the then general adoption of the "tension-bar" type of back plate for this work, the spring receiving an initial tension from the scroll irons, the idea being that easier riding resulted on a light vehicle. The effect on the back plate of this initial tension was to endeavour to part it—by straightening it out—from the succeeding plates, and the function of the shackle was to hold it on to these plates. An additional feature was that (weakening) studs, preventing lateral movement, could be omitted in the back plate, as had it not been for the provision of the shackle, such studs (with their potential fracturing features) would have been needed, as the tension bar type of scroll iron is not arranged to hold the second plate, as usually happens when vertical hangers are employed. Modern drawings frequently duplicate the lateral check on shackled springs, by insisting on (quite unnecessary) studs and slits in addition to the shackles. Railway spring shackles are frequently designed of fancy shapes, but are quite as effective made out of 1 in. by $\frac{1}{2}$ in. flat. Their use is fast decaying, but when employed, they are generally fastened and arranged as shown in Fig. 140. The rolled eye form of attachment is sometimes used in heavy automobile work.

One disadvantage of clips on railway springs—which are of long life compared with automobile springs—lies in the fact that when tight fitted (which is generally insisted on, for no real necessity in modern suspension with its vertical hangers) they gradually abrade the back plate until it breaks. The soft material of the clips collects and embeds hard dirt, and then has a grinding action on the hardened and tempered spring plate. The distinction in railway work "to clip" and "not to clip" is frequently based on the classification of the spring as an "engine" or a "carriage"—if the former,

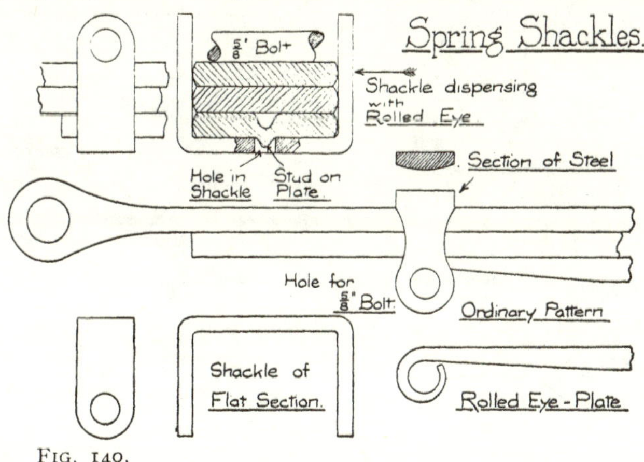

FIG. 140.

no clips, if the latter, clips must be fitted—and on occasion, springs for the same service, of the same length, have been noted, with no clips for the motor coach springs (engine) and clips on the trailer coach springs (carriage). After a few visits to the repair depôts, however, these latter shackles generally disappear, and leave an empty rolled eye.

For clips in general, standard flat sections are as good as anything, providing they are arranged according to the strength of the spring. Three standards will be found satisfactory for all springs, light (say to $2\frac{1}{2}$ ins. wide) of section 1 in. $\times$ $\frac{3}{16}$ in.; and medium (say $2\frac{1}{2}$ ins. to 4 ins.) of section $1\frac{1}{4}$ ins. $\times$ $\frac{1}{4}$ in.; and heavy (say 4 ins. upwards) of $1\frac{1}{2}$ ins.

× $\frac{5}{16}$ in. American standards are somewhat as follows :—up to $2\frac{1}{2}$ ins. wide, $\frac{3}{4}$ in. × $\frac{1}{8}$ in. ; $2\frac{1}{4}$ ins. to 3 ins., 1 in. × $\frac{1}{4}$ in. ; over 3 ins., $1\frac{1}{4}$ ins. × $\frac{1}{4}$ in. ; with inside width $\frac{1}{32}$ in. more than the nominal spring width.

The high-production manufacture of clips can be carried out best on machines (hereafter illustrated) of the American rolled-eye pattern, with compound movements. They can, however, be made very cheaply and efficiently with the ordinary cutting up, spearing, and punching tackle of the usual spring forging shop. For the bending, however, some form of fly or other press is required, as the stroke of the usual forging machines is not sufficient for deep clips, which

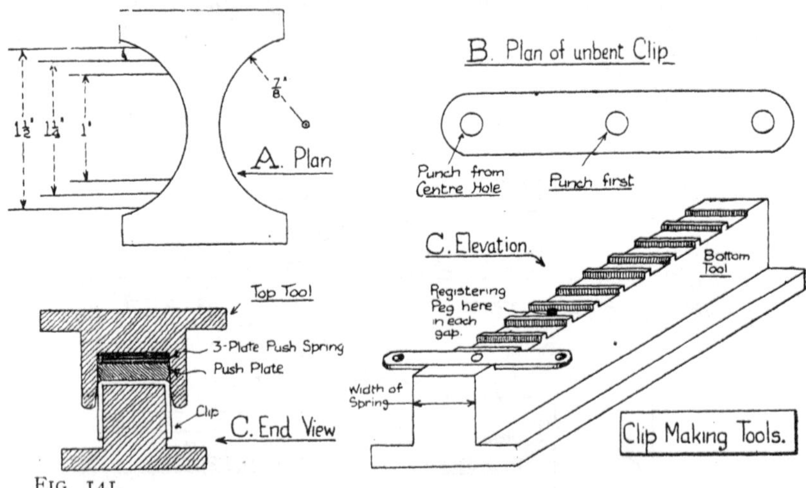

FIG. 141.

may have 6 in. or 7 in. legs. A set of tools for clip making is shown in Fig. 141, in which

A is a combined cutting and spearing punch.

B is the punching arrangement, all holes in one clip being the same size.

C is the bending tool—holding 10 clips, and taking all lengths.

Different tools are needed for each width.

The material used for clips should be good quality soft steel, which will stand cold bending—tensile say, 25 to 30 tons.

On occasion, rebound clips are supplemented, or replaced, by "rebound plates," of the type shown in Fig. 142. The custom is not far-reaching—it appears to have commenced in the U.S.A., then touched this country, and has now reached the Continent. The few American and British designers employing the device are mostly dropping it, and probably in due course, Continental designers will follow the same line. For these special plates to be of any value, they must be fitted with considerable nip, which should bear a relation-

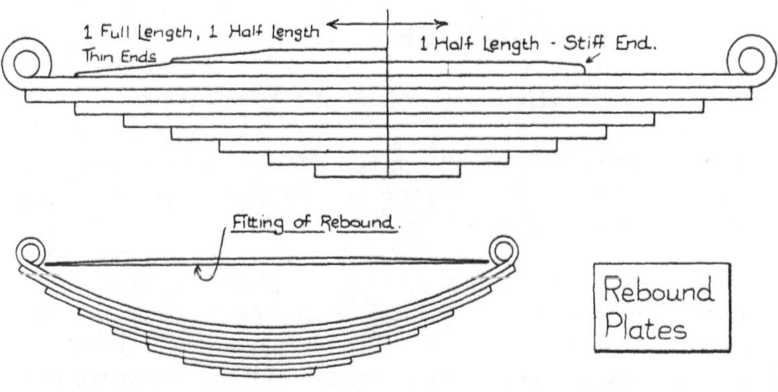

FIG. 142.

ship to the maximum working (or test) point of the spring. For instance, if the spring may go straight, the rebounds should be made straight, and pulled down. The "best" form consists of two or three stepped plates with long drawn points. Sometimes, however, only a half-plate, with stiff ends is arranged for, which is worse than useless, and tends to cause early failure of the back plate, for the same reason as badly fitted clips. If a spring is well clipped, rebound plates are redundant.

CHAPTER XXXII

FORGING SHOP MACHINERY

A BRIEF resumé will be now given of the various types of machines in use for the purpose of "forging" spring plates to any required pattern. As before indicated, there are two schools of forging practice, namely, those preferring the single operation machine, and those preferring the multiple operation machine. Generally speaking, it is impossible to dogmatise upon the better forms, as it is entirely a matter of trade conditions as to which pattern is likely to prove the more economical in working. For repair shops, undoubtedly the multiple operation machine is likely to be the better type, as the use of it results in lower first cost, and a minimum of driving arrangements. Production work in the U.S.A. is generally carried out on single-purpose machines, or at the most double-purpose machines, owing to the simplification of spring design, whereas here, and on the Continent, from three- to seven-operation machines are common.

In Fig. 143 is illustrated a typical British pattern of single-head press, generally used for hot work. This machine will be arranged with either nibbing tools, stud and slit tools, or spearing tools, as may be required, and it can be fitted up fairly readily from one to the other according to the orders of the moment. The stroke of the head is fixed, usually to about 3 ins., but the depth of the punch or nib can be readily graduated by the insertion of packing pieces into the slider by the uplifting of the vertical lid—the varying packs being introduced between the eccentric pin die, and the inside of the head. The machine is fitted with a claw clutch, which can be arranged to be held down so as to render the motion continuous.

An American single-head press for similar medium work is shown in Fig. 144. This machine is of the geared pattern, and has a standard stroke of $1\frac{1}{2}$ ins., which can, however, be readily altered within limits by the changing of the eccentric on the front end of the driving shaft. A depth

FIG. 143. SINGLE HEADED PRESS. CAMS LTD., SHEFFIELD.

adjustment for the punch is provided by means of the screwed locknut shown just below the driving eccentric—this device obviating the need for loose packings. A pin clutch of special pattern is used, operated by the treadle, which can be arranged for single stroke or continuous movement of the press head. Machines of this general pattern are in extensive use for spring manufacture in the U.S.A.—where the operations are performed by unskilled labour, and the machines set up by the shop fitters. In Europe, semi-skilled labour is more employed in the forging shops, the men setting up and working their own machines. Accordingly, for this

FIG. 144. SINGLE HEADED PRESS. TOLEDO MACHINE AND TOOL CO., TOLEDO, U.S.A.

practice, the loose packing strips are better for punch adjustment than the screwed connecting rod—as the type of labour is better suited with relatively simple devices of this nature, strips of all sizes being always to hand.

Another American single-head press is illustrated by Fig. 145. This has a wide head, which will take five sets of spearing tools of varying shapes, and a shear blade set. The arrangement and design of the length gauges will be noticed, the cutting being worked from the centre hole of the plate, which is one of the advantages present with this form of centre fastening. The registering pin is sometimes fitted to a sliding block on the length rod, so that the plate can be

FIG. 145. POINT TRIMMING PRESS. COULTER & MCKENZIE, BRIDGEPORT, U.S.A.

put on quite clear of the shear or trimming blades, and when in correct position, advanced under the press head for cutting. An arrangement of this sort is a certain check on accident to the plate or operator. but somewhat slows down the general operation.

A heavy type of press, American design, particularly for railway springs, is illustrated in Fig. 146. This is designed to cut cold plates up to 6 ins. $\times \frac{1}{2}$ in., which is the maximum engine spring size of the U.S.A., and punch out the large slots (hot) which are required for the insertion of the suspension hangers. It will be noted that the shear blade

FIG. 146. SHEAR AND PUNCH PRESS. COULTER & McKENZIE, BRIDGEPORT, U.S.A.

portion of the head has provided for it a guard to prevent the bar kicking upwards when being out—the prevention of this assisting in the production of square cuts. A nibbing attachment can be quickly fitted in the place of the slot punch, so that for the forging of railway springs, to American design, no other machine but this is necessary.

A double machine, of English design, for cutting up and centring plates, is shown in Fig. 147. This is of the usual eccentric driven ram pattern, arranged for the actions to come on alternately, so as to equalize the driving power required. The side and end gauge stops for the punching operation are clearly shown. As usual in English designs,

FIG. 147. SHEARING AND PUNCHING MACHINE. CRAVEN BROS. LTD., MANCHESTER.

no throw-out motions are fitted—great strength and simplicity being aimed at, and achieved. Machines of this pattern will cut steel up to 6 ins. × $\frac{3}{4}$ in., and punch up to $\frac{5}{8}$ in. diameter through $\frac{5}{8}$ in. plate, cold. The limiting equation for cold punching is " diameter of hole = thickness of plate." When the hole is less than the plate thickness, hot punching is a necessity, as cold punches will not survive long on a small hole through a thick plate—the high stresses set up breaking the punches very quickly. A hole ½ in. diameter through ordinary 0·50 per cent. carbon spring steel, ½ in. thick, requires a pressure of about 35 tons to shear it through, equal to about 180 tons per square inch if the punch were in dead compression.

Fig. 148. Double Arm Eccentric Press. Engel & Biermeyer, Hagen, Germany.

A Continental pattern of double-head forging machine is illustrated by Fig. 148. As will be seen, a geared drive is provided to an eccentric shaft, with inner and outer slides. Any variety of tool can be fitted to these heads, but shears are generally attached to the inside. Stop motions of simple design are arranged on each of the slides.

A heavy triple-headed machine, of English design, is shown in Fig. 149, as set up for studding, slitting, and spearing. This machine forms, in conjunction with No. 147, the full layout for a heavy spring as regards press tools. Substantial castings, and large bearing surfaces characterize the general arrangement of the machine, which is, as usual in British practice, not provided with stop motions. Three strong gauge rods of identical design are supplied as shown. In the foreground are various tools for spearing and "diminishing" plate ends.

FIG. 149. TRIPLE FORGING MACHINE. CRAVEN BROS. LTD., MANCHESTER.

Another English design of triple-headed machine is illustrated in Fig. 150, which is intended to forge cold plates of water hardening steel up to 6 ins. × $\frac{5}{8}$ in. In this design, the first head is arranged for a pair of shear blades, the second carries studding and slitting tools, and the third includes the spearing tools. Other tools can be introduced as desired, as the clearance of punchings is provided for under all the heads by means of the channels cored through the base, which can be seen. No stop motions are provided. A pressure of about 180 tons can be exerted through any of the rams, and the drive is through double helical gearing, which is a particularly good form for heavy machines of this pattern.

It will be noticed from the picture that large bolts, of special steel, run vertically through the machine, firmly clamping together the underside of the base, and the top of the main frame—with a view to stiffening the casting, and preventing breakage of the same by any untoward incident that might, for instance, occur under the spearing head, and throw an abnormal stress on the frame. In heavy tools of this sort,

FIG. 150. TRIPLE FORGING MACHINE. F. BREARLEY, BINGLEY, YORKS.

it is advisable to introduce some form of breaking box, as it is by no means unknown for spring shop machines to have their heads pushed off through accidents in handling the spring plate, or tools, or through movements of the latter, causing thereby the downward movement of the press head to strike the equivalent of a solid block, with resulting disaster at some point. Too much attention cannot be given to the security of the fastening means provided for the punches and beds. More ingenuity seems to be lavished on this important feature by American makers than by European makers—

the reason presumably lying in the differences of labour available for the forging shop.

A good heavy machine for spearing work is a valuable asset to any spring plate forging shop, as it can be used for other purposes besides its nominal function. For instance, it will additionally diminish or slot plate ends, for the reception of hangers. The latter operations are sometimes (in this country) carried out in the machine shop, but with a spearing machine in good condition, with well-made beds and punches, a thoroughly workmanlike job can be made by punching, the spring only requiring trimming afterwards by filing or

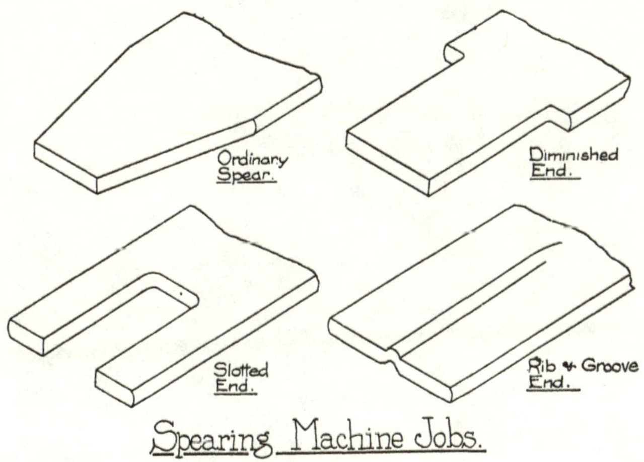

FIG. 151.

grinding. Diminished-end, and slot work is invariably done hot, unless the plates are very thin. Certain odd jobs can also be done on spring buckles, but to safely carry these out, the ram should have a stop motion fitted. The pressing of end ribs and grooves, to take the place of studs and slits, can also be carried out under a machine of this sort. Fig. 151 pictures the above mentioned work.

If plate ends have to be drawn, or tapered in thickness, this operation can be carried out before any centring is done, so that the latter is worked backwards from the drawn and trimmed plate. It is generally, however, done in the reverse order, so as to provide a definite gauge point for all operations by means of the centre hole. Fig. 152 shows the type of eccentric rolls generally employed for this work, which must of course, be done hot, and a small coal, gas, or oil furnace, has to be provided adjacent to the rolls. This particular machine (of English design) has a bottom eccentric roll, which is driven, and a top cylindrical roll, which runs free. The plate end is caught between the two as the eccentric roll revolves, and this extends the length by means of a tapering thickness. Should a long draw be necessary, the ends are thickened somewhat, and the plate width reduced, by the

FIG. 152. ECCENTRIC ROLLS. CRAVEN BROS., LTD., MANCHESTER.

use of the squeezer—this procedure preventing the plate splaying sideways as it is thinned by the rolls. Railway springs do not as a rule, include very long draws—which are more common on automobile springs of light design.

An American pattern of eccentric rolls is shown by Fig. 153. There are several divergencies from English practice here, the chief being the provision of special removable die blocks on the eccentric roll ; and the horizontal action of the squeezer, which obviates the necessity for turning the plate through a rightangle. The American automobile spring plants are, however, rapidly reducing drawn point work to a minimum—and it can be anticipated that in a very brief time, hardly any springs of this class will be manufactured there. Rolls of this type are sometimes used for forming ribs and grooves in the ends of plates.

FIG. 153. ECCENTRIC ROLLS. J. T. RYERSON & SON, CHICAGO.

A multiple operation machine—six in all—is illustrated by Fig. 154. This example is of Continental manufacture, and includes four press heads, which can be supplied with tools as desired, plus eccentric rolls and squeezing dies, the whole being driven by one through shaft. Each head is fitted with a stop motion, and a bar crossing the whole machine renders it also possible to stop instantly the driving shaft by throwing out the main clutch. These features have their good points, as too often with the strictly English pattern, machines cannot be stopped instantly, and in the event of some tool accident, even if the belt goes off, the machine continues to run and cause more damage until the flywheel energy has spent itself.

Fig. 155 shows a maximum of multiple operations, seven in all, embodied in the one machine, of English design. This particular pattern is very popular in the automobile spring shops of this country, and its compactness and simplicity

FIG. 154. SIX-OPERATION FORGING MACHINE. SOCIÉTÉ LIEGÉ-LONGDOZ, BELGIUM.

justify to a large extent the repeated demand. In succession, the operations for the machine are

(1) Cut off the plate by shears on the right. Cold.

(2) Centre hole, by punch under the shear blades. Cold.

(3) Heat the plate end, and squeeze in dies fixed to top of press head on the left, No. 1.

(4) Taper thickness in eccentric rolls.

(5) Trim or spear, on head. No. 1. Hot.

(6) Slit, on head No. 2. Hot.

(7) Stud, on head No. 3. Hot.

No stop motions are anywhere provided. It will be noted that the bottom shear blade is the moving member in this case, the slide being in one with the underneath punch slide. The standard machine of this multiple-pattern is made for steel up to and including 3 ins. × ½ in., and a heavier type takes up to 5 ins. × ⅝ in.

It will be noted that with machines as Figs. 154 and 155, owing to the low build, certain operations have to be carried out by a man or lad in a pit adjacent. No objection can be seriously raised to this if hot plates are being worked through, but such arrangement cannot be recommended for cold work,

155. SEVEN-OPERATION FORGING MACHINE. F. BREARLEY, BINGLEY, YORKS.

as in the event of any accident occurring to the tools or spring plate, the worker cannot remove himself with sufficient celerity to be certain of avoiding injury.

For springs including a "saw and bead" instead of a stud and slit a special machine is required, as shown in Fig. 156, which is an American tool. The heated bar is first beaded under the punch, and then placed on the carrying table adjacent to the slitting saw. This table can be moved upwards to any desired extent by a treadle, which is then operated, and the plate thus forced against the saw, which slits it until the table reaches the stop determining its travel, and therefore the depth of the slit. The plate is then placed under a wider saw, or milling cutter, which clears off the fraize from the slitting saw.

The foregoing brief resumé of world practice in spring shop forging machinery will give an idea of the trend of design in the three great schools of British, Continental, and American practice. British practice is characterized through-

FIG. 156. BEADING AND SLOTTING MACHINE. COULTER & MCKENZIE, BRIDGEPORT, U.S.A.

out by machines of indubitable solidity and simplicity, deficient in stop motions, and not conspicuously wonderful in fastening means for tool attachments. One reason for the latter point is that many spring makers prefer to make and arrange their own tool fastenings, according to their particular practice—and they frequently insist on machines being supplied without any tool attachments. Continental machines are not as heavily built as the British type, but are replete with stop gears. Tool attachments do not receive any more attention than in this country, owing to the spring makers having their own ideas on the subject. American machines are generally much lighter than those built here or on the Continent—owing to the well-known American axiom of running a machine full speed for five years or thereabouts, and then scrapping it, and buying the latest improved design. The single-headed type is the most popular in the U.S.A., and nearly all are clutch operated. A maximum attention has been given to stop gauges, punches and beds, etc. Most American machines will work with little or no foundation, whereas in Europe, fairly substantial concrete beds are invariably used. Each school has evolved machines most suitable to its own particular class of labour and trade, and in such circumstances, comparisons, as such, are worse than useless—the remarks in connection with each particular type having been made with a view only of calling attention to leading type characteristics, and not with the idea of indicating that any one pattern is superior to any other.

CHAPTER XXXIII

BACK PLATES—PLAIN AND FORGED

THE back or top plate of the spring (known in American as the " master " plate and in French speaking countries as the " mistress " plate) is without doubt the most important plate of all, as from the design point of view it includes the suspension arrangements to the vehicle, and in the manufacture, it is the first plate worked upon, and from it all the others are fitted. Very great divergencies occur in the designs of back plates, as will be evident hereafter.

Every country follows, for its railway requirements, one of the three great schools of railway practice, namely British (meaning English), American (meaning the U.S.A.) or Continental (meaning that of Belgium, France, and Germany). The countries following British practice comprise chiefly the Argentine, Australia, India, South Africa, Egypt and the various Crown Colonies. Those following American practice include Canada, Mexico, and certain Central and South American States. Those following Continental practice are those included in the Continent of Europe, and the various subject states of these countries, such as Algeria, the Dutch East Indies, etc. Certain areas, such as China, Japan, Chili, etc., are rather a melange of all three varieties, according to which class of trader or financier has been in the ascendant. Automobiles invariably have back plates of the one type only, namely, the " rolled eye."

With this pertinent digression, types of back plates, under the headings of these three schools of design, will now be examined and commented on, with a prelude as regards their broad methods of manufacture. These divide into six (each type in Fig. 157) as follows—

(1) Plain. The plate just as cut off—a small end turn may be required, which is done by the spring fitter.

(2) Forged. Hook ends, or similar arrangements, which are made under a steam-hammer or forging machine.

(3) Rolled Eyes. The eye may be completely machine made, or more or less hand-made with a steam hammer and anvil, or made by a hand rolling machine.

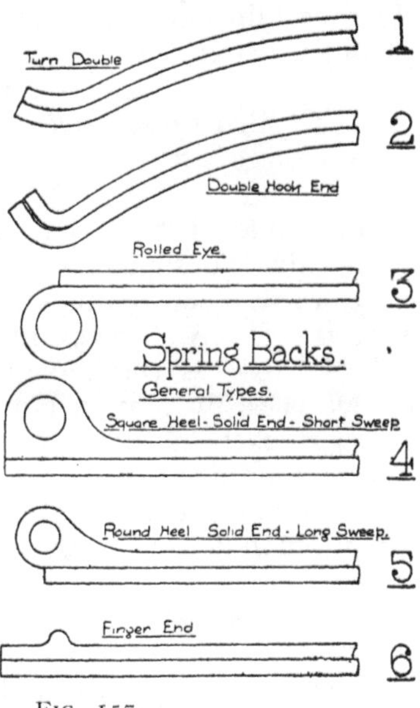

FIG. 157.

(4) Welded. Made always under the steam-hammer. Known here as "solid-ends."

(5) Drawn. "Solid-ends" drawn out from a billet under a steam-hammer or similar machine.

(6) Jumped. Enlarged ends, which are jumped or thickened up from the rolled plate.

The word "forged" in spring shop practice is generally restricted in meaning to work done in the spring plate forging shop, and backs made under the steam hammer are known as "smithed."

The plain end is rarely used in locomotive work here, but Continental designed narrow-gauge engines are frequently fitted with such type. It is a very much employed pattern in the States, in conjunction with suitable end washers to take the suspension rigging. No use is made of the plain end in coaching stock, but it is frequently fitted to four-wheel and bogie wagons of the British pattern. Also, certain heavy road vehicles with radius rods and chain drives use this type, in addition to tramway vehicles of various descriptions.

The forged end, with single or double hooks, or pressed ribs, is not included for engine work here or on the Continent, but America uses to a certain extent the pressed rib type. Forged ends are not much employed on coaching stock, but the hook pattern is now standard in this country for wagons, and is also much used in " British practice " areas. A proportion of heavy road wagons are fitted with spring top plates of this type.

Rolled-eye backs, for some abstruse reason, have never been very popular with railway designers here, but there is an increasing tendency towards their employment. There is little use for them in the U.S.A., except for bogie bolster springs, and four-wheel cabooses (brake-vans) but on the Continent the rolled-eye is easily first in popularity, nearly all coaching and wagon stock being fitted with spring backs of this pattern, and a fair number are used for engine (non-driving) springs. Throughout the world, this type of back plate is employed for automobile work to the almost complete exclusion of all other patterns.

The welded back, or " solid-end," is standard British design for engine springs. It has various forms, which are later illustrated. British type coaching and wagon stock springs are also very largely made with this type of back. Neither American nor Continental practice favours the welded back, and it is very little used in countries following these designs.

The drawn back plate has not yet come much into use. Primarily it replaces the welded solid-end, and does not therefore seriously pertain to American or Continental practice.

The jumped end back plate is a common manufacture in America and on the Continent for engine springs, and to a certain extent for British type wagon work.

The separate manufacture of plain end backs is negligible,

as it is usually done by the spring fitter working to a template for the ends when he bends the plate.

Forged backs are easy to make, requiring only relatively simple tools, and under a light hammer, or a forging machine, they can be manufactured with great accuracy at a high speed. An arrangement for making hook ends is shown in Fig. 162—A following.

Rolled-eye backs are an extremely useful type. They can be made under the steam hammer, with many or few forming tools in conjunction with the anvil, or they can be made at one stroke under heavy forging machines, or with a series of strokes under lighter ones.

One great advantage of the foregoing three types of back plates is that the same quality of steel can be used for them as for the manufacture of the succeeding plates. The processes of making involve heats very much lower than the welding heats required on solid-end backs, and not only that, but with proper tackle, they can be made in one heat. There is not, therefore, the risk of damaging the steel structure that there is with the high welding heats needed on the solid-end back type as usually manufactured.

Numerous varieties of the plain and forged types are shown in Fig. 158, and the following describes them in detail.

Nos. 4 and 6 show respectively a "single-turn" and "double-turn," at one time invariably used for British type wagon work. So far as this country is concerned, they are now practically obsolete in railway work as there is nothing to prevent the spring coming altogether adrift from the wagon, should the latter have been derailed, and re-railing is being done by crane power. Backs of this pattern, are, nevertheless included in most heavy British steam wagon designs. No. 10 shows the R.C.H. standard type of "hook-end"—the hook being on the second plate—and this, with the safety bolt through the shoe, holds the spring to the wagon frame in the event of accident necessitating the lifting from above of the body. In other countries following British practice, with lighter traffics, and more leisurely methods, the disadvantage in this respect of the "turn" end does not appear to have been seriously felt, and such type is still very largely made. This pattern of hook-end is frequently used without the plain top plate. No. 11 shows a "double hook-end" which is not obviously much better than No. 10. Nos. 14 and 15 show respectively a "loop-eye" and "duck-egg," both at one time largely used, but being now practically obsolete as regards new stock. No. 15 would seem to be the better from the point of view of bearing on the shoe. One incidental advantage of these ends is, that for a given design of wagon, with a simple spring shoe, the camber of the spring can be reduced about 2 ins., as compared with 4—6—10—11, which certainly gives a generally better spring.

All the above types are for use in connection with spring shoes fastened direct on the wagon under-frame, or bogie frame, as the case may be. A few typical shoes are shown, Nos. 1, 2, 3 representing a plain pressed shoe, No. 5 a very neat pattern, also stamped out and pressed to shape, which is an alternative device for reducing camber to the " loop " pattern of end; Nos. 7, 8, 9 show the R.C.H. standard cast-iron shoe, and No. 16 gives an end view of a simple type used for side bearing springs on plate frame bogies, being merely a piece of steel flat bent over.

Generally speaking, it will be obvious that if the spring weight carrying points are " shoes " the springs for such purpose must have a positive camber, and this is the case, excepting a few Continental examples provided with specially deep shoes, which allow of a straight spring.

Nos. 12 and 13 give interesting examples of what can be done in the way of efficient and economical spring design for relatively light narrow gauge stock. No. 12 shows an upturned hook end for a side bearing spring, the pin being shackled with tension links, and the spring itself being arranged to be straight with the light vehicle weight, and taking a reverse camber when loaded. No. 13 shows an inverted bogie spring, with the hanger riveted to the hook end and bent over to fit on the pin on the equalizer—the hanger it will be observed can be made from ordinary flat steel.

No. 17 shows a transatlantic form of bolster spring. British and Continental practice favours rolled-eyes for these springs (which will be shown later) with intermediate washers and through bolts. This design is perhaps somewhat more artistic, but loses a lot in cost in comparison with the type shown, more particularly on elliptics with a small number of plates where difficulties frequently occur in hooping. Bolster springs cannot well come adrift, except the bogie becomes a wreck, as they are well held between the bolster and spring plank, and most bogies have safety hangers over these latter. With this type of bolster spring the ends are turned inwards

SPRING BACK PLATES.
PLAIN AND FORGED.
1
2
3
4
5
6
7
8
9
10
11
12
13
14
15
16
17
18
19
20
21
22
23
24
25
26
27
28
29
30
31
32
33
34
35
36
37
38
39
40
41
42
43

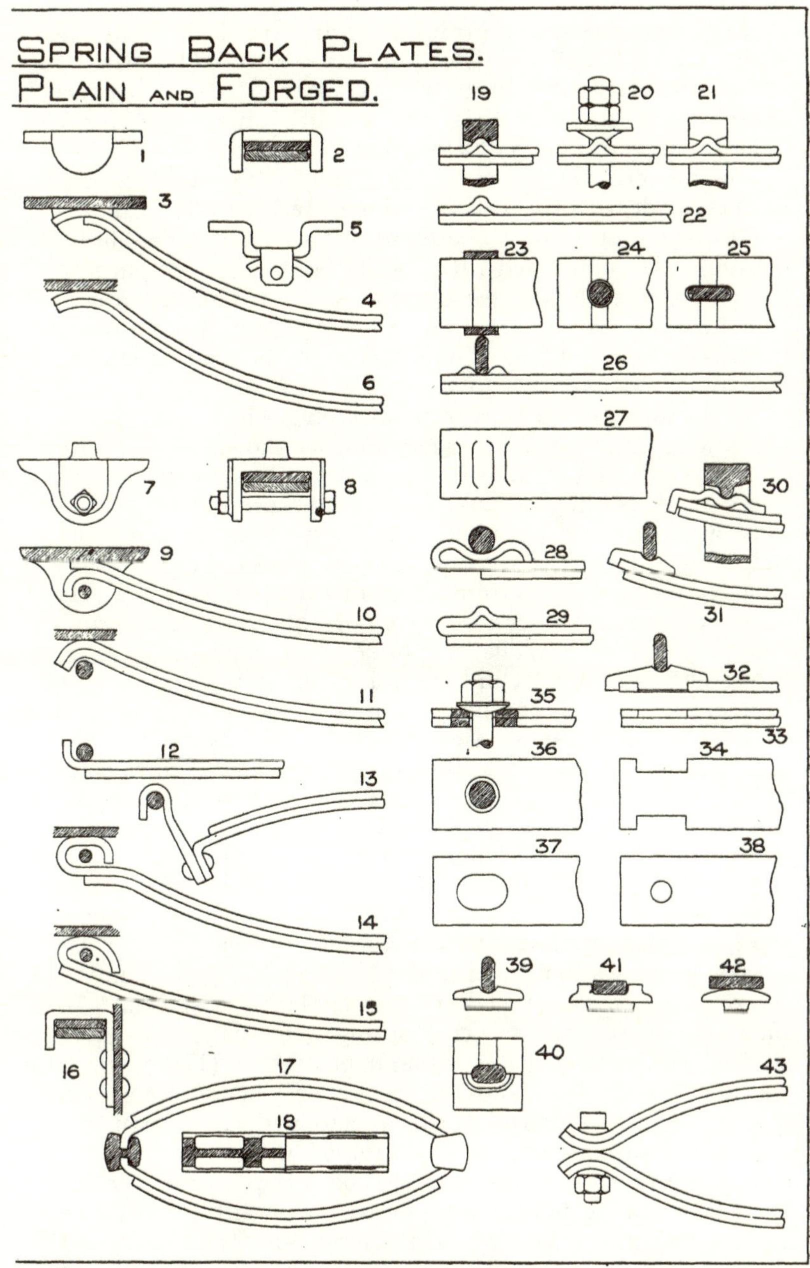

FIG. 158.

slightly, and the whole group of springs rests in a casting, either steel or malleable iron, at either end. This casting, by intermediate ribs, keeps the necessary division between the elliptics, and No. 18 shows a half-section, with inside flange removed, and half elevation.

This pattern of bolster spring is invariably used on locomotive tenders in the U.S.A., which are generally triples or quadruples, comprising therefore six or eight semi-elliptic sections. It is regarded with special favour for this purpose, as the heavy tenders in use are somewhat hard on the springs, and if a " section " fails, it is a very simple matter to renew it during the few hours the engine is in the roundhouse for attention in between its journeys.

No. 22 illustrates a type of pressed end, which is adaptable to various suspension arrangements, as shown in elevation in 19, 20, 21, and in plan in 23, 24, 25. Nos. 19–23 show a box hanger, needing no hole through the spring plates, 20–24 show a washer and adjusting nuts, and 21–25 show a solid-end hanger form. This type of end would seem to have much to recommend it from all points of view, as it is a very simple press job, and can be done hot under ordinary spring-making plant, such as shearing or spearing machines.

Nos. 26 (elevation) and 27 (plan) show a pressed end, having two ribs extending partially across, and forming a depression instead of an elevation, which lends itself to the American gib fastening. Nos. 28 and 29 show " turnover " ends, giving respectively any desired depression or elevation, and are alternatives to Nos. 22 and 26, without appearing to offer any serious advantages over the simpler forms.

A variety of adaptations for absolutely plain back plates are shown in the following sketches. No. 30 is a plain plate, unholed, with a box-shackle suspension, and a loose pressed plate washer to take the latter. No. 31 is similar, but the washer is stamped steel or malleable cast iron. In both these cases, the working spring must have a positive camber in order to keep the washer in position. These washers can also be used for through hangers. No. 34 shows a plan of a shaped plate, of which No. 33 shows the elevation and No. 32 the same with cast or stamped washer in position. This pattern can be worked straight or with negative camber. The plate can be punched to shape under any spearing or similar machine. No. 35 (elevation) and No. 36 (plan) illustrates a plate with a round countersunk hole, and spherical washer taking the weight on to the back. Nos. 37 and 38 show holes in plain plates suitable for the washers Nos. 39, 41, 42. No. 39 (elevation) and No. 40 (plan), with No. 41, are suitable for through hangers, and would insert into No. 37 plate. No. 42 washer is for a rigid box hanger, and needs accordingly only the small registering pap as shown, for plate No. 38.

The " turn " arrangement, illustrated in No. 43, is very suitable for single elliptics, which are sometimes used instead of side bearing springs of the ordinary pattern, and provides a simple and efficient end security. (Used only in loco. work).

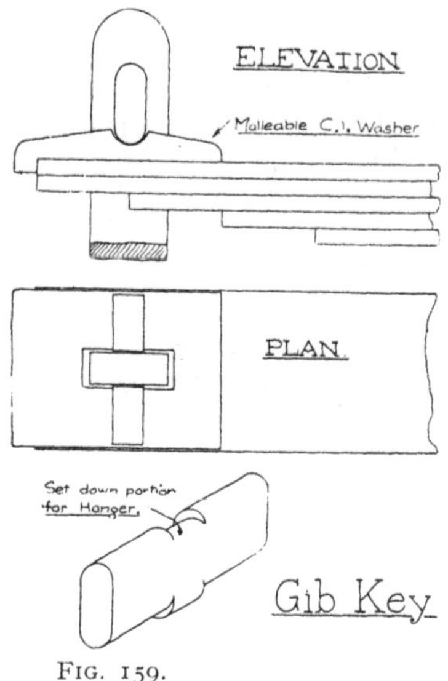

FIG. 159.

It is worthy of note that the arrangements shown from Nos. 17 to 42 are predominantly American. No. 30 is very largely used there for positive camber springs, and Nos. 37–39 for negative or straight cambers. Of their simplicity and efficiency there can be no question, and the obvious cheapness and general soundness of the designs are worth carefully noting. Whilst here stamped steel washers would be favoured American builders have a decided tendency towards malleable cast iron, due, perhaps to the fact that the U.S.A. is the leading producer of excellent quality malleable castings.

The weakest point in the stamped steel washer seems to be that they are frequently made from odd ends of spring steel plates. This is really too hard for the job, unless annealed, and judging from observation, numbers of these washers must be quenched after stamping, as it is no uncommon thing to see them broken in half under the gib cotter, thus letting this cotter down on to the back plate. The loss of this washer, and the resulting effect throughout the equalized spring rigging universal in America, cannot be regarded as very happy, as the spring kicks over to make up this loss of height, which kick makes itself felt at adjacent springs by pitching them out of the horizontal.

Fig. 159 shows the standard gib fastening of American practice. Its simplicity becomes the more obvious when it is contrasted with the pin fastening necessitated by the solid end back, as this type has to have drilled holes in the back, with a turned pin, which is, in its turn in its simplest form, drilled for a split or taper pin.

CHAPTER XXXIV

BACK PLATES—ROLLED EYE

ROLLED-EYE backs from many points of view present the same general advantages as the plain and forged backs previously reviewed. The standard quality of spring steel can be used, and—when machine made—they are comparatively cheap and an excellent job. The eye can be well formed to practically exact size and only requires reamering to reasonable limit gauges to make an excellent hole, whereas the solid-end eye has to be drilled from the solid, and then (in welded backs) the weight comes on to the iron of the eye, whereas in the rolled-eye, it is taken by spring steel.

Various types of rolled-eye backs will now be described, with the assistance of the illustrations in Fig. 160.

No. 1 shows the "perfect" eye, correctly scarfed, and rolled to a completely circular form. No. 2 shows an eye, equally good practically, in which the scarfing is omitted. No. 3 is an unclosed type, needing no scarfing. No. 4 shows an eye made from material jumped at the end, the metal round the eye being thicker than the body of the plate. In Nos. 5–6 the second plate is shown rolled more or less round any of the foregoing types of eye. It is a moot point as to whether any serious advantage is derived from this in the case of railway springs. Certain useful features present themselves in road vehicle work justifying this particular type, but generally, it is not recommended for normal railway springs. No. 7 shows a very small eye, which is for enclosing in a box hanger, with merely a safety pin or screw in the ends of the eye, instead of a bearing pin as in the other patterns. No. 8 is an eye of exceptional size, frequently 2 ins. to 2½ ins. dia., as used in Continental practice, generally for engine and tender buffing springs. With this size, there is a certain justification in rolling the second plate around, and it will be noticed that it is also stiffened by having two laps of metal around the pin. The only point that can be claimed in favour of such an eye is that if it is absolutely essential to have a 2 in. to 2½ in. bearing pin, this form is a great deal better than a solid end back. No. 9 shows a non-scarfed form, correctly drawn for simple hand manufacture. The No. 1 type is always acceptable in lieu of No. 9 as this latter does not lend itself to machine making, as does No. 1. No. 10 shows an eye in rib and groove steel, with the rib and groove included. This is sometimes insisted on, but from the British and American point of view is not very useful and complicates the manufacture, as some tools require to be split longitudinally to free the rib of the completed eye. Accordingly it is preferred to make such eyes as shown in No. 11 by grinding off the rib a sufficient distance to permit of the ready extraction of the finished back. This special objection to the rib and groove rolled eye does not obtain with usual Continental manufacturing methods. No. 12 shows a normal rolled eye studded in the hole. This is good practice if it is impossible to enclose the second plate with the shackles, as the stud distortion is in its correct theoretical position, away from the areas of maximum stress. No. 13 shows an elliptic spring, with " he " and " she " ends. The type illustrated is one frequently desired, to the great tribulation of manufacturers, as it is not possible to drive the hoop over the " she " end, which is deeper than the inside hoop depth. One end of such back plates has therefore to be left more or less straightened out as shown, and after the hoop has been put on, this end has to be heated up and curved round the " he " back. Cases like these specially lend themselves to the plain bent-over end, previously illustrated. Nos. 14–15 show a normal elliptic, with fastening for a bolster group. The intermediate washers are sometimes plain, cut from rolled bar and drilled, or similarly made, and at other times are malleable cast-iron washers, shaped with a groove. No. 16 is a very much stiffened eye, which takes the bearing pin pressure on the full thickness of the plate instead of partially on the scarfed edge. This shows a lateral check not to be recommended, consisting of a small pin riveted into the back plate. It has a certain advantage on long springs, as it can supplant a side-check shackle, at the expense of lengthening

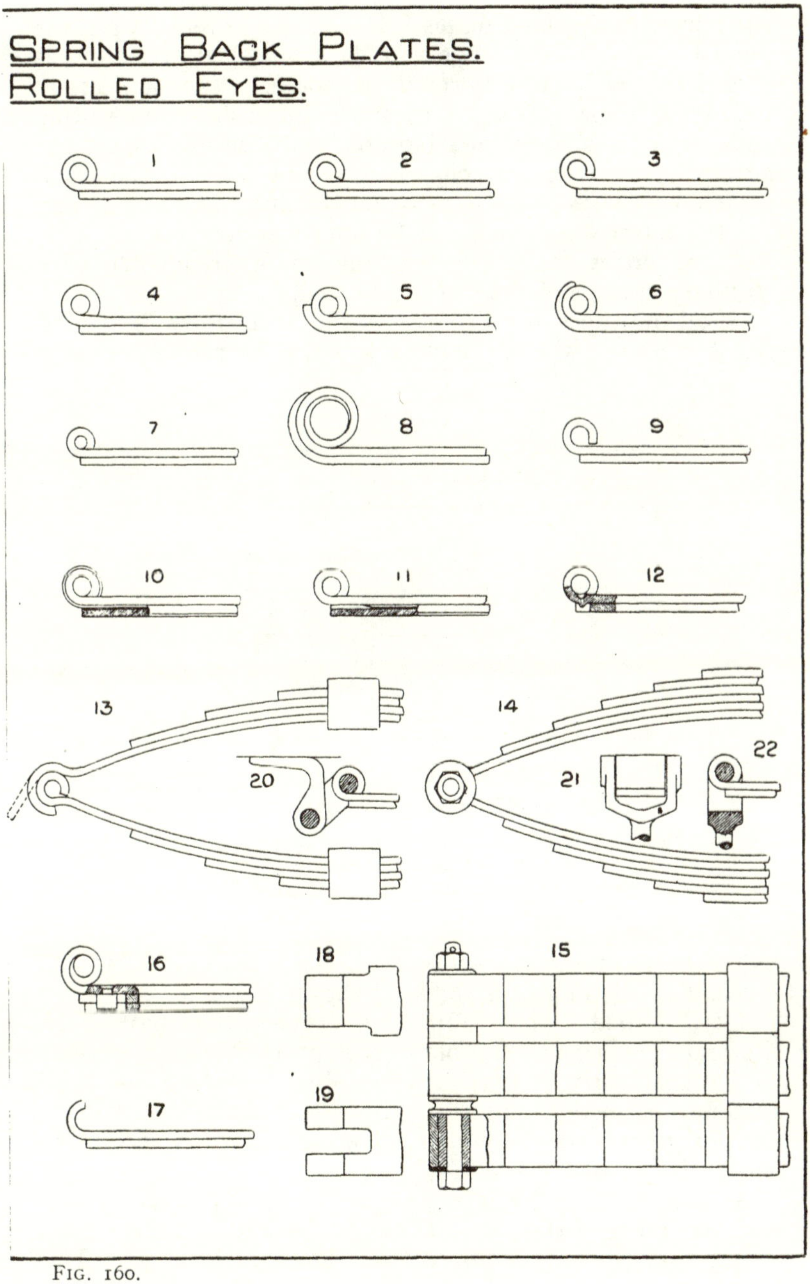

SPRING BACK PLATES.
ROLLED EYES.

FIG. 160.

the normally shackled plates. No. 17 is a cross between the plain forged hook end type and the true rolled eye, and has been included because normally it would be easiest made in rolled eye tools. Nos. 18–19 show diminished and slotted ends, which are sometimes required in rolled eye backs, but generally, work of this character is not needed, the usual attachment to the vehicle being on the lines shown in 20 and 22, the former being general for four-wheeled wagon stock, and the latter for bogie coaching stock, frequently with auxiliary springs of the coil type fitted.

Of all the types of rolled-eye or quasi-rolled-eye illustrated No. 3 is, on the whole, the most practical for railway springs,

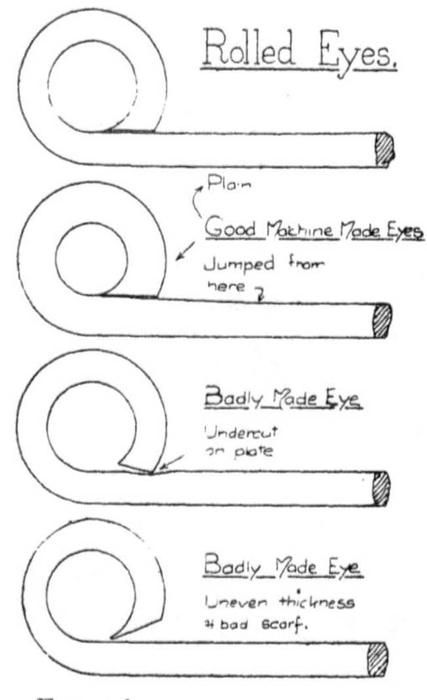

FIG. 161.

as there is no thinned scarf present, and it is in every way as efficient as the more perfect form shown in No. 1.

The "forged" and "rolled-eye" types that have been illustrated present no trouble to the spring fitter as a general rule, particularly if they are machine made, as the back remains reasonably square, and without twist or "wind" and all types, except the one or two special ones noted, are relatively cheaply made.

A hand-made rolled eye designed with a perfect eye, is not the most perfect of backs, as certain manufacturing defects can present themselves, as shown in Fig. 161. The thickness may be uneven or the back plate may be undercut at the bottom of the scarf, which is a serious defect, hastening breakage. The machine-made rolled eye is, in all ways, the best for all classes of work, as it can be made a very perfect job, and the eye can be reamered out within very fine limits, if considered necessary. Moreover, the cost bears no comparison with that of the "solid-end" back.

Machines for the manufacture of rolled eyes have reached a very high development in the U.S.A., where, by the use of the latest patterns, it is possible to have a continuous pro-

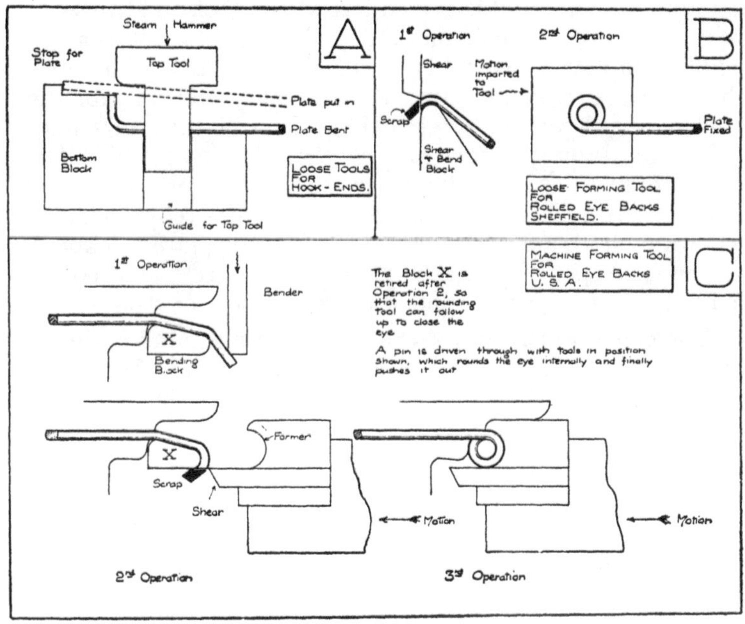

FIG. 162.

duction of 200 eyes per hour (on say 3 ins. × ½ in.) of very good form and finish. The automobile trade has been responsible for the perfection of this process, as the American railroads make little use of the perfect, or scarfed, eye. (Fig. 160/1.) In this country, up to recent years, rolled eyes for the railway trade were made by the spring smiths, and rolled eyes for the automobile trade partly by local methods, and partly by American machinery. On the Continent, simple hand machines are generally used—as will be described.

Contrasting methods of modern rolled eye manufacture are shown diagrammatically in Fig. 162. In this, Sketch B indicates a process in use in this country—the plate end being heated, and then being partly bent and scarfed on a special block under any forging machine. The second operation consists in either continuing the heat, or re-heating, and placing the bent end sideways in a solid tool of the form shown. Motion then being imparted to the tool, the back end of the plate being solid against a stop, and the plate held down to prevent its buckling, the eye rolls itself round into the required form, and is forced out of the tool by a pressure stroke sideways. This method of manufacture can be usefully employed wherever a machine is available with either a direct stroke, or convertible stroke, sufficient for the travel required for the solid tool. Heavy forging machines of the bull-dozer pattern can be usefully employed. The method shown has the advantages of cheap tools, and loose tools, no setting being wanted when it is desired to change over from one to another dimension of eye or plate thickness. The speed of operation can be made to produce 50 to 60 eyes per hour, of very perfect finish, on the largest of sizes.

The general principle of the American machines is shown in Sketch C. The plates are sheared off to desired length, heated one end, and this end introduced to the machine, which

(1) Bends the plate over a die (X), and immediately afterwards

(2) Follows with a shear blade, which gives a bottom curve, and continues by scarfing the end.

(3) The scrolling dies roll up the eye, which then has

(4) A pin forced through to " size " it and completely round it.

The Coulter and McKenzie Machine, for maximum size of 6 ins. × $\frac{1}{2}$ in., is illustrated by Fig. 163—this instance showing a self-contained electric drive. The heated bar end is placed under the gripper, on the left of the picture, the cold end being against a stop. This gripper is worked by a toggle mechanism and provides a very firm hold on the plate which prevents it pushing back, or buckling. When in position under this gripper, the treadle is depressed, which starts the cycle of operations by throwing into gear the special driving clutch. This first causes the toggle movement, which brings down the gripper, and the instant this has occurred, the bending lever and scarfing blade operate. The machine then stops, the gripper is released, and the plate is removed to the second position, on the right, but still under the gripper, which is sufficiently wide for this purpose. Another treadle depression, and the scrolling tools form the eye, the plate being again released. It is this time removed to the position outside the gripper, on the extreme right, and a third treadle movement brings up the outside finishing tools for the eye, and forces the sizing pin in from the side,

FIG. 163. 6 IN. ROLLED EYE MACHINE. COULTER & MCKENZIE, BRIDGEPORT, U.S.A.

afterwards withdrawing it, and retiring the tools. The eye is then complete, and for railway work, where springs are not bushed, and quite satisfactory with $\frac{1}{32}$ in. clearance between pin and eye, the job can be considered finished. It is often considered finished for automobile work, as bushes can be made a light driving fit, and accordingly introduced with no distortion resulting on the finished bush bore.

A development of this general principle for lighter work is illustrated by the machine shown in Fig. 164, which is known as the "single-stroke" machine. In this pattern—made to take up to and including 3 ins. × $\frac{3}{8}$ in.—the dies are moved relative to the plate, which remains always fixed under the gripper. The bending lever is clearly shown adjacent to this gripper, which is opensided in this machine. It is stated that 10 eyes per minute have been made by this process, which is possible, but obviously, it cannot be kept up by any operator, and a fair figure for continuous working would be three per minute.

FIG. 164. SINGLE STROKE ROLLED EYE MACHINE. COULTER & MCKENZIE, BRIDGEPORT, U.S.A.

It is obvious that in order to justify the use of machines of the two foregoing patterns, the demand for springs must be very high, and large quantities must be forthcoming of any given pattern, so as to reduce to a minimum the time lost in setting up the machine, and also the expense of different tools. Such machinery is admirably adapted to the U.S.A. trade, but it is an open point whether it is likely to be quite as useful in this country or the Continent. The labour aspect is particularly difficult, as it certainly does not pay to set up for less than 200 back plates, and as the majority of individual orders are for less than 200, clearly hand labour will have to be resorted to for the oddment work, which makes for complications as regards the adjustment of prices.

In Continental practice, rolled eyes are used almost exclusively for railway carriage and wagon stock, and they are generally made on a small hand machine, of the type illustrated by Fig. 165. This pattern has many advantages at all times, as the only detail required for varying sized eyes

FIG. 165.—HAND MACHINE FOR ROLLED EYES. ENGEL & BIERMEYER, HAGEN, GERMANY.

is the central pin. The large roller at the back (recessed for rib and groove steel) is adjusted up to the plate, and the swinging frame taken to its extreme position towards the right of the machine. The hand lever is forced down, and this movement, by the interposition of an eccentric in the slide, causes it to grip firmly the scarfed end of the plate, which can then be pulled round the pin. To completely close, probably a second pull will be needed. The sizing pin is on centres, and when the eye is finished, the top hand wheel is run back, releasing the pin, which can then be driven out from the plate. There is a tendency for the grip to mark the outside of the eye, but this is a matter of no moment except to the hyper-critical. It is claimed that 40 and 50 eyes per hour can be made with this machine—which is obviously a very cheap article, and requires no tools. Normally, the eyes made will be of the patterns shown in Fig. 160 (2—3—9) but perfect eyes can be produced if a scarf and bend is first made on a forging machine, as Fig. 162—B.

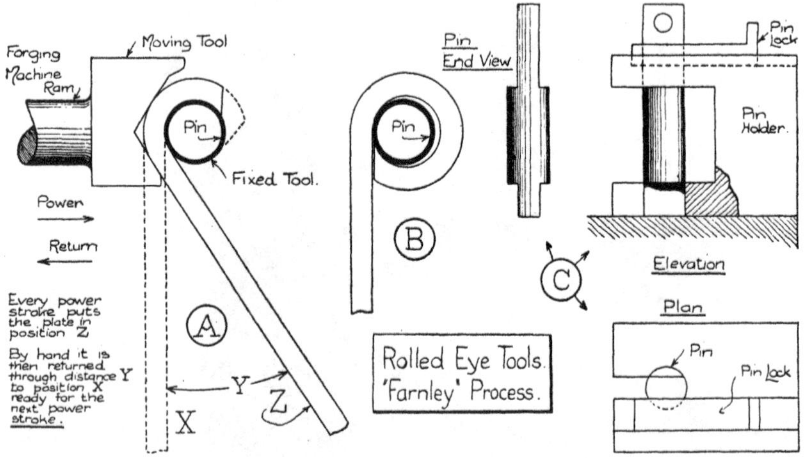

FIG. 166.

Another machine process for rolled eye manufacture is illustrated in Fig. 166, which is somewhat quicker than the foregoing, and also eliminates the heavy manual labour required. A large range of sizes can be made with one pair of front and back tools, the eye diameter being determined by varying pins. Adjustment of the tools is readily made. Scarfing and bending, if required, has to be done on a forging machine. It will be noted that with this process no gripping device is required for either the plate or the eye.

In the case of hand-made eyes, it has been customary here to " jump " the material in the plate end before forming the rolled-eye ; in its minimum form, to prevent any reduction of section due to the rolling, and in its maximum form, to considerably thicken and " strengthen " the eye. The only machine process which will safely and effectively " jump " the material is the Sheffield method of Fig. 162/B. the thicken-

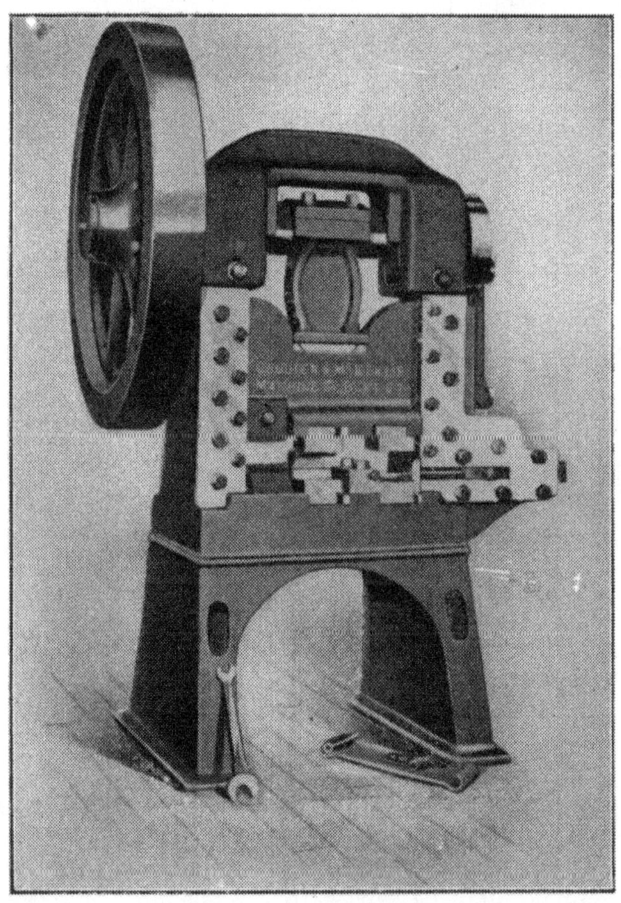

FIG. 167. EYE AND WRAPPER MACHINE. COULTER & McKENZIE, BRIDGEPORT, U.S.A.

ing being done in the tool. Generally speaking, the jumping of the material in such a vital part is not too good a process, particularly in the case of hard alloy steels, and when such are jumped by light tackle, serious damage may be done to the steel owing to the great heat necessary. Many rolled eye breakages can be traced to this procedure, and it would be found no disadvantage if it were entirely prohibited, as the eye made from the " as-rolled " section is always of ample strength. If stiffening is desired, it is preferable to introduce this by the support of " wrappers," which act additionally as safety devices if the main eye should fail.

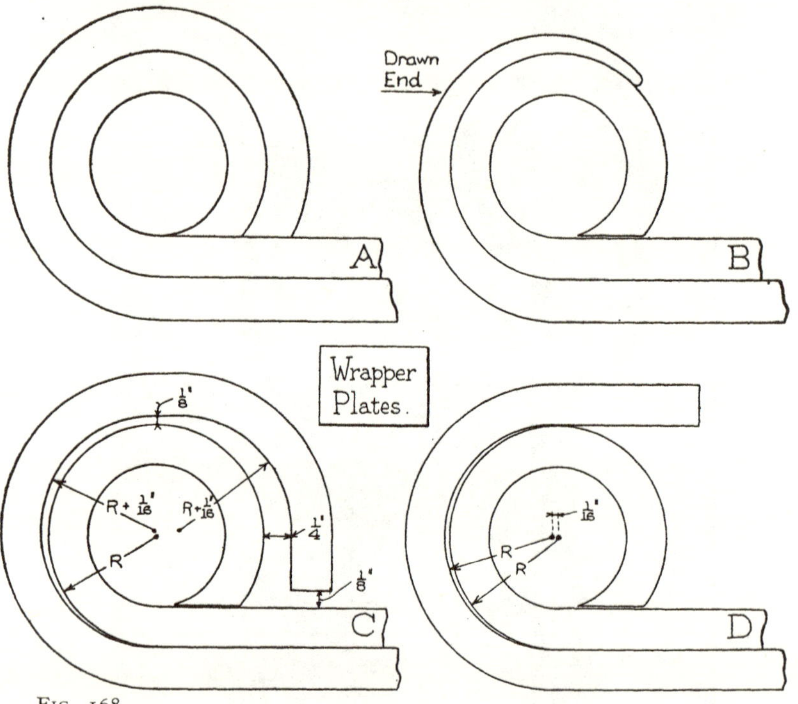

Fig. 168.

In the high production machine process, the steel must be rolled within fine limits, or alternatively, sorted out within fine limits ; otherwise trouble will be experienced by the plate (if thin) slipping through the gripper, or conversely (if thick), damage may be caused to the machine. Precaution in this direction is not necessary in the Sheffield method, or the Continental method.

The scarfing of plates needed to produce the " perfect " eye, can be accomplished by the use of eccentric rolls, but the method of simultaneous scarfing and partially bending is preferable, as the first small bend thus obtained is of great assistance in the final manufacture.

For the formation of " eyes and wrappers " such as are required for railway car bolster springs (finished plates shown in the foreground of the picture) American makers generally use a heavy machine of the type shown in Fig. 167. Wrappers are sometimes thinned by the eccentric rolls, and this machine includes a shear blade set for trimming the ends. A plate, thus trimmed, is passed to the die on the right of these blades, which die starts the bend, and also presses the plate with the necessary " kink" generally designed in bolster springs. It is then passed to the second die, which finishes the eye forming, and simultaneously, the sizing pin at the side, worked by rack and pinion movement, is driven through.

It is not unusual in the automobile trade to bring the second plate of the spring around the eye, as has been previously referred to. This is by no means a bad practice, as if the main eye fails, the spring cannot get adrift from the chassis, and the vehicle can be taken home, as the roll of the second plate will hold the broken eye approximately in position. It is generally frequently expected that the second plate should be a perfect fit to the back plate, and to carry out this idea, the wrapper has to be rolled round the back plate after the spring is otherwise finished. A perfect fit of this sort, however, retards considerably the freedom of the plate movement, and the back plate is in permanent tension when under load, with a reacting tendency to break off the inside of the eye of the wrapper. It is better practice to leave clearance between the two eyes, so that the movement of the plates can take place properly, particularly as the safety function of the arrangement is by this means in no way impaired. Fig. 168 includes some aspects of eye-wrapping, as follows :—

A shows a wrapper and rolled eye as generally drawn, and expected. As already pointed out, this form locks the plates, and whilst not particularly objectionable on thin plates for lightly loaded springs (as indicated in Chapter XIV, Fig. 49) is not very good practice for thick plates, heavily loaded springs, and probably high cambers.

B shows a form with a " drawn end " wrapper. This is of no particular value—the drawn end itself being a very much weaker item than the rolled eye. The chief attraction in the supply of this type is the comparative ease with which it can be hammered on after the spring is fitted.

C This is one U.S.A. standard form, showing specified clearances. There is no doubt regarding the practical nature of this wrapper, as it is an enclosing security means, whilst allowing for the movement of the plates. It can be easily machine made as a rolled eye, and flattened to suit.

Fig. 169. Wrapping Machine. Usines de Braine-le-Comte, Belgium.

D An alternative suggested form, not difficult to make, and providing for excellent security.

Types C and D can be made complete as separate plates, and put on the spring as such. Types A and B have to be included in the spring with straight ends, which have to be heated (this heating involving more or less of the adjacent spring·plates) to be bent round as shown. Type A is, in cooling, liable to contract in and throw out of shape the bearing eye. Wrappers of any pattern should not be used for dumb iron bearings, but should only be arranged on springs with suspension hangers.

In Fig. 169 is shown a Continental machine, very similar to their rolled-eye machine, for wrapping, or forming eyes. In this case, the plate or plates swing on the carrier against the adjustable back block, which is capable of adjustment to give any desired " kink " for bolster spring plates. The central pin is raised by means of the front hand lever, and does not entirely remove from the machine with the spring plate, as with the pattern shown in Fig. 165.

FIG. 170. EYE WRAPPING MACHINE. COULTER & MCKENZIE, BRIDGEPORT, U.S.A.

The illustration, Fig. 170, shows an American pattern of machine for wrapping plates round eyes. This will also make rolled eyes along the same lines as the previous types of Figs. 165 and 169. It is more generally limited, however, to wrapping.

CHAPTER XXXV

BACK PLATES—SOLID-END

THE oldest method of fastening suspension arrangements to a spring is that involving the employment of a "solid-end back," otherwise, a form which has enlarged ends, nominally solid with the spring plate. Horse-vehicle springs still maintain to some degree the type, but for the more modern automobile, it is used only in rare instances. In railway work, with few exceptions, British practice only continues the use of this pattern, but the trend here now is towards its elimination in favour of cheaper and simpler forms. This back, outside British practice countries, is almost entirely confined to locomotive springs, with the exception of the "half-moon" type, which is essentially a wagon or buffer pattern. Large numbers of solid-end backs of the "square-heel," "bevel-heel," "round-heel," "tension-bar," "pendant," and "half-moon" patterns are made here, however, for carriage and wagon stock for India and the Colonies. There is an increasing tendency just now to substitute rolled eyes for the first five types, owing to their lower cost, and it is certainly a wise course of action, as a good machine-rolled eye is a generally safer back in service and a much easier job to handle in manufacture.

One incidental disadvantage with solid-end backs of the heavy type, with a long sweep, is in the point that they are not operative as spring plates owing to the length of the sweep. This is best indicated by the sketches in Fig. 171. In these circumstances, it follows that the spring plate ends fitting under these heads have to be straight, and cannot take the true arc—in fact, the whole spring shape is incorrect.

The result of this is that the spring action is crowded up in the middle of the spring, due to the non-operation of the flat plates under the stiff solid end, which leads to the spring failing in service long before it would have done had it been fitted with clip or similar type ends.

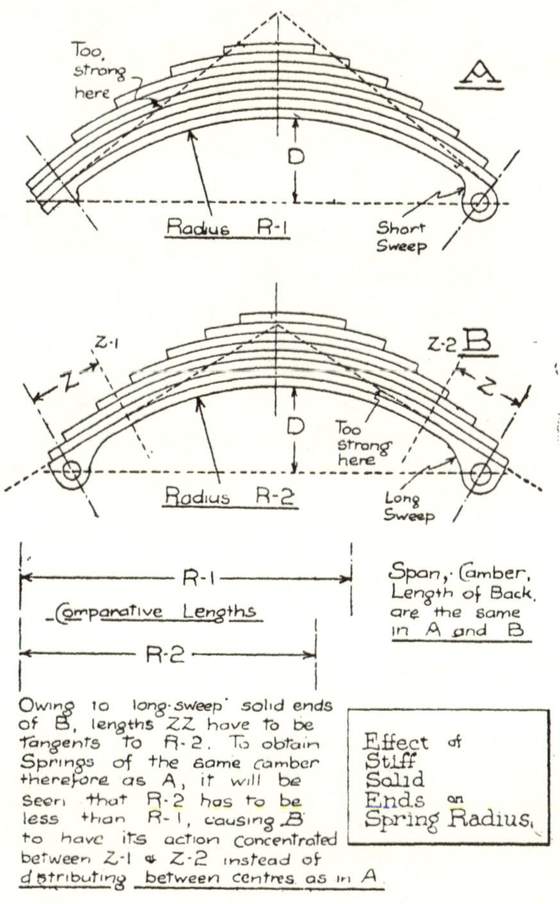

FIG. 171.

Fig. 172 includes a number of patterns of solid-end back, with methods of end finish, briefly touched on as below. The titles allotted to each are more or less local, and are frequently different in different shops.

Spring Back Plates. Solid Ends.

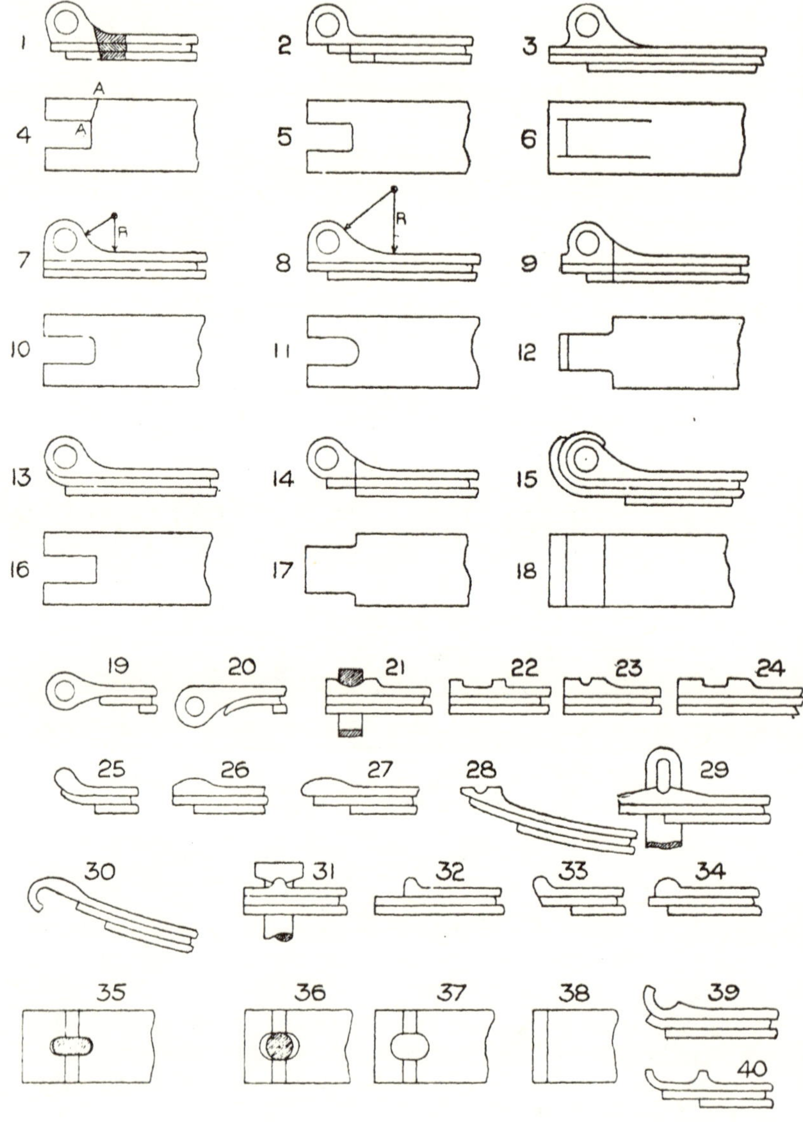

Fig. 172.

No. 1 is a " bevel-heel " generally designed for a heavily cambered spring for no special reason except that the end of the back is thereby brought parallel with a vertical hanger. The manufacturing cost is thereby incremented. The slot here is also inclined, and it will be noticed that it is square cornered in plan, No. 4, which is very objectionable indeed, and should be strictly avoided, as fractures frequently occur over the line marked AA. No. 2 is a " square-heel, short radius," which is rather difficult to make but reduces the stiff length of the back. It will be noticed that the succeeding plates are very carefully spaced so that the theoretical design is very closely adhered to. Obviously, therefore, this back should not be of the (*a*) method of manufacture, (*d*) would be very much better. Plan No. 5 shows rounded corners to the slot. No. 3 is a " lug " back, with the solid end machined away at each side (owing to small clearance spaces on the engine), as indicated on No. 6. No. 7 shows a " square-heel " which is the most used type of any for locomotives, including pinned hangers, and No. 10 shows a usual slot in plan. It is not uncommon, however, to use " solid-ends " without slotting. suspension hangers being on each side. No. 7 is a " short-sweep " or radius (marked R) and is more practical for the spring-maker than the long sweep with the similar head shown in No. 8, as will be shown later. The plan of this, No. 11, shows the slot with rounded end, which is a good form, but requires more material removing than the slot with rounded corners. No. 9 shows a " lug-end " back, " diminished-end " finish, the solid-end and plates adjacent being cut away at the sides to allow of a tension hanger with narrow jaw. It should be noted that such diminishing should be arranged with round corners as shown in No. 12.

Nos. 13 and 14 illustrate respectively a " round-heel short-sweep " and " round-heel long sweep "—No. 15 showing the latter with plates wrapped round. Nos. 16, 17 and 18 shows probable plans of such backs. These three patterns are favoured to a very limited extent in automobile practice, but are invariably left full width for the suspension. No. 16 with the square corner is to be avoided. No. 13 is not a very practical method of finishing off the second plate, which it will be noted is curled slightly round the back. No. 14 is in all ways the best finish for the second or " eye " plate. No. 15 is a particularly objectionable form from the manufacturers' standpoint, and its value is rather dubious. If the second and third plates are wrapped round a good fit when the spring is light, the back will not be free to extend properly when loaded, owing to its being locked by the wrapping. Usually, therefore, such wrapped plates are not turned round a specially good fit, so they are therefore not very valuable as a strengthening device. The only useful purposes they serve is in the event of the head breaking, they act as a check to keep the fractured end in position.

Nos. 19 and 20 are known respectively as a "tension-bar" and "pendant-head." The former was universal for coaching stock suspension with the old horizontal adjustable suspension. The suspension gear is now nearly obsolete, but the type of head remains. The advantage of the "pendant-head" lies in the fact that it is possible to bring the whole spring nearer to the bottom of the underframe, owing to the head being included in the overall spring height instead of being additional thereto, as in the case of the other solid-ends illustrated.

Nos. 21, 22 and 24 show American type ends for strap hangers, and 23, 28 and 29 show similar practice for sword, gib, or box hangers. No. 29 is perhaps the most practical type, as it suits all styles of camber. No. 28 is used chiefly for inverted locomotive bogie springs. No. 34 is another American type, known as a "bulb" end. Nos. 39 and 40 are peculiar constructions sometimes used for transverse truck springs on U.S.A. engines. These patterns are all made either by "jumping up" or "dabbing on."

No. 25 is a "claw" end, particularly popular in this country at one time for the overhung springs on leading or trailing radial axleboxes. Nos. 26 and 27 are "half-moon" backs, which can readily be jumped. No. 30 is a half-moon back with hook-end, acting as the safety device to retain the spring to the wagon.

Nos. 31, 32 and 33 represent the bulk of spring ends used in Continental locomotive practice, and they are a very efficient type. No. 31 is shown as a "rib-end" or "duck-bill" and No. 32 as a "knuckle" end. Plans Nos. 36, 37 and 38 show the usual end finishes, in No. 37 it will be noticed that the hanger is held by the plates only being slotted, the back No. 32, being simplified by having only an open slot. No. 35 plan is the ordinary American end with single hanger.

The welded, or "solid-end" back, can be made by several methods, enumerated below—

a. Bending the plate as required, and filling in with iron bar at a welding heat. This is most generally the practice in this country, and is regarded as a very sound commercial method.

b. Making a separate stamped end, and welding it on to the plate, 3 ins. or so from the end. This method is now nearly obsolete here.

c. Building up the end with thin stampings, heating the lot, and finishing in shaped tools. Used to a certain extent in the U.S.A. for small springs, although it would lend itself to many large types. The blanks forming the head are dead soft steel.

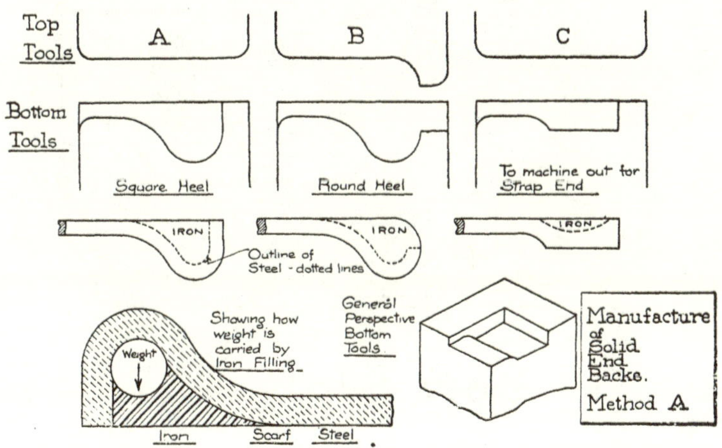

FIG. 173.

d. "Dabbing on" iron as required on to the top of the plate and finishing all off at a welding heat under the stamping tools. Chiefly American practice.

e. Jumping the metal up at the end until the required shape, regulated by dies, is obtained. This can be done with a steam hammer for relatively small enlargements, but for heavy heads, a strong forging machine is necessary.

f. Drawing down the middle of the plate, and stamping to shape the ends—starting with a billet of larger section than the end.

Method (*a*) is illustrated by Fig. 173. The end of the plate is frequently jumped to make up for the reduction in thickness due to the "hooking." Fixed tools are generally provided for the work, *i.e.* tools actually attached to the steam hammer tup and anvil respectively. For small quantities, however, loose tools are employed, but these are not as satisfactory. Owing to the great diversity of design, the number of tools on hand in the average Sheffield spring smith's shop is beyond reasonable limits, and, if solid-end backs are going to survive, it is most necessary that an attempt be made in conjunction with the railway authorities to standardize the sizes, which should not be a difficult job. As invariably happens, the worst sinners in the respect of divers patterns, are the customers

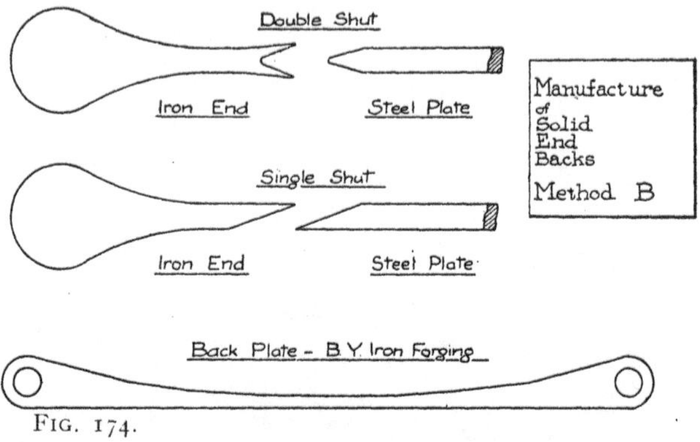

FIG. 174.

ordering four or six springs, and great difficulties are experienced in attempting to meet all requirements. Standard models should be decided on, and drawings of the same circulated to the various locomotive builders, etc., so that their designs can be made to accord with existing tools and patterns.

It will be noted from Fig. 173 that the weak point of this type of solid end lies in the fact that in most cases the suspension pin takes its bearing on the iron part of the head, which wears elliptical in service. It could be arranged for the steel to be on the underside, but this would necessitate bringing the scarf to the topside, and every smith cannot finish a scarf to be presentable when in such obvious position. Badly smithed backs, with the scarf on the underside as is usual, are particularly objectionable to the spring fitter, as the job of fitting a second plate to a solid-end back with a bad scarf and curved underside is one of the worst in the trade.

Fig. 174 shows the separate stamped end with the plate tongued therein. The defect of such manufacture is obvious. This type was contemporary with the complete best York-

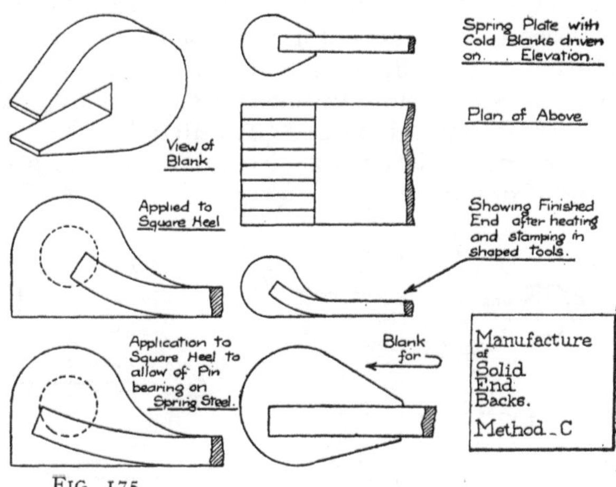

FIG. 175.

shire iron back plate, where the plate acted merely as a pack between the end suspension gear and the body of the spring, and was, of course, negligible from the spring point of view.

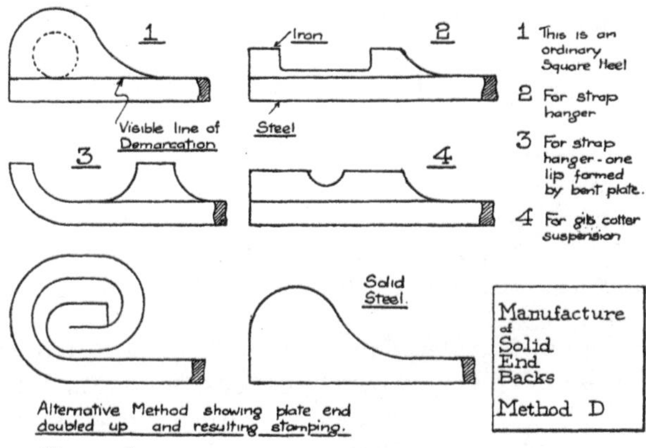

FIG. 176.

Fig. 175 shows the method of building up an end with small blanks. There would certainly appear to be much in this method to recommend it, and its possibilities for standard English backs are sketched.

Fig. 176 illustrates the "dabbing on" process, which is generally practised in the U.S.A. for solid-ends. The line of demarcation between plate and end is generally fairly visible. In view, however, of the bulk of their solid-ends being for strap and "sword" hangers, this method is by no means unsatisfactory. For repair work, ends of the types 2—3—4 can, in these days, be built up on stock plates by electric welding—which is one advantage presented by them. An alternative for large solid-ends is also shown on this sketch, which is used occasionally.

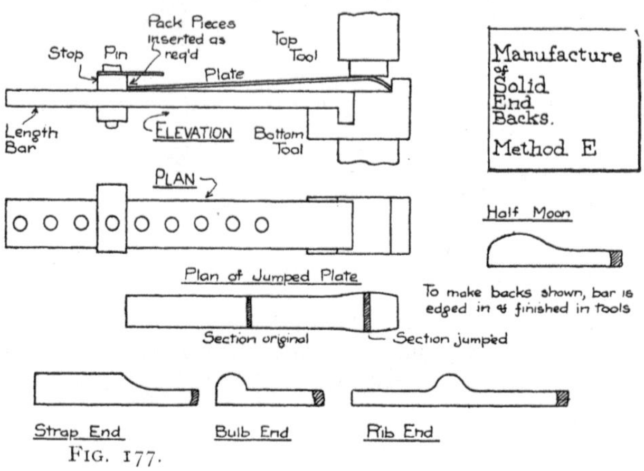

FIG. 177.

Fig. 177 shows jumped ends, and steam hammer jumping tools. For a small back, jumping is not objectionable. On a large head, however, the structure of the steel is not improved from the strict metallurgical aspect, but practically, this would not appear to be of the highest importance for this particular job. A point of greater moment is the risk of overheating the steel on its surfaces which renders it liable to rapid failure.

Fig. 178 gives a general idea of the "drawing from the solid" process. This is relatively expensive, and is not very popular. Unless great care is taken with the edging of such a back, so as to obtain the same radius edge as the succeeding spring plates, the appearance is bad (a very great consideration from the British engineer's point of view), and, more important, extremely great care is required to obtain the same thickness throughout the plate, such as is obtained in the rolled bar, the non-uniformity of the thickness being a very serious point to reckon with as regards the stressing of the plate.

It has so far been found impossible to design a more expensive back plate than the solid end type, if attention is going to be paid to the excellence of manufacture. Wide plates, say above 5 ins., are particularly awkward to handle by any of the methods sketched. A well-made, large, solid-end back is really an extremely fine example of smith work, but this is a doubtful justification for the perpetuation of such an expensive and not entirely satisfactory design. One immediate disadvantage presents itself in welded backs, as, in order to ensure safe welding, the back should be of a different quality from the plates (that is, if a high efficiency steel is to be used in the spring). Generally speaking, back steel

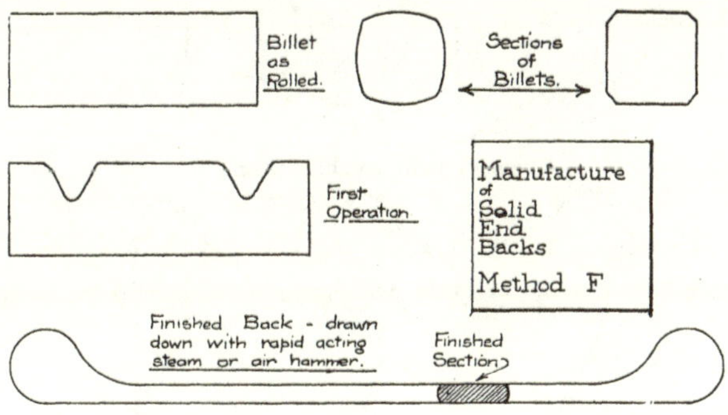

FIG. 178.

is 0·10 to 0·15 per cent. carbon below plate steel, and it should never exceed 0·45 per cent. carbon total. This means keeping many additional sections of special back quality in stock. The fact of welding being required in certain manufacturing methods introduces an element of risk which is not present with the other types of "plain and forged" and "rolled-eye" backs already referred to. The practical thickness for the manufacture of solid-ends is another consideration as such should be limited between $\frac{7}{16}$ in. and $\frac{5}{8}$ in., whereas the other types are satisfactory for all thicknesses.

Certain types of horsed road vehicles still retain a solid end form of back, particularly for full elliptic springs, such as are in common use on light coaches and buggies, running in Australia, South America, etc., where there is a lack of made roads, and a need for high flexibility (or low periodicity) which can be most practically obtained from the full-elliptic with very thin plates. Solid ends for this type of spring are made by a modification of Method C, stamped clips being clinched by machine or hand hammer, on to the sides of the cold plate, which is then raised to a welding heat, and finished under some form of shaping tool. Types of these heads, as made, are illustrated in Fig. 179, together with the tool block used for the finishing process. Power presses are generally employed in the U.S.A. for this welding job, and an illustration of one is

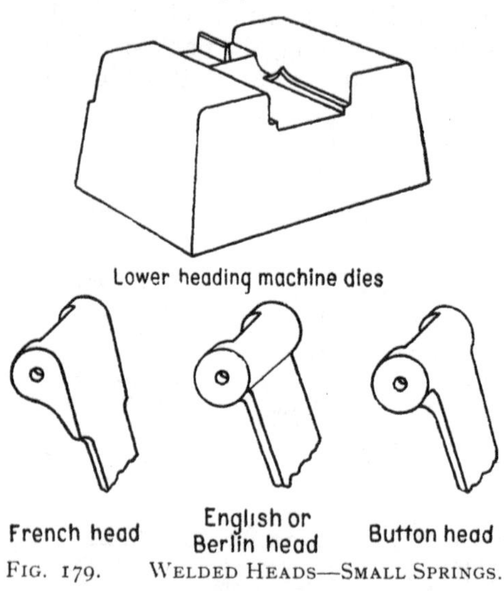

FIG. 179. WELDED HEADS—SMALL SPRINGS.

given in Fig. 180. It is simply a heavy forging machine type, and can be equally usefully employed for ribbing ends, etc., but where installed, is generally kept constantly at work on solid-ends for these light springs, an enormous export trade in which is done from America to all parts of the less civilised world.

For the efficient jumping of backs into solid end forms, a better tool is required than the usual steam hammer of the spring trade, and on the Continent, where the rib-end back is very usual practice, a machine of the " bull-dozer " pattern is employed, as illustrated in Fig. 181. The spring plate, with one heated end, is placed in the geared saddle, which includes a gripping device, and is set from the stop in the tail of the frame. This saddle is advanced as required by the hand-wheel, and the jumping tool is put in motion by the closely adjacent lever. The pattern illustrated will work steel up to 5 ins. wide, has a stroke of 2 ins., and exerts a maximum pressure of 100 tons.

Solid-end backs of the English pinned pattern can be made by a similar process, but as might be expected, a very heavy

FIG. 180. HEADING MACHINE. COULTER & MCKENZIE, BRIDGEPORT, U.S.A.

machine is required, exerting up to 250 tons pressure. Such a back might require to be made with a 2¼ in. head from a ½ in. thick plate, and then requires a preliminary roughing process in the jumping tools before it can be reheated for the final shaping. On the whole, the finish of jumped backs off machines of the type illustrated, is not as good as the finish of similar backs made by skilled spring smiths.

For the manufacture by the various processes indicated, nothing out of the ordinary run of power hammers or drop stamps is required, apart from the machines illustrated. Steam or pneumatic hammers of 10 to 20 cwts. capacity, or drop stamps of 5 cwts. upwards, will be found to meet nearly all requirements.

It is sometimes stipulated in Continental specifications that not less than four heats are to be employed in the manufacture of the usual " rib-end " back, with a view of ensuring that the steel is not overheated so as to allow of the back being made in one or two heats.

Fig. 181. Jumping Machine for Rib-End Backs. Engel & Biermeyer, Hagen, Germany.

CHAPTER XXXVI

THE MACHINING OF BACK PLATES

According to the general spring and suspension design, so more or less machining, and more or less cost, is involved in the final manufacture of a spring back plate. The general trend of such requirement is covered by the following remarks :—

Automobile practice generally uses only rolled eyes, with side hangers (or shackles—auto. trade term) and the finishing of these is confined to reamering the bore of the eye (sometimes) and grinding the sides of the eyes (according to specified requirements). Machining, as such, is therefore non-existent on springs for this trade.

British railway practice involves so many solid-end backs, that a very large amount of machining is involved. With few rare cases, all solid-ends require drilling for the suspension pin ; they are frequently slotted or diminished for hangers, with all classes of springs, loco. carriage, and wagon ; and sometimes finished at the sides of the solid end with light machining as distinct from heavy diminishing. The material to be removed to permit of the insertion of hangers frequently covers three or more plates following the back plate, and as, until quite recent years, the punching of hanger slots was objected to by the purchasers, machining from the solid, not unusually on hardened and tempered plates, became fairly standard practice. Fortunately, the latter requirement is becoming less pronounced, and accordingly, considerable punching of slots is now resorted to on the plain plates, although it is not practicable to punch the large solid end for obvious reasons.

Continental railway practice, using largely "rib-end" backs for engine stock, requiring no pin holes, includes a minimum of machining. The "rib-end" can be punched readily, and the same punch will clear out the following plates, a different bed only being wanted if perfect work is to be done. The rolled-eye backs used for carriage and wagon stock have generally outside shackles to the suspension hanger, and require no slots. When inside hangers are used, the slot can be punched before the eye is rolled.

American railway practice entirely misses back plate machining. All hanger slots are punched, and as, apart from bolsters, few laminated springs are in use on passenger and freight cars, the general simplicity of spring work is at a maximum.

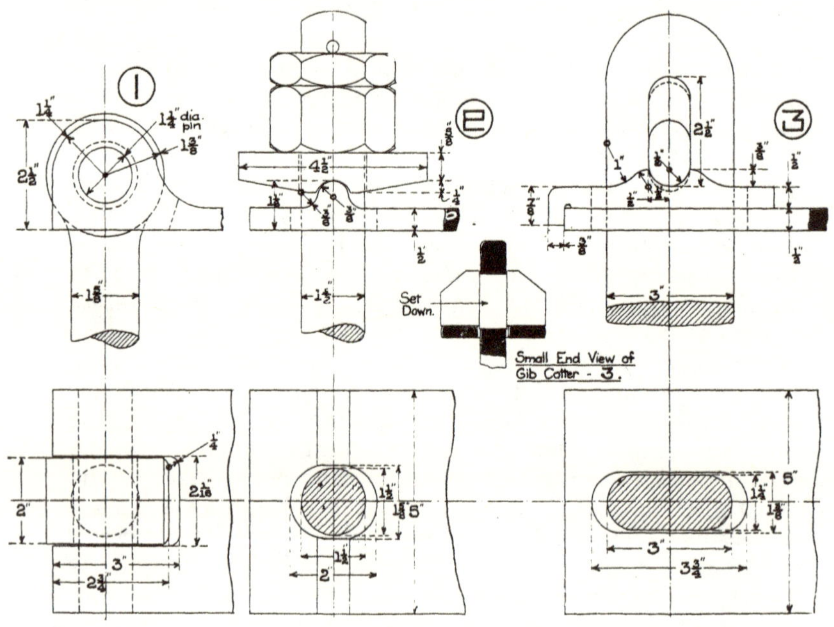

FIG. 182. FORMS OF BACK PLATE ENDS.

In Fig. 182 are illustrated the typical locomotive suspension ends of the three practices, with remarks as follows :—

(1) Illustrates the standard British form. It is thoroughly sound as regards design, and indicates an apparently good, solid, workmanlike job, with the drilled eyes, machined ends, and turned pins. It also carries a cost commensurate with the workmanship. The solid end back is usually made by welding iron on to the looped steel plate, which sometimes causes the

latter to be overheated, and the steel spoilt. The inherently weak back is then further weakened by the machined-in jaws to permit of the introduction of the hanger. When these jaws are finished with sharp corners at the closed end, breakages are invited. Frequently, to add yet another potential failing point, large nibs are hammered in close to the weld, to fit into slits on the second plate—an idea which is quite consistent with the necessity of studding and slitting all plain section plates to check lateral play, but which has no merit but consistency, as in the majority of cases tension hangers are used, which themselves preserve in alignment the full length top plates. Lastly, the turned pin in service, " collars " at the two junctions of plate-hanger-plate, and is difficult to knock out, when it will generally be found that the eye in the spring back is elliptical, and the back accordingly useless for further service.

(2) This shows a largely used Continental pattern, with a small solid-end, called here a " rib-end." This is generally specified to be made by machine-jumping the plate, and to prevent risk of overheating it is also frequently specified that not less than four heats must be employed. Such successive heatings de-carburise the surfaces of the back ends, but as there is ample material there, and the bending moment range is zero to a very small amount, the matter is not of importance. The bearing on here is usually a stamped steel, and case-hardened, rocking washer, which cannot rust in or damage in service. The hangers sometimes work through punched " oval " holes in the plate, but not infrequently box links are used, which leave a full width bearing. No machining at all is introduced on the back plate as a consequence.

(3) American practice, showing one type of loose washer end. The numbers of patterns of loose washers are many, and they are variously made as bent steel plate, malleable iron castings, steel castings, and drop forgings. When Continental practice indulges in these, the washers are always specified to be steel stampings. This form of back plate counts merely as a top spring leaf, as there is no need for it to travel anywhere from the cutting-up shop except straight to the spring fitters. No special steel quality is required, as in the British type, wherein lower carbon steel is specified, and used, to reduce the risk of burning, thereby necessitating two qualities of steel to be made, rolled, kept separate, heat treated accordingly, and finally introduced into the same

spring. The attachment to the spring rigging is simplicity itself, merely a stamped " gib cotter " being used, which fits into a punched link. No machining anywhere, all the spring plates having " slots " which are punched.

The ends just described are all drawn and dimensioned suitably for the 5 in. × ½ in. springs.

FIG. 183. TWO SPINDLE DRILL. JOHN HETHERINGTON & SONS, MANCHESTER.

As will have been seen, the machining of back plates is an important subject in British practice and to effectively cope with all types that come along a well-equipped machine shop is needed. For solid-end backs, stiff drilling machines are required, and double spindle machines are preferable. Such are not always used for drilling simultaneously both ends of the one plate, but more frequently for drilling one end each of two plates. To perform the simultaneous drilling of the two solid-ends accurately, with spindles set to the exact centre distances is, of course, the object to be aimed at, but owing to the impossibility of obtaining smith work within extremely fine limits, this is not always attained. Generally speaking, the backs are smithed within limits of plus and minus $\frac{1}{16}$ in., and the holes are drilled in the centres of the

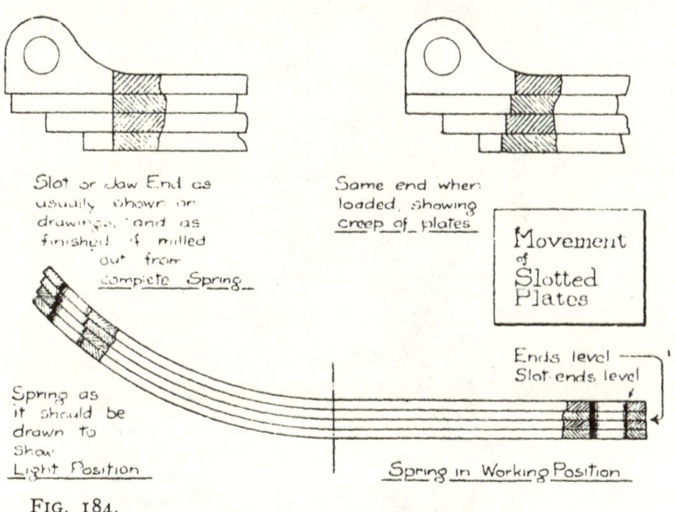

FIG. 184.

ends. Fig. 183 shows a double spindle machine suitable for this class of work, with self-acting feeds, and capable of drilling 2 in. dia. holes, and taking backs up to 8 ft. long. A single spindle radial drill of a good type can be used, and could be arranged to take several ends, the operator removing a drilled end, and replacing it, whilst another end is being drilled, thus speeding up the work. A well made rolled-eye has barely $\frac{1}{64}$ in. of material (in thickness) left in for finishing, in fact, as a rule, reamering is a needless refinement for carriage and wagon springs, but if required, any good sensitive drill will easily deal with such work.

Diminishing and slotting (open and closed) of solid end backs is carried out with either slotting or milling machines. Closed slots, of the Continental pattern, on rib-end backs, can be punched out, but unless a quantity is required, it is not worth while making tools for this job, and drilling centrally is resorted to, the oblong slot being finished with shaped tools in the slotting machine. Jaw-ends of the British type can be slotted completely out, and diminished ends can be finished in the same way with double tools.

Some firms diminish or slot the back and succeeding plates as required, after the spring has been fitted and hardened

FIG. 185. HEAVY SLOTTING MACHINE. J. BUTLER & CO., LTD., HALIFAX.

and tempered. Others, and this is the better manufacturing process, punch the plate slots, mill the back slots, and then fit the spring, all work being thus done with the flat plates and flat back. The result, practically, is more satisfactory, although the appearance of the spring (light) is not as good. The explanation is given best by Fig. 184. From this it will be noted that the spring (light) when machined as finished, has, of course, a level surface over the inside end of the slotted plates. When this spring is loaded, the plates alter in length from the neutral axis to the surfaces, which push them into an irregular line at the ends of the slots. It has been known for springs so machined to be perfect to drawing, but when on the engine, the extension of the plates has cut into the hangers to such a degree as to prevent the spring setting to the load. By the simpler method, *i.e.* punching out the plates, and machining the back, before

FIG. 186.—SINGLE PUNCHER SLOTTER. J. MUIR & CO., LTD., MANCHESTER.

fitting, allowance can be made for this alteration in plate lengths, so that when the normal load is being carried the plate ends in the slot are level. A very objectionable process was sometimes resorted to, of (permanently) softening the whole of the spring at the ends in order to machine the slots. This is now less frequently done owing to the improvement with recent years of cutting tools.

One type of slotter suitable for this class of work is shown in Fig. 185. Two types of jigs are required, one for square cut jaws, and the other for angled jaws, as No. 1, Fig. 184, this latter being made adjustable for all normal angles. Fig. 186 shows a heavy puncher-slotter, which will work with a 2 in. wide tool, and acts thus as a good test for proving the efficiency of the welding.

All classes of end slotting or diminishing can be economically performed on a machine of this class, which is greatly used in this country for spring work, having now achieved

FIG. 187. HEAVY PLANO-MILLER. KENDALL & GENT, LTD., MANCHESTER.

a high development. Relieving tools are sometimes employed, by means of which the broad cutter misses the back plate on the upward stroke. These are not, however, very popular in the spring shops, as such numerous dimensions have to be worked to, that it is considered better—and with reason—to retain solid tools, and amend them by smithing as required.

Fig. 187 shows a plano-miller of suitable pattern for jaw and diminished ends, which is frequently preferred to the slotter. The arguments pro and con are that the slotter is the better machine for oddment work, and the plano-miller for quantity work. The slotter is more easily set up for lots of fours and tens, but stops more often for tool changing on thousands than does the plano-miller. The slotter tools are the cheaper and more easily amended, but per unit weight of tool steel, actually used, the plano-miller

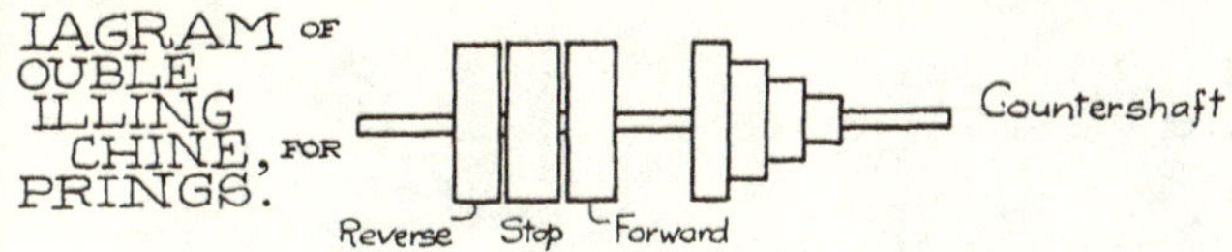

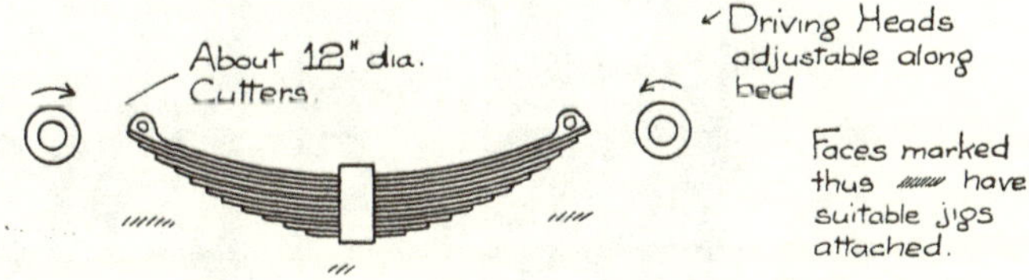

FIG. 188.

will do more work, owing largely to the fact that tool lubricant is invariably employed on the milling machine, and not on the slotting machine. It can be used on the latter, but gives no adequate return as a rule for the cost and mess involved. A point in the plano-miller for quantity work is that several cutters can be fitted, and the number of backs allocated to suit.

A double-ended milling machine, for working on the complete spring is illustrated by the diagram Fig. 188. This type is not used to any extent, and is really a survival of the days when all engine springs here were milled in their finished state, with softened ends. As pointed out previously, the finish of the spring with no load looked particularly good, but when loaded, all the plates staggered themselves outwards, and the cleanly milled slot bottom disappeared.

Machining in its true sense is carried out to a certain extent on forged heads for "coach-work." Fig. 189 shows these heads as they come from the machine (Fig. 224) and also as finally finished on a special double-milling machine, which operates on both sides simultaneously. The "female" end for the head shown requires only reamering to make it a good fitting job.

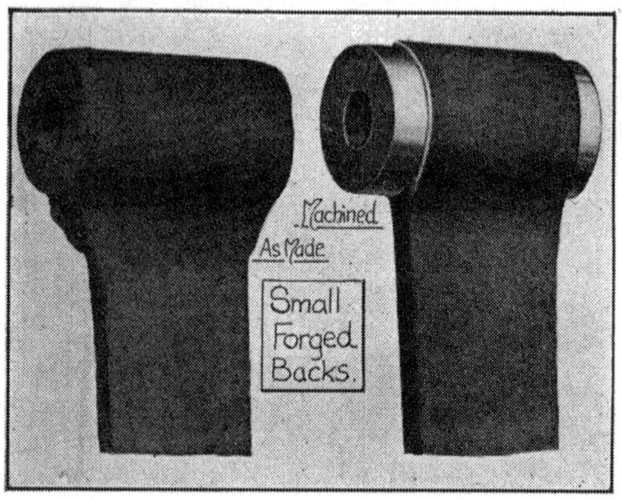

Fig. 189.

CHAPTER XXXVII

SPRING FITTING AND HEAT TREATMENT

THE term "spring-fitting" includes the heating, curving, quenching, tempering and adjusting of the plates composing the laminated spring, in order to produce a relatively perfect job. Spring-fitting at its best is undoubtedly seen only in the Sheffield district, which is the home of the British trade. The spring-fitting of the Continent is extremely good, but the spring shops there do not have to handle the problems of making a good spring from a bad design, as has to be done in the British shops; Continental springs, particularly for railway service, being invariably perfectly designed. American spring work throughout presents the same advantage—their designs being generally excellent from the spring-fitters' point of view, with the exception undernoted. Fig. 190 gives examples respectively of a British, Continental, and American, engine-spring. No account is taken of the skin stress, etc., for the loading, as this does not seriously interest the fitter, whose considerations are chiefly applied (apart from the steel quality) to the length and finish of the plates, and the type of back plate. It will be observed that the British spring has a long sweep solid-end back, drawn points, a long "short" plate, and several plates nearly full length, also three thicknesses of plates. The Continental spring has a small rib-end back, square points, a short bottom plate, two plates full length, and all plates of one thickness. The American spring has a plain back, (for loose clips,) square spear points, a short bottom plate, four plates full length, and one thickness of plate. Of the three, the British spring will take roughly three times the amount of fitting necessary for the Continental design. The latter "fits" itself almost automatically. The one awkward feature of the American design is the large number of full-length plates, making the spring far too strong at the ends. Unless this is corrected in the fitting, the spring becomes a bad shape under load taking it nearly or quite " straight."

Whilst the term " laminated spring " generally visualises the semi-elliptic form, numerous other more or less complicated shapes are in use—with a greater or lesser demand according to their complexity. The art of fitting these numerous shapes has to be additional to the art of dealing with the inherent defects in the design or practical incidents which may arise in manufacture. With the decline of the horse-drawn vehicle, the most awkward shapes are disappearing,

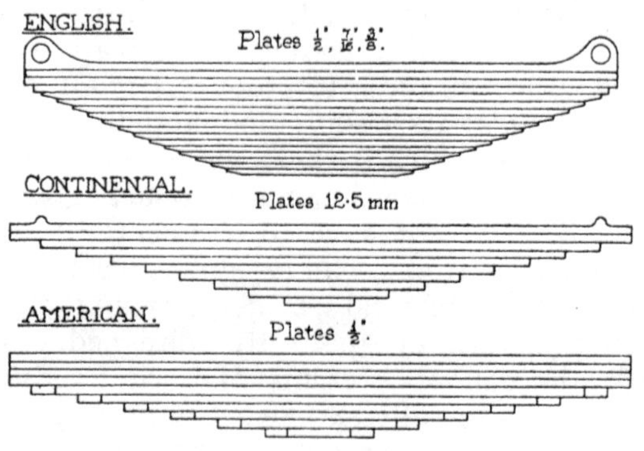

FIG. 190.

and it is pleasing to note that the automobile trade has nearly entirely settled down to the use of the semi-elliptic, chiefly direct, and occasionally as a cantilever. The only other pattern that counts in this trade is the true cantilever, or quarter elliptic. The bulk of railway springs are also of the semi-elliptic or full elliptic forms. A brief review of some of the patterns manufactured follows (see Fig. 191).

A. Semi-elliptic. Straight as made. Negative camber when working.

This is in request for small-wheeled locomotives having underhung springs, where the clearance from the rail is too small to permit of any positive camber.

B. Semi-elliptic. Reverse camber as made. More reverse camber when working.

This is a rare type, but will probably be remembered by many as the characteristic of the Belpaire locomotives of the Belgian State Railways—which included numerous types—and the spring pattern existed until the period (about 1900) when British influence asserted itself, and the railway authorities bought the five " Dunalastairs." The reason for the design was a general combination of considerations embracing the Belgian permanent way of the period, outside frames, overhung spring rigging, etc. The one thing needful was a spring type for heavy engines to stand up on a bad road, and therefore specially thin plates were resorted to. This practice meant a very large number of plates, and many springs were 10 ins. and 11 ins. deep at the buckle. All were overhung, and had a positive camber been included, the lengths of the hangers would have been abnormal, as the suspension point was underneath the deep outside frame. Negative camber " as made " was therefore introduced, which enabled the spring hangers to be substantially shortened, and reduced the overall height, which in the case of these engines, owing to other features in the design, was a very necessary item. Except in circumstances of this description, there is no particular justification for the type—which, however, works quite well. The design being abnormal, however, increases the fitting cost. With designs of this form, very special attention must be given to the rounding of the bottom inside of the hoop, otherwise trouble will be experienced with the short plates.

Type C. Semi-elliptic. Positive camber as made. Less positive camber, or straight, or reversed, when working.

This is far and away the most used for all purposes, railway and road. Automobile practice generally tries to run with a straight or nearly straight spring. Continental and American railway practice uses the type straight, or practically straight, for all rolling stock design where it is employed. British practice tended to rather high cambers, for locomotive springs, but is now approaching the " straight when working " design. Coaching stock is always run on straight springs, but wagon stock, owing to the large use of spring shoes instead of spring hangers, includes fairly high cambers. The adaptability and practical manufacturing considerations relative to this pattern of spring fully justify its extended employment.

D. Semi-elliptic. Double sweep. Positive cambers as made. Less positive camber when working.

This pattern dates back to the early days of the heavy spring, then desired chiefly for locomotives. It was thought to

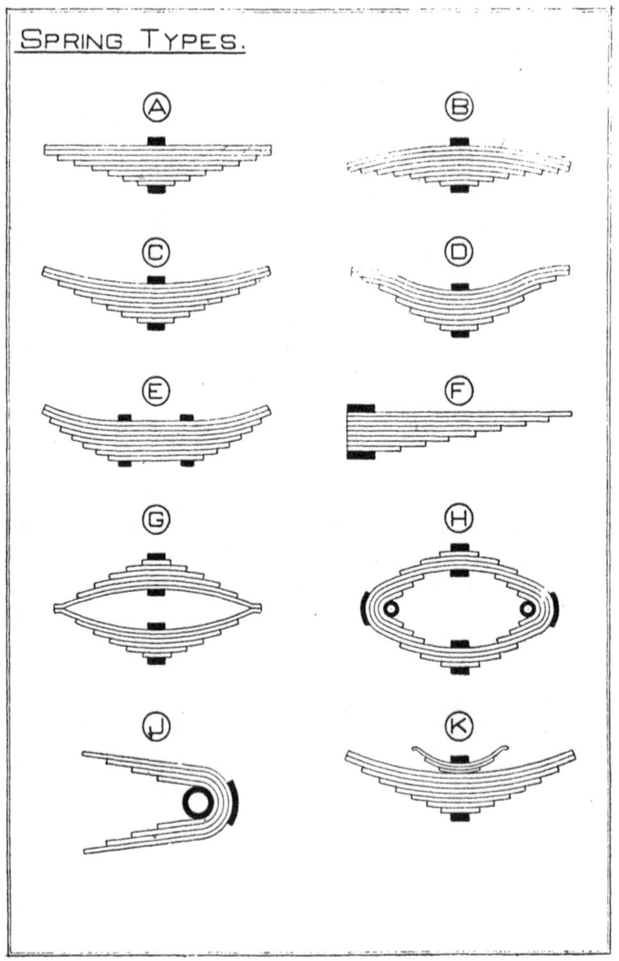

FIG. 191.

simplify hanger bearings, by giving a horizontal top with a well cambered spring. Gradually the type is falling into desuetude so far as this country and the Continent is concerned, but it is still largely used in the U.S.A. for heavy horsed vehicles and automobiles (trucks). The one thing in its favour is that when badly overloaded, the span is shortened to such a great degree that the spring is nearly unbreakable because it cannot deflect, and the vehicle runs practically springless. A study of loaded road vehicles, of the heavy patterns, in cities of the U.S.A. indicates the high wisdom present which insists on the retention of this type. It is not considered a very satisfactory spring from the point of view of high-class workmanship, as the plates " gape " badly in the vicinity of the junction of the two radii if they are perfectly fitted, and it is difficult to fit this type in any way which will ensure continuous close plate contact under all deflections.

E. Double-cantilever. Positive camber as made. Less positive camber when working.

A type in occasional use as a buffing spring for railway work, the object being to obtain a relatively light spring of considerable power, the span having to be the distance apart of the fixed centres of the buffers on the vehicle.

F. Quarter-elliptic, or cantilever. Generally with positive camber as made, and straight when working, but used with other camber possibilities.

This pattern is rare in railway practice, but employed in an increasing degree for light patterns of automobiles, as it cheapens substantially the cost of the suspension arrangements. It is also used to no little degree for tramway work. It is quite an efficient spring, but not as easy for hand fitting as the semi-elliptic pattern, owing to its being unbalanced. When this type is used, care should be taken to employ very substantial clamping devices on the end "encastre."

G. Full-elliptic, or bolster. Always positive camber when fitted.

This spring, generally grouped, is in universal employment for the bolster springs of railway coaching stock vehicles, but to only a limited degree for locomotive work; the elliptic bearing springs for the coupled wheels of engines, and for rigid axle tenders, on the (late) L. & Y.R. being one outstanding example; and the bolster springs for U.S.A. bogie tenders being another. Very little use is made of the type in automobile work, but for horsed vehicles—particularly in rough countries—it is one of the most used types.

H. " Elliptic " alternative, or endless-back bolster. Always positive camber as made.

This spring is really a combination of two semi-elliptics and two double cantilevers of the " U " pattern, of type " J " following. It is not in very common use, as it is an expensive spring to make, and it is more than doubtful whether the increased cost over the ordinary full elliptic is justified by any superior qualities that can be discovered.

J. Double cantilever, " U " pattern.

This is not often employed, but it is not a bad type of spring should such necessity arise as to insist on its employment for any special work.

K. Semi-elliptic, with auxiliary plates. Alway positive camber as made, and when working.

A pattern which is in fairly extended use for railway work, for both bearing and buffing and drawbar springs. The shorter bumper spring will, of course, take up a share of the load when the main spring has deflected to a certain point. In this country it is employed on rigid wheel base coaching stock, such as parcels vans ; and in America for full elliptics on passenger car bogies. This arrangement is quite satisfactory if the vehicle runs always " light " or " fully loaded." Under conditions of intermediate loadings, it has disadvantages.

A further miscellaneous assortment of shapes is shown on Fig. 192, and briefly referred to as follows :—

L. This is a four-plate door check spring for railway wagons, and from a cost point of view not to be recommended. A single-plate spring, properly designed and heat treated, is just as efficient for this purpose—and neither type assist in reducing maintenance expenses on doors.

M. Another type of door check, which is really a controller —the door being connected to the spring through the rolled eye—which also acts as a stop when the door has dropped to its approximately vertical position.

N. This is the usual British type of wagon door check—and consists of two plates ½ in. thick. As may be imagined, its effect on a heavy door causes severe stresses in the timber and ironwork.

O. A special pattern of multiple plate spring as made for use with small power hammers.

P. One type of many as used for horsed vehicles. In this trade, the number of shapes, and arrangements of semi-elliptics and cantilevers is very great, and the spring maker has cause to be grateful for the fact that automobile designers have shown no inclination to follow the practice of earlier road vehicles.

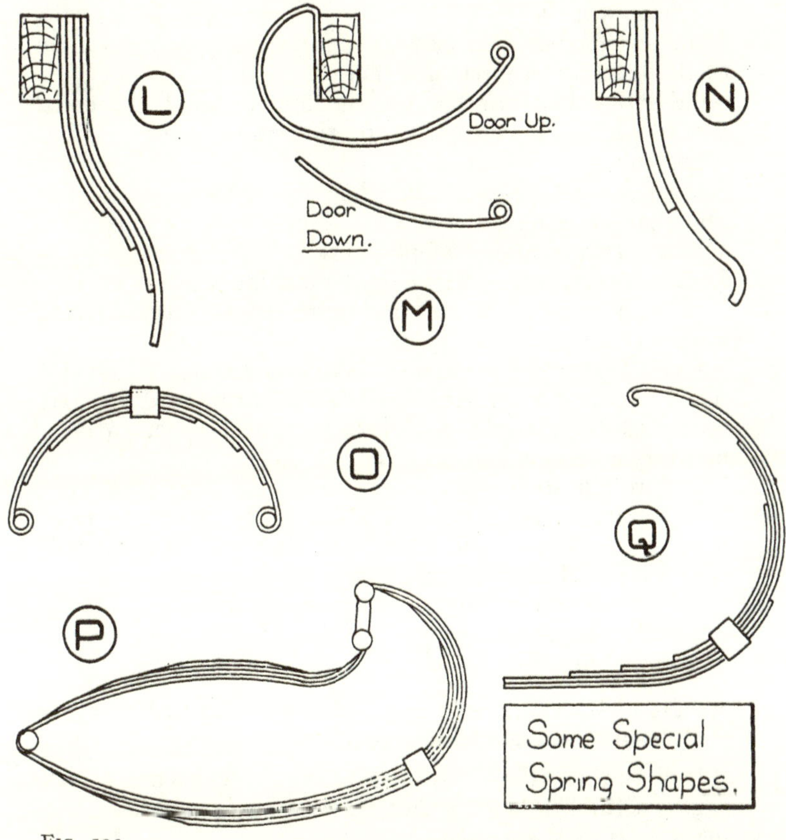

FIG. 192.

Q. This shows the " C " or " Cee " spring largely favoured for horsed vehicles on bad roads. The great advantage of this pattern lies in the fact that an initial tension on the spring can be provided by means of a leather strap and ratchet gearing, which makes for easy riding on an unloaded vehicle.

Necessarily, as the semi-elliptic spring is the general world-standard, the following expositions regarding spring fitting are mainly limited to this type, other types being abnormalities. The chief considerations rendering this standard pattern difficult to fit, are as follows :—

1. When the positive ratio (camber : back length) is more than 1 : 8 or less than 1 : 16. (Applying to hand and machine work.)
2. Long sweep solid-end backs. (Hand work only, as none of this pattern are machine made.)
3. Solid-end backs not flat on the underside at the ends, due to bad hammer tools or inferior smith work. (Hand work only.)
4. Drawn points, particularly when the drawn part of the one plate overlaps the start of the draw on the plate above. (Hand and machine work.)
5. Long short plates. (Hand and machine work.)
6. Too many top plates of the same length. (Hand and machine work.)
7. Varying plate thicknesses. (Hand and machine work.)
8. Abnormal test requirements. (This feature is dying out owing to the general use of rational test specifications, but was a serious consideration at one time.)
9. Badly rolled steel—having too much or too little concave ; varying from nominal thicknesses ; varying from nominal widths ; including laterally bent (dog-legged) bars, etc. (Hand and machine work.)
10. Steel of un-uniform quality, assuming the supply to be generally in accordance with the specification limits. (Hand and machine work.)

Opposite considerations which cause a spring to become a relatively easy manufacturing job, are :—

1. When the positive ratio (camber : back length) lies between the limits 1 : 8 and 1 : 16.
2. Plain or rolled-eye backs.
3. Square or spear points.
4. Short (correctly designed) " short " plates.
5. One or two full length plates only after the back plate.
6. Plate thicknesses uniform throughout the springs.
7. Reasonable test requirements—for instance, British Standard Specification.

8. Steel rolled well to gauge, clean and straight, with suitable concave for the design.
9. Continuously uniform material (that is, varying within minimum limits of analysis.)

If the above conditions are fulfilled in designs, the spring

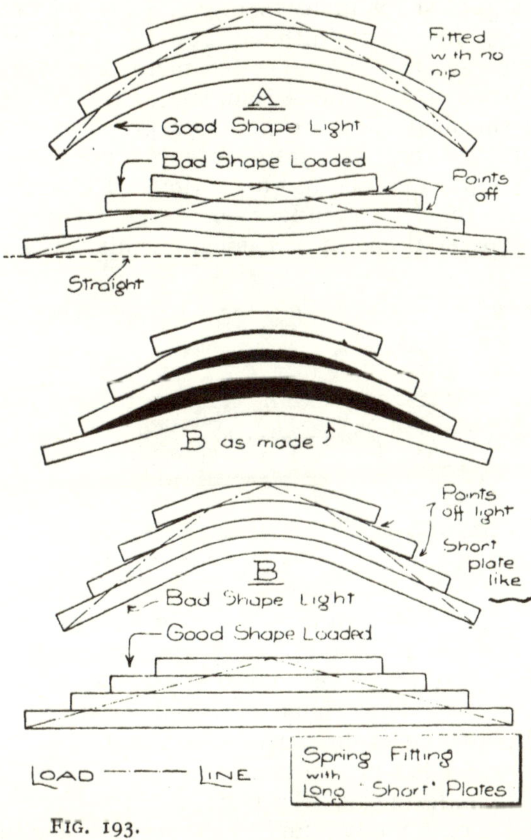

FIG. 193.

is a comparatively easy manufacturing proposition either by hand or by machine. The fulfilment of these is a necessity for the successful economic operation of any machine process, which accounts for American springs being invariably designed in accordance with the above requirements; and machine manufacture will not be commercially possible in this country until designers agree to strict adherence to these conditions.

With " short " plates dimensioned much longer than the correct length (a very frequent fault) it is necessary to badly shape the plate in order to ensure the points being on when under load. The choice in spring manufacture with a bad design is between making a perfect shape light, for a bad shape under load, or making a distorted shape light, which will set to a good shape under load. The latter is invariably, and logically, adhered to, although, owing to the amount of " nip " involved, the spring is far more severely stressed under these latter conditions than under the former. Nowhere has this bad shape to be more pronounced than in long " short " plates, and from two to four plates above, according to the number in the spring. The line of load through a spring has already been dwelt upon and when superfluous material has to be carried outside this automatic

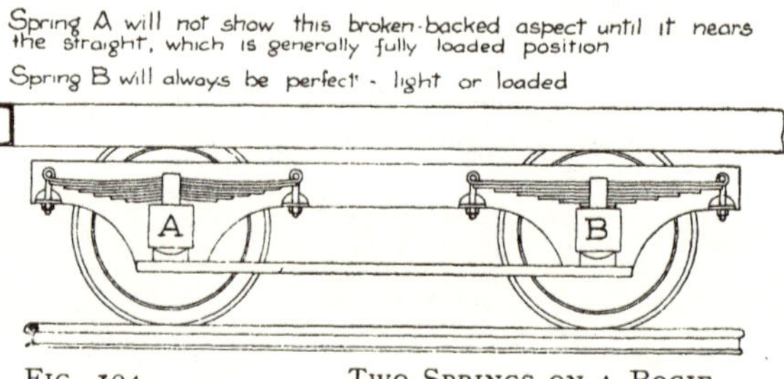

FIG. 194. TWO SPRINGS ON A BOGIE.

load line, it has to be shaped accordingly to keep up appearances. The result is best shown diagrammatically in Fig. 193, which shows that the " short " plate is as far away removed from its correct shape as it well could be. All this labour and plate distortion can be very simply avoided, as it is invariably avoided in American and Continental designs, by employing uniform thickness plates, with the length of the short plate approximately equal to the straight length of the spring, divided by the total number of plates, a 40-in. spring of ten plates therefore having a 4 in. short plate, or perhaps one slightly longer to provide for practical hooping. From the purely commercial point of view, such correct design is worth adoption, owing to the saving of weight, and consequently, cost, as numerous springs can be re-designed with 10 to 20 per cent. less steel, when the short plate is made of a length approaching the theoretically correct dimension.

The possibilities which occur with flat coaching stock springs badly designed, and difficult to fit, are best illustrated in Fig. 194. It will be observed that with the numerous horizontal lines present in the elevation of the coach and bogies, a "straight" spring that is not straight is at once noticed, and placed to the credit of the spring maker—which is the reverse of honest, as the fault lies entirely with the design, and the spring shop will have done its best to fit it so that it will be straight—an endeavour not always possible of perfect attainment. It might be enquired why, if the spring was obviously badly designed, the spring manufacturer did not take up the matter with the rolling stock builders with a view to obtaining a practical amendment, but, owing to considerable experience in such attempts, manufacturers in this country have reason to be always chary of raising any points of design with the designers. In this respect, British purchasing concerns might with advantage copy the practice usual in America, where spring makers are supplied with a skeleton drawing showing essentials, such as length of back plate, type, end slots, width of hoop, etc., thickness of plates, and working load, and then left to themselves to determine the number of plates, with the lengths of the short and intermediate plates. (See Fig. 93.)

It will now be of advantage to digress from the actual spring to a few remarks on the heat treatment of steel, since the whole object of the hardening and tempering of spring plates is to raise their elastic limit, and consequently, the allowable working stress. A spring could be designed and made out of dead soft steel, of 0·10 per cent. carbon, elastic limit about 12 tons per square inch, but it would require to be four times as heavy as one made from 0·55 per cent. Carbon steel, hardened and tempered, which has an elastic limit of about 55 tons per square inch (about 24 tons unhardened). Numerous theories have been advanced to account for steel, when heat treated along certain lines, improving in its physical characteristics. These belong properly to metallurgical works. What can be visibly described as the result of varying heat treatments, are the different fractures of the bar. Fig. 195 shows four fractures of standard water hardening quality spring steel, each of which has been obtained in the manner hereafter described. Each plate was heated edgeways, so that one edge was raised to the extreme temperature of 1200° C., when steel of this carbon, 0·50 to 0·60 per

Fig. 195—1.

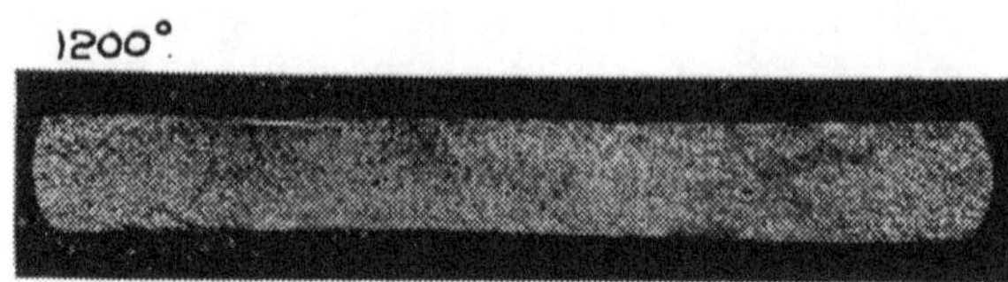

Fig. 195—2.

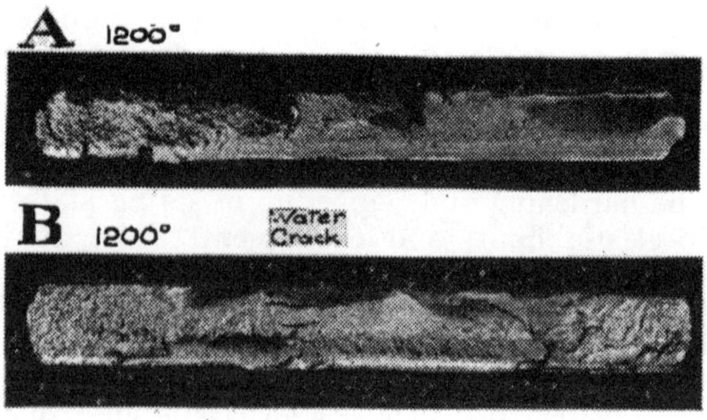

Fig. 195—3.

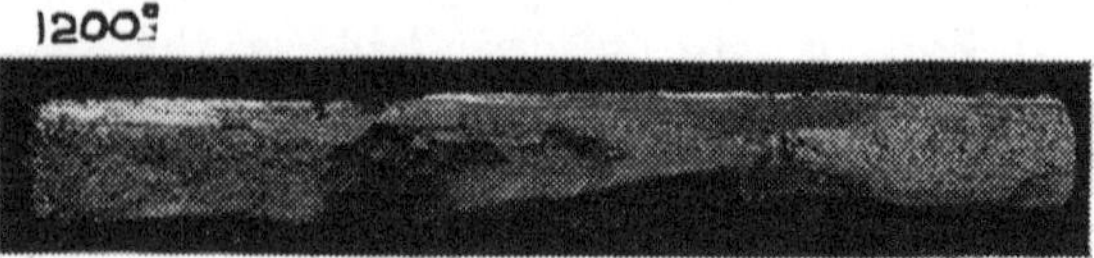

Fig. 195—4.

cent., commences to burn or " frizzle," the other edge being outside the fire. The temperature therefore ranged from 1200° at the burnt edge to 700° or lower at the outer edge, caused by radiation. The middle of the plate would be somewhere between these temperatures.

Fig. 195/1 shows the fracture of a piece which has been burnt at one edge and has been in the neighbourhood of 700° at the other edge, the hot piece being then plunged into cold water. The burnt edge, readily noticeable, by its irregularity, is on the left of the picture, and it will be observed that the fracture here is extremely coarse or crystalline, which grades away into the extremely fine crystal line structure of the cooler edge. Owing to the heat of the piece when quenched it has cracked in various directions. Such cracks are known as " water cracks " and one is very obvious along the bottom edge, at a position of about 1000°.

Fig. 195/2 shows a piece which has been fractured after heating as before described, the burnt edge being 1200°. It has then been allowed to cool until this edge was about 900°, before being water quenched. It is to be noted that no water cracks appear in this piece. Three distinct types of fracture are here visible, namely, at the burnt edge a coarse crystalline structure, in the middle a very fine structure, and at the cooler edge a small crystalline or granular structure. This edge has been under the " recalescence point " or change point when cooled, and the fracture accordingly is the same as that which would have been obtained from the bar as rolled. The centre structure is the correct fracture of the hardened and untempered plate.

Fig. 195/3 shows two fractures of a piece heated as before, one edge 1200°, water quenched and tempered at 450°. Dealing with the picture A, the part adjacent to the burnt edge, on the left, shows crystalline, indicating thereby that the steel, by overheating, has been ruined, and is not susceptible to being brought back by normal heat treatment. As a matter of fact, even this structure can be partially restored interiorly by known methods, but as the outside surfaces are hopelessly decarburized, it is a useless proposition to attempt restoration from a spring making aspect. From the midway point onwards to the cooler edge, although the variation of temperature at quenching would have been between 950° and 700°, there is no apparent difference in the fracture, which would be described as " fibrous "—a term not strictly correct, except for puddled iron, but generally accepted and understood in the trade. This is a reasonable structure for the finally treated spring plate. Owing to the high temperature of the overheated side when quenched, however, the piece is badly water cracked, and the second illustration of this sample (B) shows a characteristic water crack running from the surface one quarter through the thickness. Another feature that will be noticed in this side of the fracture is that the cooler edge appears crystalline, whereas 2 in. away (in length) it was, as shown on the right-hand of (A), " fibrous." A careful examination of this shows that it was due to the sharp blows necessary to detach the piece—and this part of the section has snapped suddenly as the result of the remainder having broken. From the original piece it can be seen that the crystals here are more cohesive than those in the previous structures explained. The same thing can be made to occur with good quality wrought iron, a sudden blow through a sharp nick resulting in a square crystalline fracture, whereas a blunt nick in the same piece, and a gradual bend, would have opened up a beautifully fibrous structure only.

Fig. 195/4 shows the fracture of a piece heated as before, with one edge to 1200°, then allowed to cool to 800° before being water quenched, and finally tempered at 450°. A crystalline structure is still visible in the overheated portion, but the middle is obviously in an extremely " fibrous " state, the ideal in fact, showing that the quenching temperature of this part was very close to the change point, that is, approximating 700°. This middle has bent, and would have bent through an angle of 90°, but for the retardation of such action due to the over-hot and under-hot sides respectively. The structure on the right shows that the quenching and tempering has had no effect on the original, owing to the low temperature, about 400°, which this portion was when put in the water. From such fractures the whole of the treatment can be reasonably well gauged. The experienced hand fitter, however, does not desire to see fractures, and he bases the treatment he gives the various steels almost entirely by the " feel " of his peaning hammer when he is adjusting the plates after tempering.

The " recalescence point " or change point has been referred to in connection with these fractures. This is the critical point in the cooling-off of steel—its antithesis being the "decalescence point," occurring during the heating-up of steel. These points may be simply defined as temperatures at which occur a definite change in the rate of cooling or heating, otherwise, of the rate of radiation or absorption of heat. A piece of steel cooling from, say, 900°, will fall in temperature uniformly until it reaches about 700°, when it not only ceases to cool, but actually becomes momentarily hotter, after which it proceeds again to cool at a uniform rate

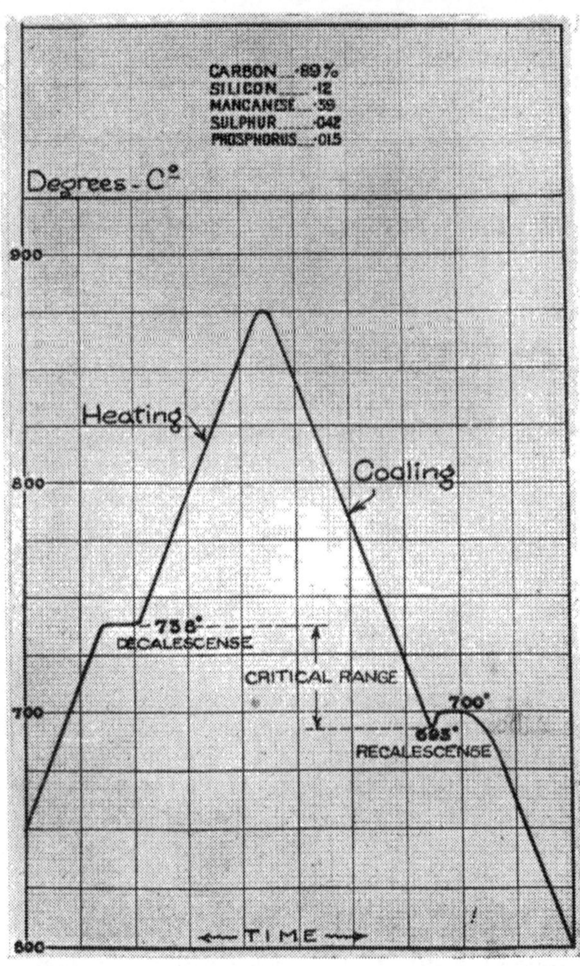

Fig. 196. Steel Heating and Cooling Curves.

to the temperature of the surrounding atmosphere. Equally, when heating-up, at about 730° heat is absorbed for a period without the steel becoming visibly hotter. On normal carbon steels these temperatures are between 700° and 730°, the only element in such steels seriously affecting them being manganese. With the percentage of this element below 0·50 per cent. the normal carbon steel range applies, but with an increase of manganese to 1·00 per cent. the recalescence point is sensibly lowered, being then in the neighbourhood of 670°. The decalescence point is not seriously affected. The critical points for ordinary spring steels will therefore be about 680° and 730°.

A typical approximate curve for the heating and cooling of an oil-hardening spring steel is shown by Fig. 196. Metallurgical science recognises for varying steels more than one "critical point" but these additional points are not far removed from one another. The total temperature distance covered by these points is known as "the critical range in heating" or "the critical range in cooling" as the case may be.

The phenomena presented at the "recalescence" and "decalescence" points are generally taken to indicate changes in the internal structure of the steel, and matter dealing further with this can be found in any book on steel metallurgy. When looked into closely, the subject will be found extremely complex, as steel comes under neither of the well-known definitions or categories of "mechanical mixture" and "chemical compound," it being a combination of both. Definite, however, are the facts of these critical points occurring, with corresponding critical ranges; and unless steel is heated above the decalescence point (say 730°) and quenched above the recalescence point (say 680°) it will not harden.

The spring fitter has therefore to watch that his steel is about 900° before withdrawal from the furnace, and not less than 700° when plunged into the cooling medium, the necessary shaping of the plates having to take place between these temperatures. The temperature should not exceed 900°, because (1) the growth of crystals in the structure, gradually enlarging until they reach the size indicating overheated steel (Fig. 195/1), steadily deteriorates the steel and (2) the extra heat increases greatly the rate of decarburization of the surfaces, thereby reducing their elastic limit. It is however, thought that a "smoky" furnace checks this latter possibility.

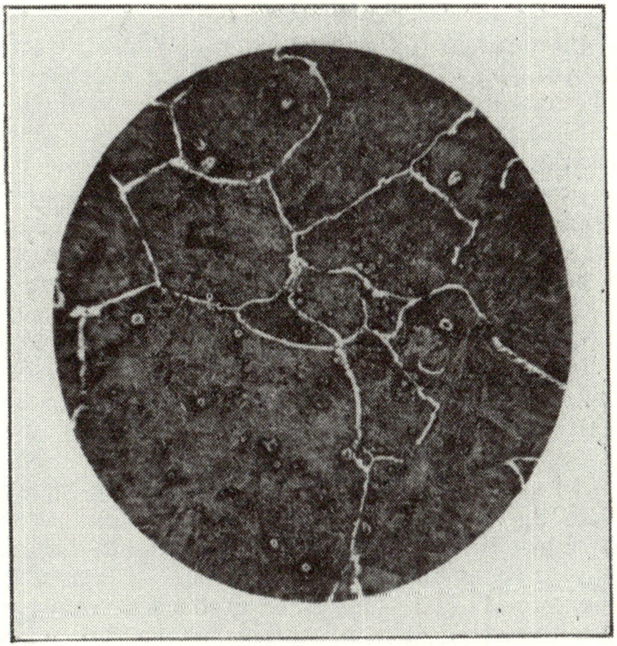

Fig. 197—A. As Rolled × 50.

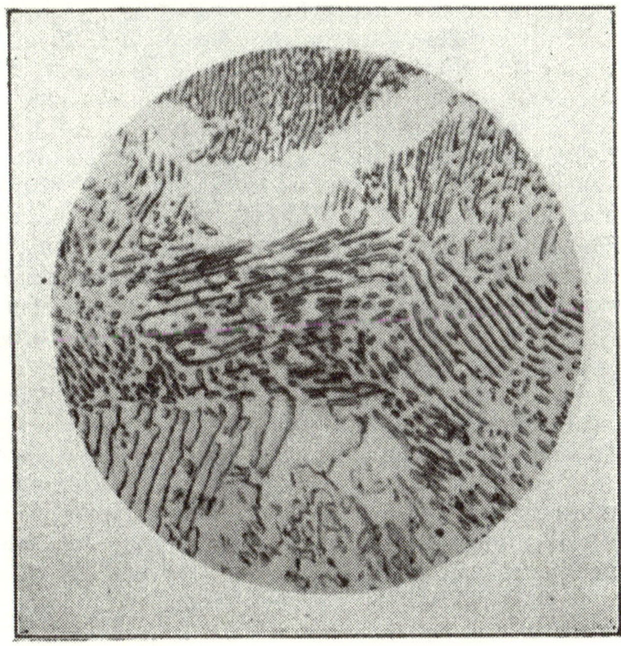

Fig. 197—B. Pearlite × 1500.

As an interesting supplement to these photos of fractures, Figs. 197, (A and B), 198, and 199, are given—these being micro-photographs of spring steel bars under various conditions Fig. 197—A shows, at a magnification of 50 diameters, the appearance of the structure of the spring steel bar as rolled. Broadly speaking the white areas consist of practically pure iron and are known microscopically as " ferrite." The dark areas (which are called " pearlite ") consist of a mixture of iron and carbide of iron arranged in alternating plates ; the latter are very fine so that, at a magnification of 50 diameters, they cannot be seen. When examined at very much higher magnifications however, the sandwich-like structure of alternating plates becomes clearly visible as shown in Fig. 197—B, which represents one of the dark areas in Fig. 197—A, magnified 1500 diameters. The iron and carbide of iron are present in such proportions in these dark areas that the latter contain approximately 0·9 per cent. carbon. Since these pearlite areas contain all the carbon in the steel it is obvious that they will increase in area as the carbon content of the steel increases up to 0·9 per cent. Thus a sample of carbonless iron contains no pearlite ; a steel containing 0·9 per cent. carbon consists entirely of pearlite while with 0·3 per cent. carbon two-thirds of the mass consists of ferrite and one-third of pearlite.

When steel is heated up to the decalescence point, the pearlite structure changes ; the alternating plates of iron and carbide dissolve in each other forming a solid solution which is known as austenite. As the temperature increases above the decalescence point, the austenite gradually dissolves the ferrite until, at some temperature depending on the amount of carbon (approximately 780/800° C. for a steel containing ·5/·6 per cent. carbon), the whole mass consists of austenite. If the austenite be then cooled slowly, the changes mentioned above occur in reverse order ; first the ferrite begins to separate out and then at the recalescence point the austenite breaks up into pearlite. If however the austenite is rapidly cooled, as for example, by quenching in water, the changes are almost completely suppressed and a homogeneous structure which is called martensite is produced. Martensite is hard and brittle ; on being reheated to temperatures below the decalescence point, however, it gradually loses its hardness and becomes tougher. At the same time the homogeneous structure of the martensite is gradually broken down ; the carbide of iron, which has been retained in solution by the quenching, is gradually thrown out of solution in the form of exceedingly minute particles which gradually coalesce. Fig. 198 shows the appearance of a hardened and tempered spring plate at a magnification of 200 diameters ; as will be seen there is no ferrite or pearlite present, the structure consisting of fine grains of what were martensite in the quenched plate but which have now been somewhat modified by the tempering. These grains, in order to distinguish them from the untempered martensite, are referred to as " troostite."

FIG. 198. SPRING TEMPER × 200.

Referring again to the changes occurring on heating a steel, it was mentioned that at some temperature above the decalescence point the whole of the ferrite had been dissolved by the austenite so that the steel consisted entirely of this constituent. It also has an extremely fine grain at this temperature. If, however, the steel be further heated, the grains of austentite grow in size, some of the grains being absorbed by others. At first this grain growth is very slow, but at higher temperatures it becomes rapid and produces a very coarse structure in the steel. Steel with such a coarse structure is said to be " overheated " ; it is much more brittle than when it has a proper grain and when broken, shows a coarse crystalline fracture. Fig. 199 shows a spring plate which has been overheated ; it is taken at the same magnification as Fig. 197—A and comparison of the size of the grains in these two photographs shows in a very marked manner the influence of " overheating " and emphasises the importance of the control of temperatures when dealing with the manufacture of springs. With the type of spring steel

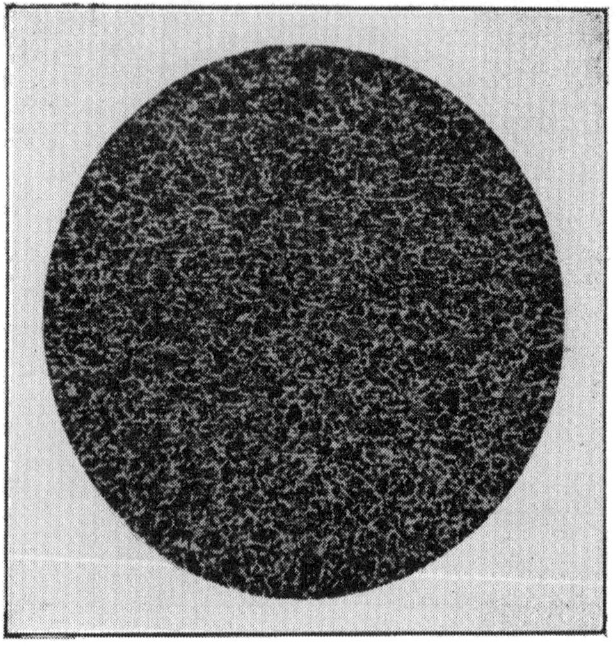

FIG. 199. BURNT × 50.

under consideration, overheating will commence to make itself felt if a temperature of about 950/1000° C. is exceeded.

Pyrometer control of the temperatures of the ordinary spring furnaces in use is impracticable, as they are generally units with from two to ten fitters working material in and out, and additionally are used for tempering as well as for hardening. With separate furnaces for such purposes, control might be used, but as with hand fitting, skilled labour is required, it is clear that the heat treatment by eye may as well be included in the training course. Pyrometer control is in fairly general use for the manufacture of automobile springs, but as a rule, the class of labour employed is not of the high skill of the heavier railway trade, and the heat treatment control, for tempering at least, becomes therefore a necessity. With machine fitted springs, however, the proposition alters itself as regards temperature controls —and this will be dealt with in its proper place.

As has been gone into, the determination of a suitable treatment for spring steels can be checked by various means, including diversities of camber or deflection tests, vibratory tests, impact tests, and transverse bending tests. A simpler method than all these, however, presents itself by the use of the hardness test of the Brinell machine. Only a small sample is required, which can be heated, quenched, tempered, and brinelled, in ten minutes or less. The author's experience convinces him that with alloy steels, the Brinell hardness should never exceed 3·10 mm. dia. for an all-round satisfactory spring, and it is better to aim for 3·20 (364). Water-hardening steels should be worked somewhat softer, for maximum hardness, and they will be more satisfactory at 3·40. In practice, with nip-fitted springs, it will be found that the plates in one spring frequently range from 3·60 on the top plate to 3·20 on the short plate, owing to the initial stressing of the latter, and the need therefore for a higher elastic limit.

The practice sometimes obtains of brinelling every plate. This does not seem of great value, as the best check of the spring being satisfactory is the final scragging test—the brinelling of each plate being more in the nature of a " talking-point " for the manufacturer than of any practical value, as there is always to be found at least one point on each plate which will be absolutely exact to any specification requirements, if the plate has received any ordinary heat treatment. In the case of automatic and controlled tempering furnaces, the brinelling of a few plates per hour is a useful check. An extremely handy machine for spring shop use for this hardness test is illustrated in Fig. 200. For continuous, and relatively rough work, the absence of cylinder, piston, valves, and gauges, renders it eminently suitable—the test pressure standard of 3000 kg. being applied by means of the hand lever at the side through a simple eccentric device in the frame. When the pressure is attained, the balance weight (calibrated manometrically) raises and locks the operating lever. For reading the impression—which for spring check purposes need not be done with the highest accuracy—either the ordinary Brinell microscope can be used, or the better miniature microscope, about 1 in. high and $\frac{3}{4}$ in. diameter, which is supplied by the makers, and has great practical advantages. However, with a certain amount of experience, the eye measurement is tolerably accurate.

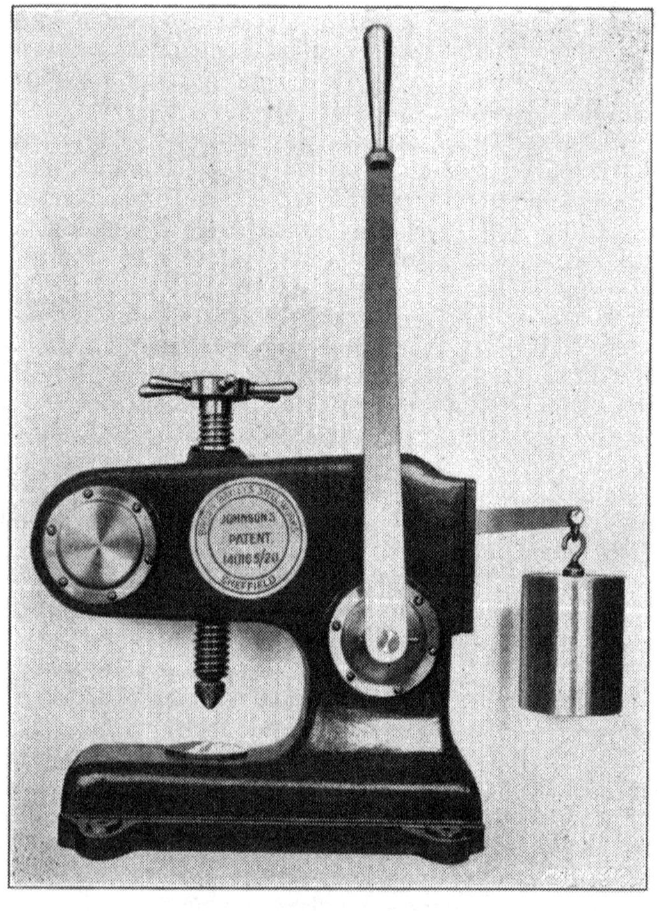

FIG. 200. BRINELL TESTING MACHINE. BROWN BAYLEY'S STEEL WORKS LTD., SHEFFIELD.

The fitted spring, bolted together if a centre hole is provided, otherwise fitted with a temporary hoop, is taken to the scrag or testing machine, and deflected through the specified test, plus the excess the spring is over the correct camber. For example, a spring desired 4 in. camber will probably come to the scrag at 4½ ins. With the test deflection at 3½ ins., the actual deflection applied will be this 3½ ins. plus the ½ in. of excess camber, or 4 ins. With this, the spring, if of good steel, well treated and having little nip, will recover to probably 4¼ ins., or ¼ in. high. Other springs of the same pattern will be therefore made by the fitter of a camber 4¼ ins. to 4⅜ ins., so that, with the scragging loss of ¼ in., they finish at 4 ins. to 4⅛ ins., which is within the limits tolerated in this country and the Continent. American practice gives an allowance of nothing under and ⅜ in. over. Locomotive and carriage springs should certainly be kept within the ⅛ in. tolerance, but for wagon work, ¼ in. is in no way objectionable. Springs coming out low under the test are returned to the fitter for resetting. On rare occasions a plate breaks and has to be replaced, but the percentage is extremely low, in the neighbourhood of 1/100 of one per cent. (1 in 10,000).

The test deflection in this country and on the Continent is generally twice the working load—in other words, the spring should be designed backwards from the permissible test deflection for the length and plate thickness, so that the working load is one half the test deflection. The formulæ in use here (British Standard) and on the Continent are not greatly different for this purpose. American practice tests to one-and-a-half times the working load. Accordingly, from the point of view on this side of the Atlantic, American springs tend to be on the "soft side," and will give way,—*i.e.* lose camber, and/or go out of shape—in heavy service, generally well before any plates break.

An illustration of "scragging" as carried out in this country is shown by Fig. 201, which shows a 12-plate R.C.H. buffing spring undergoing its test. These springs are 70 ins. (practically) between bearing points, when straight, include 12 plates, 3 ins. × ½ in., are made with 14 in. camber, and tested straight, which is nearly 30 per cent. over the B.S. test deflection of 11 ins. They are specified to carry a load of 7 tons, which, worked on ordinary calculations, gives a skin stress of 81·6 tons per square inch. Additionally, however, there is specified to be a ⅜ in. centre rivet, which needs a hole at least $\frac{7}{16}$ in. diameter, giving an actual central area reduced by this amount, which brings the nett calculated stress across the middle as 95 tons. The material used is always water-hardening steel, frequently only 0·50 per cent. carbon (Bessemer). The scrag employed is of the usual British steam type, which is, without any doubt, the most satisfactory type to use. A good scrag assists to no small degree in the effective bedding of the plates of nip-fitted springs, and as springs are frequently easier to put together with plenty of " daylight " between plates, the final " fitting " is really performed by the scrag. This is not good for the

FIG. 201. THE FINISHED SPRING BEING SCRAGGED. STEEL PEECH & TOZER, LTD., SHEFFIELD.

spring, but an easy check on workmanship of this sort can be got by finding the loss of height after the first test stroke. For instance, on the 12-plate buffer mentioned, the loss under the first stroke can vary from 4 ins. to $\frac{1}{2}$ in. according to the amount of " nip " employed by the spring fitter.

One result of machine forming, combined with the short " short " plate design, is the absence of serious nip (machine dies generally working springs to the same sweep so as to avoid various dies and the resetting of the machine). Accordingly, if the sweep of the spring, as set up on the machine, is somewhat exaggerated, as it is not corrected by fitting, the shape under load will be this—M. This shape can also be obtained by the spring middle giving way, and straightening, as may happen if a bad design causes a high central stress. Fig. 202 illustrates this particular point. Assuming a badly designed spring required to work straight, it is clear that the shape must be not a true arc, that is, if it is desired to appear straight when loaded, as usually preferred. Sketch B—1 shows the true arc and the correct sweep for such assumed bad design. It is always probable, and not infrequently happens, that in setting up the dies the sweep is exaggerated to make quite certain of the straight shape loaded. The result when working then appears as B—3. These points turn continually round the same centre—of original bad designing, which reduced down—whatever be its details—resolves itself into the sectional (developed) areas at certain positions in the spring plan being in excess of the material required, thereby causing non-uniform stressing. Expressed otherwise, it is divergence from the true rhombus plan towards the final rectangle maximum. Such relatively low stressed localities tend to lag behind in the general spring deflection, and the amount of that lag is expressed by the distance X (202, B—1) showing the difference between the true arc and the proper sweep to obtain a good loaded shape. The distance X is thus artificially placed in advance of the true arc, so that the spring comes down a good shape. Should X be exceeded, owing to not unusual errors of judgment, and the total increment be X plus Z (Z being the error) then the amount in advance of the true arc is sufficient not only to compensate for the lag due to the design, but to advance the spring in the localities including X plus Z to such an extent as to make the remedy at least as bad as the original disease had the true arc been followed.

Fig. 202 (A series) shows the partial evolution of a wrecked spring, up to the point where it is generally taken out of service. A—1 shows the spring as running light (short plate is too long, so considerable nip is introduced). A—2 shows the spring under full load. This is taken as straight for purposes of the diagrams as it brings out better the explanations, although, of course, whatever position the spring takes up under load affects in no way the general argument. In this position it is assumed that the 4th plate (or " round ") breaks, owing to numerous possible causes such as overheated steel, water cracks, insufficiently tempered, etc. The spring will not then set back to its original camber when unloaded owing to the " nip " fitted plates losing their original internal equilibrium (A—3). Consequently, when reloaded (A—4) not only is the deflection increased owing to a plate failure (in this case 25 per cent.), but owing to the loss of camber, it actually deflects a distance past the first point (A—2) equal to the camber loss plus original load deflection plus (in this case) 25 per cent. of the original load deflection. The immediate result is that the short plate (already overstressed owing to the nip) refuses to return to position when the load

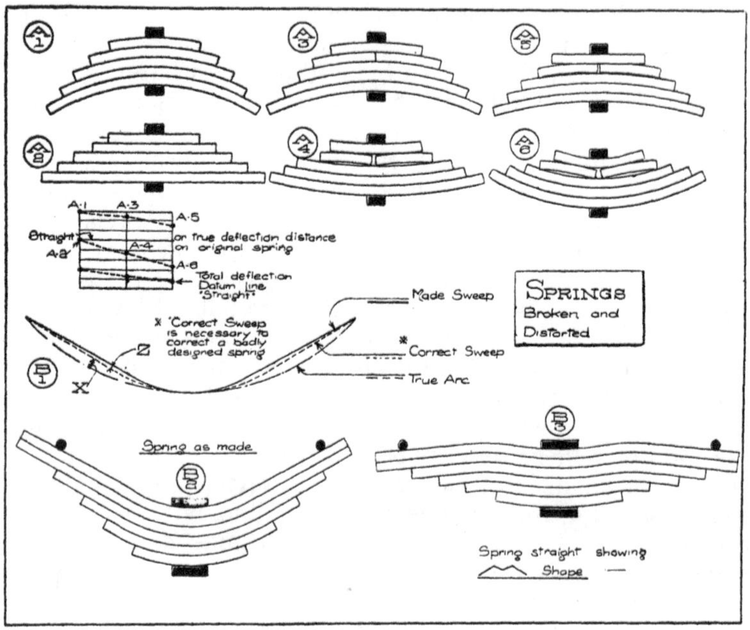

FIG. 202.

comes off, and more or less straightens out, as shown in A—5. As a result, the internal forces due to the nip, are still further disturbed, as this short plate can no longer effectively pull back the long " rounds," and more camber is lost. The resulting movement under a further reloading is as shown in A—6, where the distance travelled past the original point of deflection (A—2) is very considerable, due to the further loss of camber and the crippled condition of the short plate.

Springs with correct length short plates can clearly be taken as longer lived, as they will not lose camber to the same extent, due to a broken plate upsetting the internal equilibrium. It is not easy to detect plates broken in or near the centre, as the points usually keep on, and overstressed plates do not always break, but frequently "set"—it depends on the heat treatment they have received. "Points off," however, indicate always the probability of a broken plate being present, and the spring should be removed from service as soon as conveniently possible. American practice permits two broken plates (excluding long back plates) in engine springs before classing them as "for repair." As, however, fifteen to twenty-five plate springs are common, and they are fitted without serious nip, two plates out of action will not affect the spring to any extent.

Before concluding this general resumé of the fitting and treatment of spring steels, a few remarks might be made on the relative aspects of water and oil quenching. The whole point of quenching lies in the necessity for trapping the steel structure when it has been heated over the critical point (or points), and in order to do this, the heat contained in the spring plate must be rapidly withdrawn. The quickest commercial medium for this is brine; the next speediest water; somewhat slower is a saturated soda or similar water solution; and the slowest medium is the usual quenching oil. This latter has to be a relatively heavy oil of high flash point to prevent instant ignition. In spring work, the choice invariably lies between ordinary water, and usual quenching oils. As regards the economics of the two media, no question arises, water being comparatively costless, and oil needing not only a large capital cost, but continual running cost to make up for wastage and leakage. Additionally, if any high production is to be achieved, cooling apparatus must be installed and maintained, adding considerably to both capital and running costs. As regards the manipulation of the heated spring plates, water will ensure an absolute quench of maximum hardness, but on the other hand, with unskilled or partially skilled workers, will cause considerable trouble by providing water cracked plates, and a large amount of additional fitting work, slowing down production and increasing expense. With the employment of skilled labour, however, water cracks are almost unknown, and the method of handling the plate when quenching reduces to a minimum the fitting afterwards required. With oil-hardened material, no cracking can take place on ordinary good steels, and the need for fitting after the quench is substantially reduced. Soft plates can, however, be easily obtained, the oil being so very mild in its action. The whole aspect resolves itself into one of the "rate of removal of heat." The quick rate consequent upon the use of the water, gives rapid contraction movements to the surfaces, and the "harder" the original steel, the more rapid are the contraction movements. Plates worked through oil-quenching processes are not as clean as plates worked through the water—and in hand-fitting, water is preferred.

The great point to be aimed at to obtain the satisfactory service spring is that degree of treatment which will give the maximum "fibre growth." It would appear that with ordinary carbon steels, the softer the original quality, the better fibre can be obtained when the steel is treated to spring temper, which is one reason why in this country and the Continent, the oil-hardened carbon steel is not a standard quality. Alloy steels come into a different category, and when the spring plates are produced with 3·10 to 3·30 Brinell, the fractures of all standard alloys will be found fibrous.

CHAPTER XXXVIII

SPRING FITTING BY HAND

IN the process of " fitting " springs by hand, the spring fitter collects, or has delivered to him, the various plates, including the back plate—which are speared, holed, studded and slitted, etc.,—and his job is then to heat the plates, curve them, harden them out, temper them, and set them as required, so as to produce a reasonably well finished spring. Until comparatively recently, all classes of springs were hand-made in this country, and the Continent, the only exception being with certain standard railway springs for which a certain amount of machine work was in existence. In American practice, a small amount of hand fitting, in the generally understood sense, is performed—but the majority of skilled spring fitters there work after some class of machine former, either dealing with the treatment and setting operations, or else only rectifying machine work. Machine fitting, with other equipment to match, is in general essentially a mass production proposition, and needs large capital expenditure to devise an efficient plant. Hand fitting requires comparatively little plant, but has a corresponding low output per man. A mean can be obtained with partial machine operations, as will be described. With the numerous types of springs required here, however, and the relatively small quantities of each, machine manufacture would appear to present but little attraction, as the skilled spring fitter is a hard worker, and produces, on the whole, very good work. Unkind jibes are frequently hurled at European methods of spring-making, by ultra-scientific engineers obsessed with pyrometrical and electrical furnace control mania, but expenditure on such items can only be justified in the case of very high-production machine fitting plants, and the resulting product is not necessarily more perfect. It is possible to produce a very good spring indeed from a coal furnace, with water hardening and tempering effected by the help of a smoking stick and skilled judgment. From general considerations, one would suggest that the weakest point in

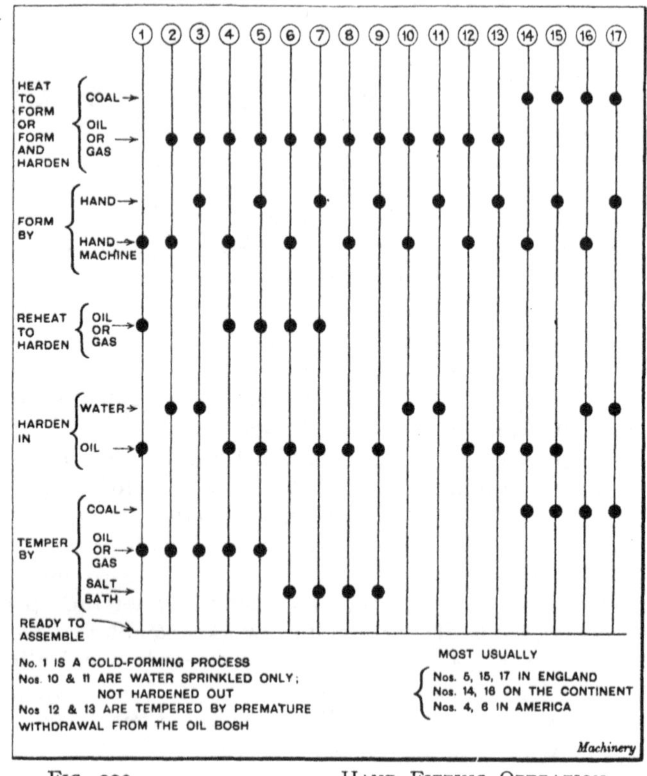

FIG. 203. HAND FITTING OPERATION.

hand-fitted practice in this country is the tendency which is not altogether the manufacturer's fault, of under-tempering oil-hardened springs (generally of the lighter varieties) whereby a very excellent Brinell reading can be obtained, to satisfy a specification, and a very good service will be given on perfect roads; but under shock, fractures, particularly with top plates, will result.

The various routing possibilities of "hand-fitting" are shown in Fig. 203, whereon is also given the more usual practices of this country, the Continent, and the U.S.A.

Some typical fitting shop layouts are shown in Fig. 204, of which A—B— and D are European, and C is American. Sketch A shows a coal-fired furnace feeding 12 fitting plates, the furnace dividing them into two equal sections, each with its own bosh. The main boshes are for water, and oil boshes

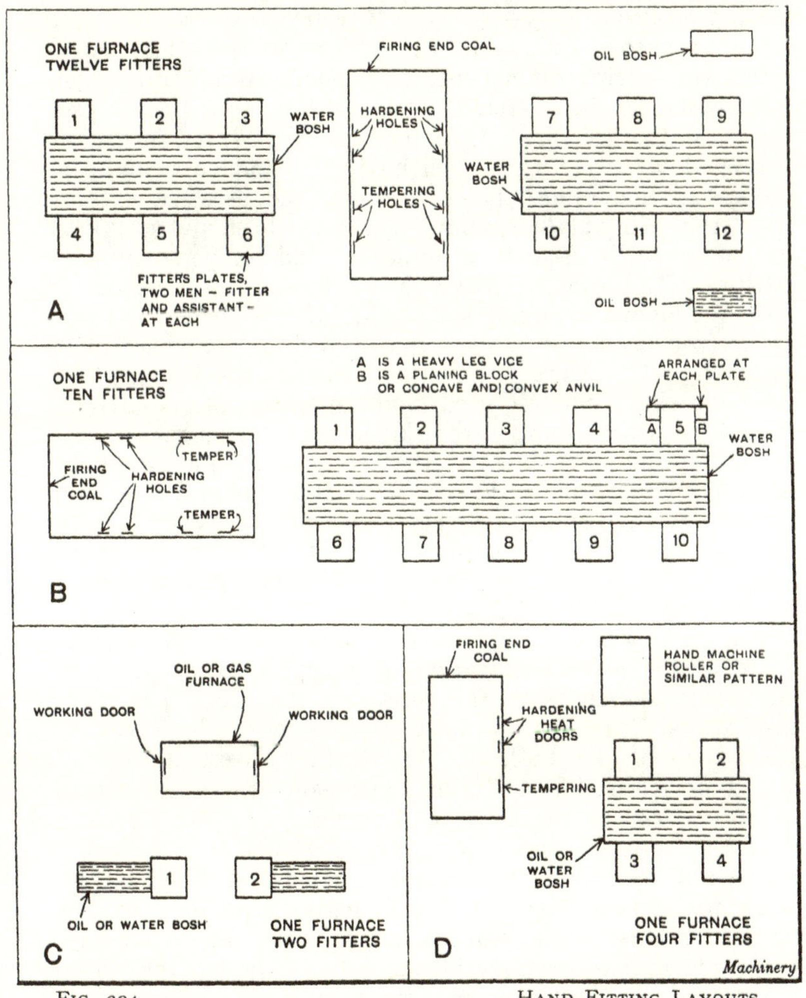

FIG. 204. HAND FITTING LAYOUTS.

as required are arranged behind as many fitters. Such oil tanks are generally water jacketed with facilities for replacing the hot water, but no general circulating cooling arrangements are fixed. Sketch B shows a similar layout, but with one furnace for 10 fitters, the furnace being at one extremity, and probably outside the main shop walls as regards its firing end. The disadvantage of both A and B is the distance which has to be traversed by the men with hot and heavy plates, particularly by the men in positions 1—4—9—12 in A, and positions 4—5—9—10 in B, which traverses can only be accomplished by passing the workers at the other plates. Otherwise, such arrangements are economical in construction and working. In Sketch C is shown the usual U.S.A. provision for hand fitting. One small oil- or gas-fired furnace is arranged for two fitters, each having an independent bosh. No special tempering " holes " are included, as when water is used, the spring plate is mounted over the bosh and sprinkled and when oil is used, it is immersed and tempered by premature withdrawal at above the oil flash point. Sketch D shows a layout including a machine of the former or roller types (which may be hand or power worked). The hot plate is taken to this, shaped, and then taken to the fitting plate, adjusted slightly as required, and hardened out. Generally speaking, in this country, the skilled fitter would rather take his hot spring plate direct to the fitting plate, and do all the shaping there, as he considers it takes little different time, and has not to be carried the greater distance necessary for the use of the machine. On the Continent, however, the roller machine is in general employment, and this type is also used to some degree in the U.S.A.

A good idea of hand-fitting practice is given by the frontispiece, which is an illustration of a typical Sheffield shop, similar to sketch B in the foregoing remarks. The coal-fired furnace (for the one bosh unit) is in the background. Each side of the bosh are five fitting plates, at each of which work the " fitter " and " viceman," the latter acting as under-hand to the fitter, and carrying the plates to and from the furnace. In the invisible foreground runs a narrow gauge railway, taking the springs when fitted direct to the scrag for testing purposes. Each fitting plate is provided on the right-hand side with a heavy leg vice, and on the left-hand side with a " peaning block." In the former the plates are curved, on the latter they are hammered to shape after tempering.

The fitter receives from the forging shop the necessary plates for the springs he has had instructions to make, and from the smiths' shop the solid end or other pattern back plates that are required. The free or light camber of the spring is sometimes specified, and in such case everything is relatively simple. Frequently, however, a loaded camber is given—when it devolves on the departmental office to estimate out the free camber necessary, which can be done with all practical accuracy by the methods previously detailed. Everything, however, turns on the spring steel rolling, and it is absolutely essential to check over, by load testing, the first spring of a new type, not with a view of seeing that the calculations agree with the spring, but with the view of ensuring that the spring agrees with the calculations, and if not, why not. Invariably this will be found to be in the thickness of the steel, or concavity which has been given.

Having the camber known, the fitter proceeds to " set " the back plate. This he does by marking out on his fitting plate a curve joining the " true camber " points. To this curve he makes a "crib " or pattern in some normal steel section, and then fits the back plate to this crib. It will be observed that the back plate as then bent is to the correct finally desired camber, with no allowance for the ultimate " nip " and " pull " which will be included in the finished spring. The object of this preliminary curving of the back plate is to ensure that the overall, or centre to centre length is correct with the desired finished camber, with the allowances made for the " sweep " of the plate according to the design of the solid ends and the spring generally. In numerous cases, owing to the design, the made spring is not the true arc of a circle, but an arc with a tangent each end, which point has been frequently emphasised.

Having now the back plate bent to the " crib," and with the confirmation of its accuracy thereby obtained, the fitter will proceed to " set " it at somewhat less than the specified free camber according to his judgment of the general fitting of the spring. Perhaps the finished camber desired is 4 in. on a 40-in. spring with twelve plates, ½ in. If the spring is normal, and the test is the standard (in this case 3½ ins.) the average fitter will set the back about 3¼ ins. camber. He will reckon this to give him a " pull " of about 1¼ ins., in other words, the fitted spring, when pulled up with the tongs, will be 3¼ ins. plus 1¼ ins. or 4½in. camber. It will be assumed that the spring will lose about $\frac{1}{4}$ in. under the test, which explains why the camber is arranged to be high before test. All these figures and allowances, are based on experience, and will vary with the ideas of different men. Admitting the constancy of the human element, variations have to be made in accordance with the composition of the steel which is being handled. Uniformity of quality is of the highest importance—steel being considered as practically uniform if the ranges of the extremes of carbon or manganese present (referring to ordinary water-hardening qualities) are within 0·05 per cent. The best practice, which is invariably

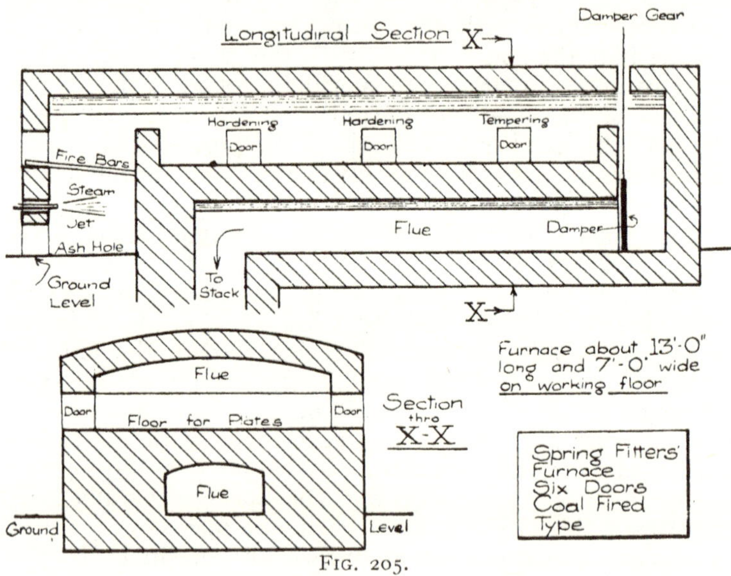

FIG. 205.

carried out in the Sheffield district, is to keep the steel bars from each cast of steel as a complete batch, and to work them out for definite lots of springs.

With the back plate set, and hardened and tempered, the fitting of the remainder of the plates can be proceeded with. It is usual in European hand fitting to work in pairs of springs, owing to the making up of the standard four-wheel wagon, which requires four side bearing, and two buffing springs. A set of bearing springs, or four, is generally given to the fitter as one job, likewise a set of buffing springs, or two. Engine and carriage work is given out in twos or fours according to the weights of the springs. The fitter, therefore, first sets his two or four backs, and then has two or four springs to be proceeded with simultaneously.

In British and Continental practice, the plates are generally heated in a coal-fired furnace, of simple construction, the ordinary Sheffield type, serving ten fitting plates, being as shown in Fig. 205. Oil has been employed here, and is a clean and·efficient fuel, but prohibitive in cost. It is likely that in the near future, producer gas will be used as a heating medium for this purpose, from a point of view of production

FIG. 206. "PINCHING-IN" SHEFFIELD.

economy, although coal has many special advantages from the men's point of view. The plates should be raised to a temperature of about 900° Centigrade—if this is exceeded there is a rapid rise in the rate of surface de-carburization, and if much below, there is a chance of the quenching taking place at too low a heat. Taken from the furnace at the correct temperature, the plate is "pinched-in" to the one which will be superimposed in the complete spring, and when fully curved, it is quenched out in the bosh, and then placed aside for tempering. This is afterwards effected by inserting the plate in one of the " cold " holes in the furnace, and raising it to about 400° C. This is a black heat, *i.e.*, a heat at which no colour shows, so the viceman judges his temperature by the " sparkle " method, consisting of rubbing the plate with an ash stick (generally an old hammer shaft) to see if the small charred fragments thus scraped off glow from the heat of the plate. If they do not, it is returned to the furnace until the desired effect is obtained. Different degrees of tempering are judged by the ease, or otherwise, with which the "sparkles" are obtained. The results got by such relatively crude methods are extremely good, in fact, one may say, with

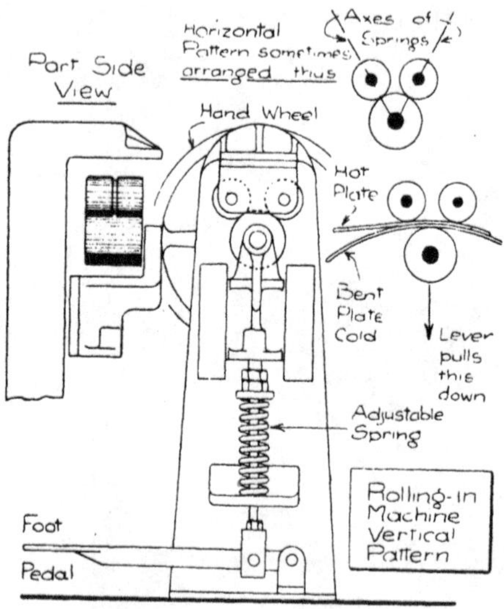

FIG. 207.

confidence, that the Sheffield-made railway spring is the best in the world, as regards treatment, workmanship and finish, and ultimate satisfaction in service.

An illustration of the process of " pinching-in " is shown in Fig. 206. One end of the plate is gripped in a heavy leg vice, whilst the other is drawn over and held down. After a rapid rough shaping, the two men work along each side with the " pinching-in tongs " which are specially made for the purpose, different sized tongs being used for different thickness of plates. In the foreground is seen a 14-plate buffing spring of the R.C.H. pattern (drawing in Chapter XVIII) showing the " nip." Actually it is more than would appear from the spring as it stands on the trestle, as the weight of the spring itself is reducing the apparent nip. The three top plates in the background, give a better idea of the real nip included. However, from the hand fitting aspect, one would call the spring shown a satisfactory job.

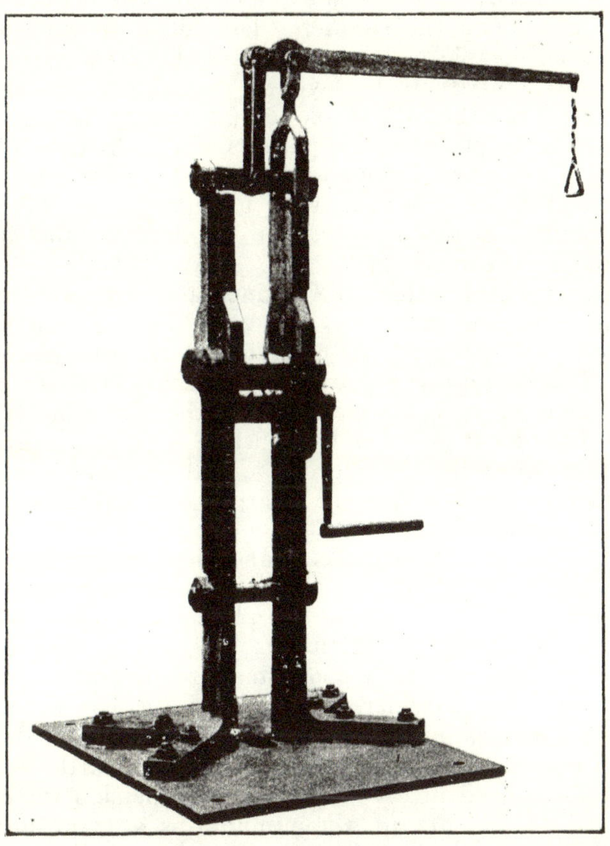

FIG. 208. PLATE ROLLING MACHINE. USINES DE BRAINE-LE-COMTE, BELGIUM.

To obviate the pinching in, various hand formers (screw-worked), and roller machines have been, and are employed. The roller type is the more usual, and is worked in both horizontal and vertical forms, the former being generally adapted for power operation, whereas the latter are usually manually operated. In Fig. 207 is shown a useful American pattern, of simple design. It is a mechanical adaptation of the hand process, the hot plate being placed on the top of the (ultimately) superimposed cold plate, and the pair together being worked through the triple rollers. Some of the power-operated patterns are fitted with arrangements which move the plates backwards and forwards a certain number of times, thereby improving the contact. The coiled spring gear is generally adjustable, so as to regulate the pressure on the plates. As pointed out, however, in this country the skilled fitter prefers handling the whole job (of curving, fitting, etc.) on conveniently placed fitting plates and attachments, to the working of a separate appliance for curving, as it is obvious that, useful as the rollers may be for eliminating the hand "pinching-in" they do not reduce the need for the fitting plate and the peaning block.

Another vertical pattern, of Continental design, and largely used there, is illustrated by Fig. 208. In this design, the top rollers are lifted for the insertion of the two plates by means of the hand lever. Any weights considered necessary can be hung on the bottom cross tie of the long roller frame.

One of the great points of hand fitting is to work at such a speed that, after the curving of the plate is completed, the temperature is approximately right for the hardening, and it can be quenched out in the oil or the water bosh so as to trap the steel structure present, and thus render the plate hard. The effect of oil-quenching, as it is very much less rapid than water, is to distort the plate to a smaller degree, and less adjustment or fitting is afterwards required to produce a spring up to a good standard of finish. Water quenching, being considerably more drastic, has the effect of warping or twisting the plate to a greater or lesser degree, in addition to altering the camber, and no little skill is employed by good fitters in the water immersion to reduce the plate movements to a minimum—as a great deal of control can be exercised in this direction. Springs which have a high ratio between length and width, say for instance, 2 ins. wide, and 50 ins. long, always give trouble with twisting, and accordingly designs should arrange for the length of a spring to be not more than 20 times the width.

When spring plates are oil quenched, tempering is frequently effected by removing them from the oil at a temperature such that the oil on the surfaces flares off. The assumption is then that the internal heat is sufficient to temper correctly the plate, due to its radiation through to the surfaces, then in air. For such assumption, it is obvious that regard should be paid to the thickness of the plates, as a temperature based on superficial oil ignition cannot give similar results on say, $\frac{1}{4}$ in. and $\frac{1}{2}$ in. thick material. As a matter of fact, with normal quenching oil, flashing at about 250° C. good results can be obtained on material $\frac{7}{16}$ in. and $\frac{1}{2}$ in. thick by this method, but below these thicknesses, the tendency is for the plates to remain too hard.

The most used tempering method with hand fitted springs is by the " stick," which should be employed for both water and oil-hardened steels. It is always used for the former, but the " flash " method more often for the latter. Best practice, however, uses the stick for both classes of work. Tempering heats for water-hardened steels are always non-luminous, and the practical method of judgment is by the " feel " or appearance of a wooden stick on the plate surface, which, as before indicated, whilst in appearance a crude method, is, in experienced hands, very reliable. The first indication obtained is a greasy feeling of the stick, as the friction of the plate reduces owing to its greater heat. The first symptom of this feeling is in the neighbourhood of 300° C. Through various grades of this the stick passes until sufficient. heat is present to cause a " smoke," which is at about 380° C. After the smoke a sparkle can be obtained with hard rubbing, and as the heat increases, numerous sparkles are obtained with light rubbing. Most water-hardening steels are tempered within this sparkle range, which is between 400° C. and 480° C. The next definite indication is the " flare " of the stick, which is at about 520° C. when the plate is just reaching luminosity, as judged in absence of external light. Oil-hardening steels should be tempered between this point and 550° C. according to material and requirements. The furnace used for tempering is invariably the same as that used for the hardening heats, the colder " holes " or door openings taking the plates for the tempering. The " stick " employed is generally an old hammer shaft, of ash or hickory. The men prefer the softer woods of birch or hazel, but these are not as a rule forthcoming for the purpose, and their standards of " smoke," " hard sparkle," " easy sparkle," " flare," etc., have to be based on available material. The charred point is used continually. It should be noted that resinous woods are quite unsuitable for gauging temperatures. By the "stick" method the good fitter can judge within 20°, which is a practical working range.

The salt-bath process of tempering spring plates is by no means as usual as the previous two methods. It is claimed that more uniform plate heating is obtained by the immersion of the hardened plates in a liquid maintained at the correct temperature, and with everything theoretically perfect, such claim would be justified, but it is possible to have salt baths with temperatures varying from point to point in the bath; and additionally, the time regulation through the bath is of vital importance. A usual mixture for tempering baths consists of nitrates of sodium and potassium, and such can be heated electrically, but external or internal gas or oil heating

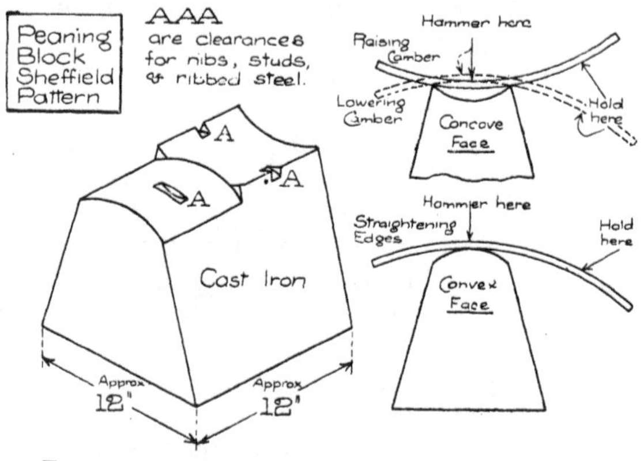

FIG. 209.

are usually preferred. This class of tempering is confined almost exclusively to springs for the light automobile trade.

The tempered spring plate, as removed from the furnace or bath, will be between 400° C. and 550° C. according to quality, and has then to be fitted, adjusted, or set, as required, to make a workmanlike job of the spring. This can be accomplished at these black heats by hammering on a special shaped anvil or "peaning block"—the Sheffield pattern of which is shown in Fig. 209. On this block, the fitter puts in the "nip" and corrects any warping which may have taken place owing to the quenching. Certain steels "run" or "warp" very badly in the water, without any particular sense of direction—two precisely similar plates of the same material may warp in different directions, due to subtle differences in plate temperatures and water temperatures. Oil quenching, being slower in action, does not tend to cause the steel to " run " to the same degree as does water quenching. This is an advantage as regards fitting, but nevertheless, any fitter would rather work on water-hardened steels than on oil-hardened steels, owing to the much higher cleanliness of the former quenching medium, and the absence of oil scale.

In Fig. No. 210 is shown a group of three fitting plates with the men at work, the first man being engaged in " pinching-

FIG. 210. SPRING FITTERS AT WORK, SHEFFIELD.

in " a special cramp bar being held over the two plates at the end. The second man is studying the form of two plates on his fitting plate, and the third man is using the peaning block. The corner of the large quenching bosh is seen by the first fitting plate.

The large water bosh in general use in this country is not one of the best patterns for perfect quenching, as only the top layer of water is used by the fitters, say, about 12 ins. deep. The boshes are frequently 6 ft. and 8 ft. deep, to allow for scale and a large bulk of water, but it will be seen that this depth is quite useless from the point of view of a large volume equalizing the temperature, as the heated water, due to continual immersion of plates, stays always at the top. As a result, towards the end of a day's work, this top layer is at a high temperature, and has frequently to be stirred and mixed with the colder bottom water in order to keep down the rising steam, which is objectionable, and also to effectively quench plates which are somewhat low in temperature. The attachment of some form of mechanical agitator to keep the water in constant circulation would be of advantage, and the deep bosh would then be amply justified.

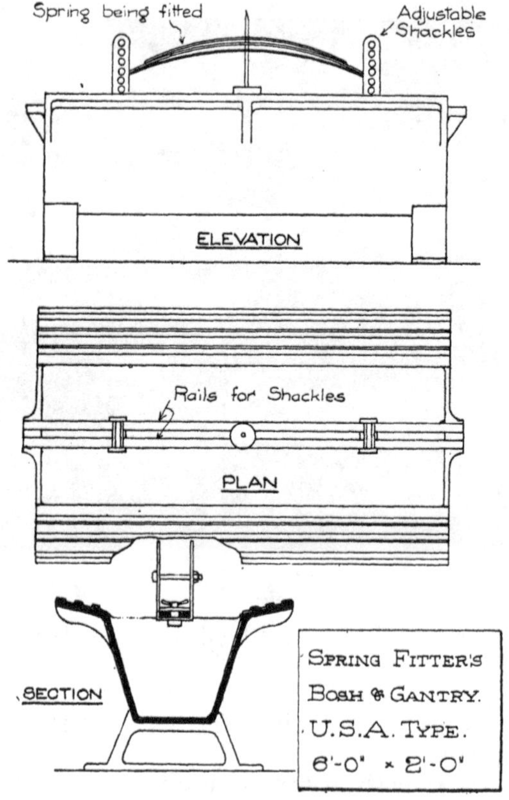

Fig. 211.

The division of the bosh in depth with an arrangement of readily removable plates, and a slow moving paddle wheel at one end, which might be worked off the make-up water supply would appear to constitute a practical arrangement.

In America, the usual type of bosh supplied is shown in Fig. 211. This has built on it the fitters' "horse" which carries the plates, and the hot plate is usually nipped to the cold one by several pairs of tongs, and then sprinkled with water, as before mentioned; or oil immersed and "flared." Each fitter is therefore an independent unit, as separate furnaces are also the rule.

It is very necessary to reduce to a minimum the need for "peaning" the plates, as every hammer blow is a potential point of breakage. With a plate requiring considerable adjustment, the blows may easily sink 1/100-in. into the surface, reducing thereby the distance of such point from the neutral axis, with a corresponding reduction in the effective width, thus increasing the skin stress on the remaining part. Unfortunately with many of the designs that have to be worked, considerable adjustment at the tempering heat is required, and the finished plates are badly hammer marked. A large proportion of this peaning is done to achieve a spring that shall be a very perfect job as regards closeness of plates, and touching of points, but the better spring would be one looking a little less perfect, but also being free from hammer marks. If the plates are of proper steel, correctly treated, and the design is reasonably sound, the load distribution will bring on all points and keep all plates tight. Bad design is responsible for most of the hammer adjustment required.

CHAPTER XXXIX

SPRING FITTING BY MACHINE

IN the manufacture of springs by the use of plate-forming machines, there are five general methods of treatment of the plate, as follows :—

1. (A) Heating the plate in an ordinary furnace. (B) Forming the plate, and permitting it to cool out. (C) Re-heating in a controlled furnace. (D) Quenching. (E) Tempering in a controlled furnace.
2. (A) Heating the plate in a controlled furnace. (B) Forming the plate and quenching it whilst held in the forming machine. (C) Tempering in a controlled furnace.
3. (A) Heating the plate in an ordinary furnace. (B) Forming the plate, and water sprinkling it whilst held between the dies, it thus being " hardened and tempered."
4. (A) Cold rolling the plate to curvature. (B) Heating in a controlled furnace. (C) Quenching. (D) Tempering in a controlled furnace.
5. (A) Heating the plate or plates in an ordinary furnace. (B) Forming under the machine. (C) Removing whilst hot, quenching in oil, and " blooming back " (tempering by the oil flare), or, with low-carbon or other steels, quenching in water, and tempering by the stick.

The first method is employed to a certain extent in the manufacture of automobile springs. It is claimed that by the controlled re-heating better results are obtained, as there is no chance of " soft spots " being present on the plate where the die fingers have been in contact. On the other hand, a double expense is involved for heating, and the first process, in an uncontrolled furnace, may overheat the steel and spoil it, particularly if a light forming machine is being worked on heavy plates.

The second method is so far only seriously employed in the U.S.A. in the high production plants, working exclusively on automobile springs, though experiments are proceeding here along these lines. As a general process, no objections can be raised, as mechanically, everything is as near perfection as is likely to be obtained for spring manufacture. It is, however, suitable only for great quantities to one standard pattern, and involves a very large capital expenditure. The

FIG. 212.—PLATE ROLLING MACHINE. A. G. VULKAN, VIENNA.

working cost per ton, for actual labour, is low—as a set-off, however, maintainance costs are high.

The third method of "spring-making" is not unusual for the lighter trade, such as springs for vehicles for animal traction. The cost of production is a minimum, and, as has been previously indicated, springs can be made quite satisfactorily from any class of steel—the service they give depends upon the design—and 0·80 carbon quality, bent cold, and used without any heat treatment, will be found excellent in service if the spring is designed to suit the elastic limit of the unhardened and untempered material.

The fourth method is employed to a limited extent for automobile springs in the U.S.A. Cold forming is, however, very general for very light plate spring work, and no treatment is afterwards given to the steel, a high carbon quality (0·80 to 1·00) being generally employed.

The fifth method is in general employment in the U.S.A. for railway spring work, and to a small extent in this country.

FIG. 213.—PLATE FORMING PRESS. ENGEL & BIERMEYER, HAGEN, GERMANY.

A very large number of machine designs for the purpose of forming and quenching, or forming only, the leaves of plate springs, are either working, scrapped, or dormant. A few will be briefly dealt with, the first illustration, Fig. 212, showing a power driven three-roller machine, not dissimilar to those employed as auxiliary to the hand fitter. The lower two rolls are driven through gear, the upper roll running free, and being adjusted for pressure by the large hand wheel. This upper roll is readily removable to allow of the curving of back plates with enlarged ends (rolled eyes, etc.). The rolls can be finely adjusted for curvature—which is set by means of a cold template bent as desired. This machine is used in Continental works, on water-hardening steels for railway springs—the plate being quenched straight from the rolls, and tempered in the ordinary way. A similar machine, of small pattern, is employed for the cold forming of plates, Method 4. The disadvantage of rolls for either hot or cold forming lies in the fact that it is necessary to carefully gauge the steel for thickness to ensure anything like uniform results. Clearly, if the rolls are set for maximum thickness, plates of lesser thicknesses will have too little curvature, and are set for the nominal thickness, thin plates are still inaccurate, and thick plates may break the machine. Very good results, are, however, claimed in the States for cold forming—which undoubtedly has its useful points.

In Fig. 213 is illustrated a forming press free from these disadvantages, but suitable for hot work only. This is a very usual type of Continental press, and is employed there for a large variety of purposes, one chief attraction being that it has a negative drive, and another that whatever excess of stock is present, whether in the form of thick spring plates, or long bolt blanks, nothing can be damaged owing to the non-fixed stroke. Dies as required are fixed to the bed and moving head, and the foot-pedal brings down the screw and ram. When pressure is released, the ram returns automatically, and knocks off when it reaches some predetermined stop. The plate is taken out after forming, and quenched and tempered in the usual way.

A forming machine of the general pattern illustrated in Fig. 214 (A and B), has been the most successful which has been employed in this country in the manufacture of springs of which quantities are desired, but has been used exclusively by certain railway spring shops for the manufacture of

" standard " bearing and buffing springs for English wagons. Both types A and B have the same general construction as regards the dies, which are bent spring steel plates of heavy section, strutted and finely adjusted by numerous set screws, etc. The top die is actuated by a power cylinder, and in its

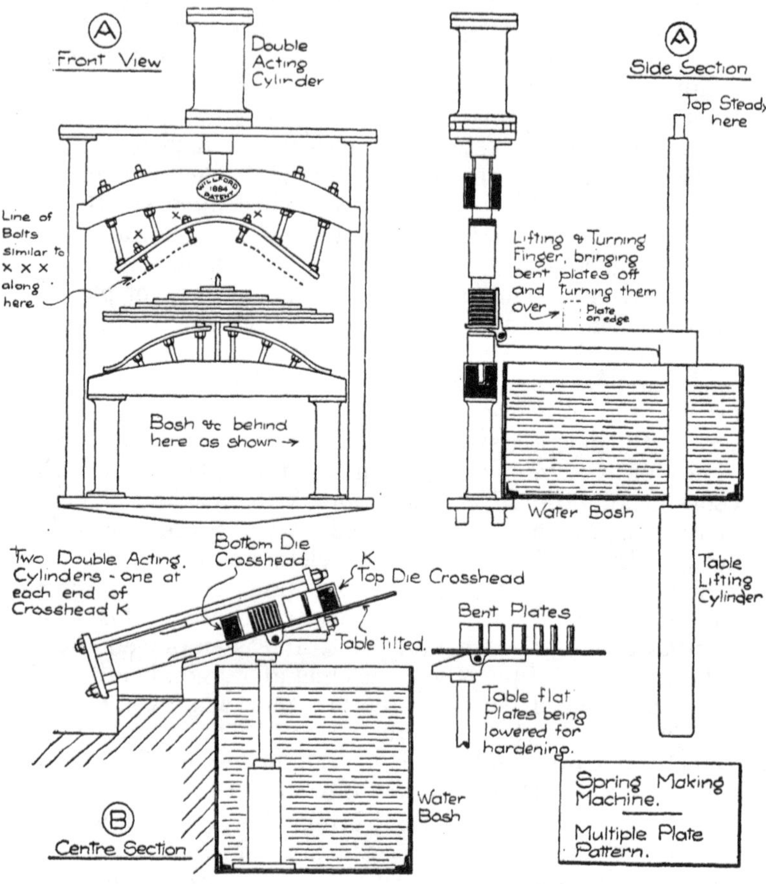

FIG. 214.

movement, the whole of the plates constituting the spring are bent to shape. In pattern A, the forming machine is vertical, and a special lifting table and equipment is provided in the bosh immediately adjacent, which lifts the formed spring off the machine, and turns it over on the table, so that the plates stand edgewise. They are then spread out, and the table and plates lowered into the quenching bosh. Pattern B is of later date; and is an ingenious improvement on the original, being, as shown, of simpler construction. The machine itself is fixed on the incline, and the table arranged to work up directly under the formed spring—being on the incline whilst the spring is being pressed, and falling into a horizontal plane when travelling downwards into the bosh, by means of the special attachment shown in the sketch.

An obvious disadvantage of any multiple plate machine, such as the above, is in the fact that the long plates are cooling whilst the others are being placed in position. A wide range

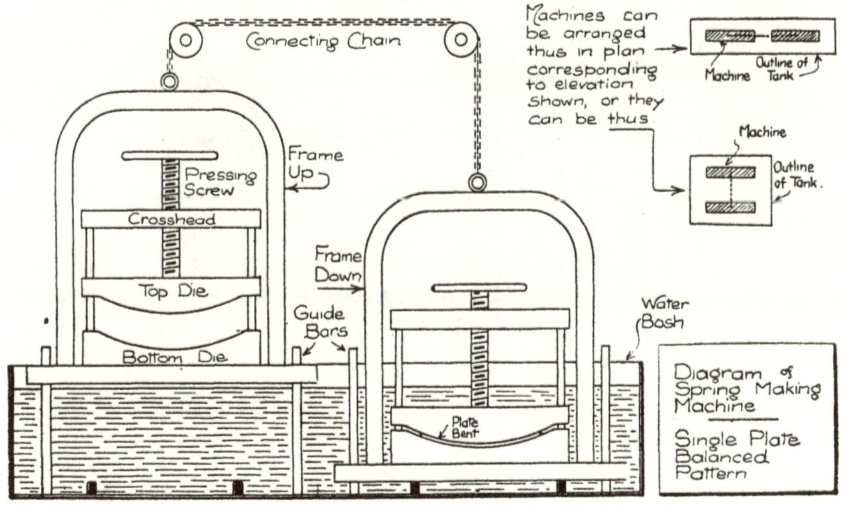

FIG. 215.

of temperature is therefore present when the spring is formed, and it probably goes into the hardening bosh with the back plate at 700°, and the short plate at 900°. The influence of the water causes, accordingly, great variations in the cold shapes of the plates, resulting in considerably more fitting being required than is wanted for similar springs hand-made throughout.

This particular machine was patented, made, and used in Sheffield, and remains the survivor of numerous designs which were patented about the same period (1870 to 1880). Most of these were multiple plate machines, and one was a balanced type—Fig. 215—one side being, with the hot spring, in the bosh, whilst the other was being loaded with plates. It is noteworthy that some of the latest American machines embody this principle.

It is, however, clear that to make any success of machine fitting, the process must handle one plate only at a time, and preferably quench off the same whilst held between the forming dies. Few machines of this type are working in the railway trade, but they are fairly common, and doing good work, in the automobile industry—particularly in the U.S.A. One argument against machines in general is the length of time taken to set up the dies. This is of no moment if orders are common for thousands of one type of spring, but where

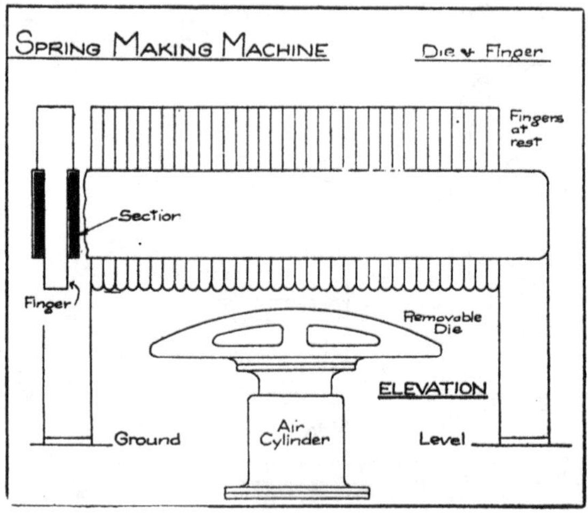

Fig. 216.

the average Sheffield shop gets one order for a thousand, it will obtain scores of orders for one hundred or less. Even in the U.S.A. where fewer designs of springs prevail, machine forming is not universal, hand fitters being retained in many of the machine shops to deal with orders for oddment lines.

Typical American spring machines are shown in following illustrations. Fig. 216 shows a pattern with a vertically moving bottom die block attached to the frame. The top former, normal position as shown, consists of a number of round-ended fingers, free to slide vertically, and taking up the position with all finger ends along one horizontal line due to gravity—the top portion of the fingers coming to rest on the cross stay. The hot plate is placed on the bottom die block, and power applied thereby raising it into contact with the top fingers, which move according to the die shape, and form the plate. When formed, it is hardened off, etc., in the usual way. Solid end back plates are usually formed by hand with this type of machine, but with the American plain back, it is obvious that every plate of a spring can be machine-formed—but all to one radius, virtually without nip. As little fitting is done, it will be appreciated that the ideal spring is being attained, and working in a circle, it is clear that the adherence to the (correct) short " short plate " by American manufacturers is as much for the ease of manufacture as for

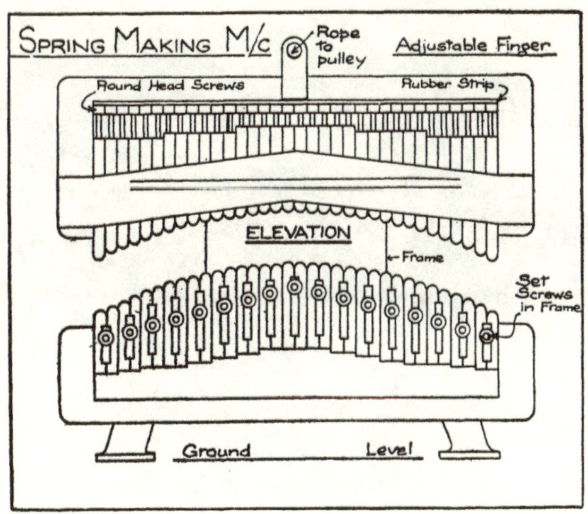

Fig. 217.

any theoretical properties possessed by such design. A certain amount of nip, just sufficient to ensure " points " being on, can be arranged by varying within limits the forming and quenching temperatures of the plate. This machine is a good type for ordinary railway shops, as one fixed die only is required, which is readily put into position, and it would pay to make such die for say, lots of 100 and upwards. A stock of dies has, of course, to be held, and these accumulate, but they are neither expensive or delicate.

Fig. 217 shows another pattern, with adjustable fingers each side of the plate. The sketch shows a vertical machine, but a similar principle is used in horizontal machines. In this case, the bottom frame is stationary, and the top balanced to stand normally as shown. It can be unbalanced and brought down by either the action of a friction pulley, or a power cylinder. The disadvantage of this type is that it only pays to set up for considerable quantities, say lots of 500 or over, as a plate has to be formed of the correct sweep, and then the top and bottom fingers set to this—which, as will be perceived, entails a considerable amount of adjustment. An advantage is that no loose dies have to be carried in stock.

FIG. 218. PLATE FORMER, FINGER TYPE. COULTER & MCKENZIE, BRIDGEPORT, U.S.A.

The makers' description of the use of this type of machine is as follows: "For bending or forming hot plates to shape desired for elliptic springs, leaving same in the machine under pressure long enough to set, using a spray of water or air to assist in cooling or setting to shape. The plates are then taken out of the machine, and tempered." A horizontal machine along the same general lines is shown in Fig. 218, which can be arranged for operation by belt, air cylinders, or hand, as required.

In Fig. 219 is illustrated another pattern, made with a horizontal table. In this, a cold back plate is used as the first die, and the arrangement of the crosshead permits of a slightly decreased radius being introduced so as to ensure the formed hot plates coming to correct camber when cold. The hot plate is placed on the back of this, and the air cylinder then operated to push it on to the chain die, the tension in which is maintained by adjustable springs. This plate is then hardened and flared off (" bloomed back ") in the oil, and returned to the table, to act as a die for the succeeding plate. The process goes on until the whole spring has been built up on the table. Alternatively, if considered preferable,

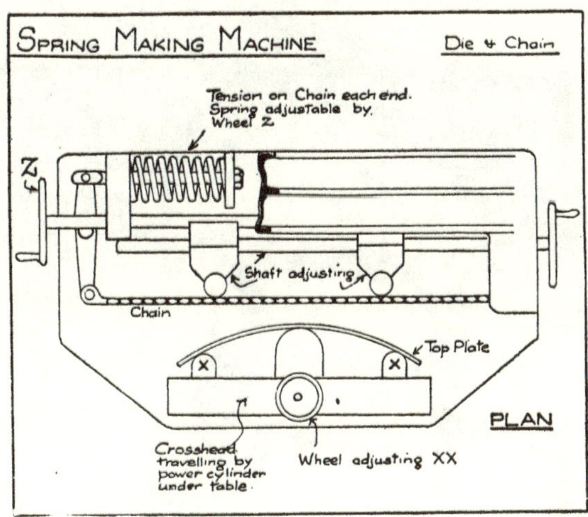

FIG. 219.

single plates can be worked through. Bearing heads are provided at the back of the chain to render it suitable for great variations in plate lengths. A photograph of this machine is shown in Fig. 220.

The illustration, Fig. 221, shows a balanced pattern of machine, arranged to drive from an overhead shaft. The throwing of a clutch causes the lowering of the one die table, and the simultaneous raising of the others ; the first having on it at the instant of the lowering movement commencing, a pair of open dies with the hot plate between ; and the second at the same instant, being oil immersed, with a formed and cold plate between the closed dies. As the first table lowers, the moving die is made to close up and form the plate by the action of the " stroke-wedges " shown. These can be removed and others substituted for stroke if necessary. The lowering continues until the whole table, with its dies and plates, are immersed. The machine stands, as shown, in a large oil tank, and it is a very efficient pattern ; one set, of two men, working both sides, from one furnace.

The machine shown in Fig. 222 is of more complicated pattern, the bottom die being a series of stressed wires—the tension on these being by means of the coil spring adjustment. These stand clear of the oil tank, and the hot plate is

FIG. 220. PLATE FORMER, CHAIN TYPE. J. T. RYERSON & SON, CHICAGO.

placed on them. The top die then comes down, being worked by an electrically driven screw, air, or other means, curves the plate against the wire, and continues its stroke by forcing down the whole of the bottom die frame into the bosh. Some machines of this pattern are fitted with electric timing devices, so that after a certain pre-determined period of so many seconds, they automatically return the wire die, carrying with it the curved plate. The main point made in favour of the timing device is that separate tempering can be avoided. This is very doubtful indeed—that is, if uniformly satisfactory springs are to be obtained. The idea is to set the timing so that the plate is brought out of the oil just sufficiently hot to

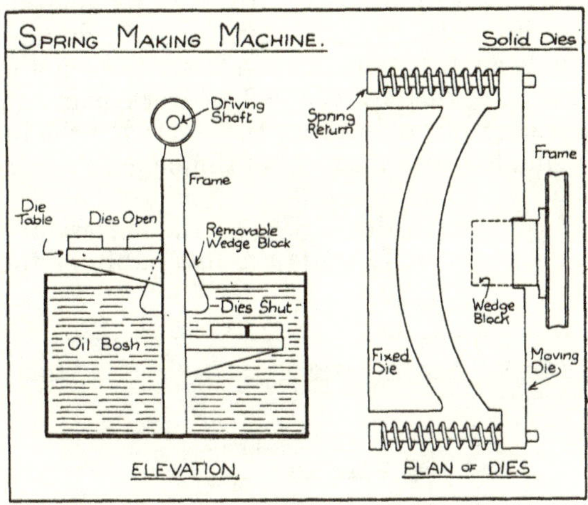

FIG. 221.

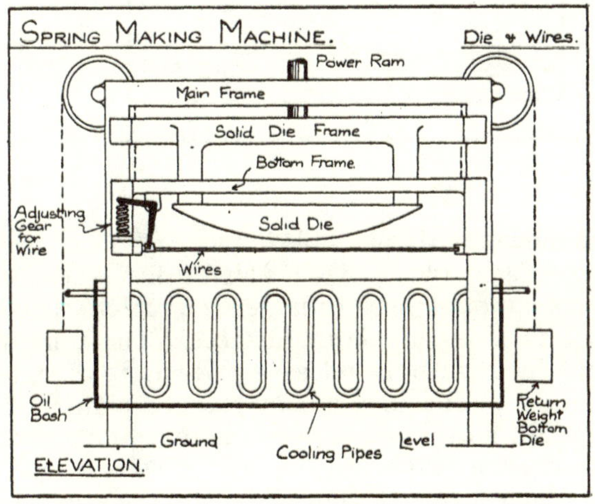

FIG. 222.

flare on its surfaces. As might be imagined, so many things are in combination to produce this temperaure, that this suggested method cannot be regarded as reliable—and it is not extensively used. It is, however, quite a sound proposition to provide an automatic time release for such a machine, as the operator is thereby enabled to work many machines without needing to trouble about the returns of the immersed plates. A simple hydraulic form of timing gear will be shown later.

The machine used in the U.S.A. on the highest production plants for the automobile trade are invariably of the rotary

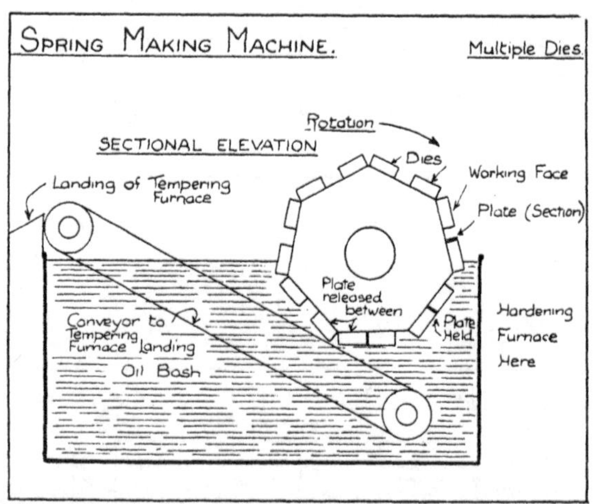

Fig. 223.

pattern, as shown by the example in Fig. 223, which is a seven-sided type. Each side carries a pair of dies, and the whole machine runs in an oil bosh of large size, replete with cooling arrangements and oil circulating devices. This tank includes a conveyor which takes the released plate up to the tempering furnace landing. Unskilled labour places a plate in the dies, presses a pedal, and dies close on, the machine rotates itself and carries the hot plate into the oil. A second face is then presented, which is given its hot plate, and the same operation occurs. At the proper point, the cold plate releases, drops into the conveyor, and is taken thereby to the top of the tank. This pattern of machine has also been worked to a certain extent in this country, but by no means on the same scale of machine design, the English types having been more or less experimental, and of cheap construction. For maximum production, no quicker method can be devised, as two men are kept continuously busy feeding the dies.

Another principle of operation is shown in the drawing Fig. 224. In this arrangement a reversing solid die and wheel are used as the forming device. The roller itself is of sufficient weight to press the plate to the die shape without external means, the whole of the power operation being on the die traverse. A clever feature in this arrangement, of great advantage, is the travelling of the plate through the

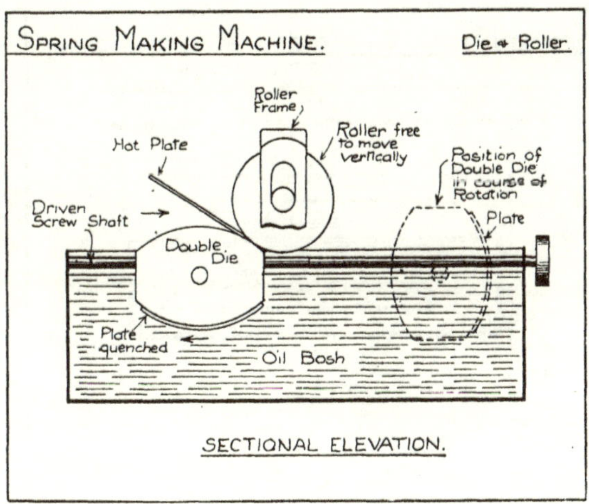

FIG. 224.

oil whilst held to the die. One awkward feature of all forming and immersing machines, is the fact that the hot plate is held stationary, between more or less solid dies, in more or less stationary oil—and the result of this, particularly pronounced with some patterns, is that the local oil is far too hot to harden the plate, and it remains accordingly soft. In the machine under consideration, the travelling of the die causes a certain oil circulation, and another good point is the fact that only one die is contacting during cooling. Whilst the die is traversing with a formed plate cooling, another plate is being shaped on its top side.

In Fig. 225 is shown a machine pattern which has been tried in this country, and, in contradistinction to all the foregoing, carried out the quenching by means of water. The top die consists of a flat plate, and the bottom die of a hardened and tempered spring plate, or series of spring plates, as required by the camber needed for the spring to be made. The steam-operated cylinder pushes down the top die on to the hot plate, and continues by thrusting both dies and plate into the water bosh. Simultaneously, water is allowed to flow into the bottom of the bosh, and travel upwards to a definite limit. The cutting off of steam from the cylinder causes the top die to return by balance weights, and the

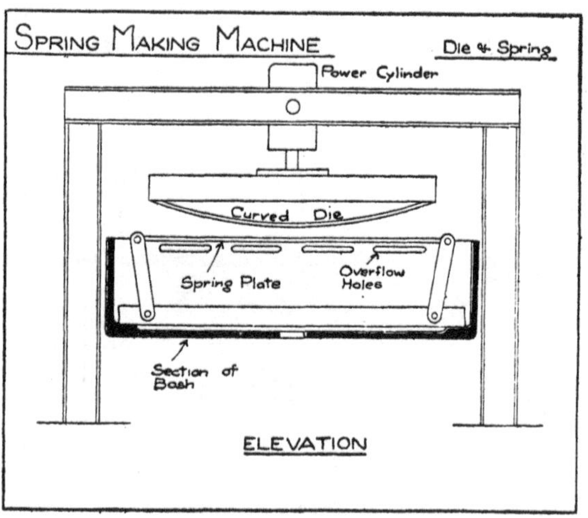

Fig. 225.

bottom die returns to its original straight shape, carrying with it the curved and hardened plate. The water at the same time, is withdrawn from the bosh. To provide for the operation by one man of a number of these machines, a timing gear is arranged of the design shown in the diagram Fig. 226. As will be seen, the device is of great simplicity, and works very well. An adjustment of the stop valve either cuts the gear out, or permits timings of a minimum of 10 seconds, which is too quick for any ordinary working. In practice, from 25 to 60 seconds is allowed, according to the weight of the plate.

This pattern of machine has its limitations when used with an unaided spring die, as very deep cambers cannot be worked thereon—no spring die being capable of taking the deflection and returning to straight. Solid dies can, however, be equally efficiently employed, with either a very thin top spring plate as a return die, or alternatively, wires or chains held at the ends by some tension device.

Fig. No. 227 shows another American machine, somewhat along the lines of the one illustrated in Fig. 222 but working

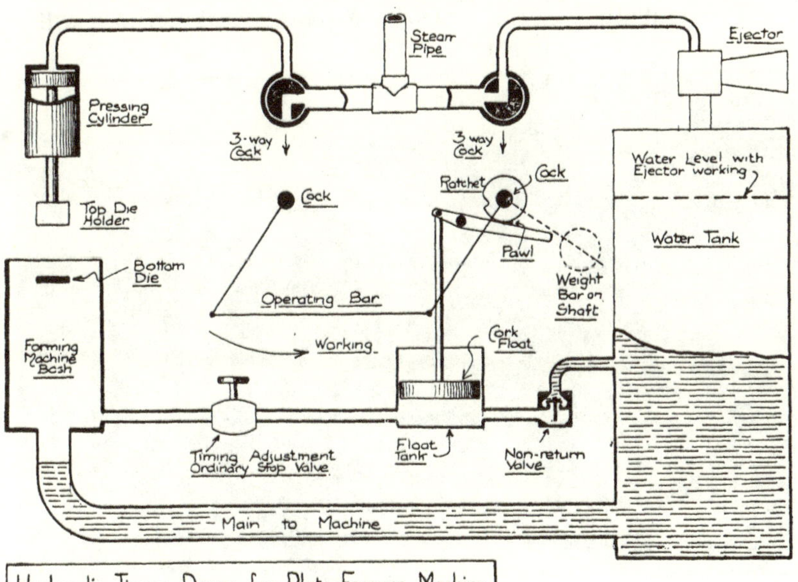

FIG. 226.

with solid dies, inverted. All patterns of machine employing solid dies have the advantage that when short plates are arrived at, two or more can be set up to be formed at the one stroke. This is not, of course, practically possible with machines employing one flexible die. It could be done, by placing a series along the flexible die, but the result would be most unsatisfactory. This photograph also shows, on the right, the chain mail heat guard at the furnace opening, and the hoods nearly always fitted in the States to clear hot air and gases out of the shop.

The machines chiefly used in the U.S.A. for railway work are those of the " solid die and finger " type, Fig. 216, and of the " die and chain " type, Fig. 219. The other patterns illustrated are limited to the automobile trade. Most " coach springs " are made on vertical finger machines, Fig. 217. Machines of the solid die or adjustable finger types are used for many spring shapes apart from the orthodox semi-elliptic. Double sweep springs can be easily formed, as can most of the special types needed for the motorcycle and horsed vehicle trade.

A careful study of the foregoing machines and methods will

FIG. 227. MACHINE FORMING, U.S.A.

indicate the difficulty, if not impossibility, of working " drawn point " plates, and the machine forming methods of the U.S.A. have been the chief factor in damning drawn point designs for the high production methods. There are existing schemes including small eccentric rollers, and similar devices, for holding drawn points up to the top forming dies, but they are not regarded as an unqualified success, chiefly because more time is occupied for setting to properly manipulate a type of spring which has no real practical value, and as every small item of this nature adds substantially to cost when mass production of automobiles is being considered, it is cut out.

No rolling device—such as the cold rolling—can form a drawn point plate ; the finger type machine can do it with more or less success by adjusting for every plate, or using packing pieces ; moving rollers or packs are required for solid dies ; and indeed every type of machine fitting includes some objection. One reason for the indifferent success obtained in this country and the Continent with machine-made springs for the automobile trade is due to the fact that it has been attempted to use U.S.A. machines to make English and Continental spring designs—English designs including drawn points on nearly all light and heavy springs as an article of faith, and Continental practice employing them to a large extent for the light (pleasure car) springs.

Most spring shops in the U.S.A. retain a certain number of skilled fitters, for rectification if the complete machine is used (forming and immersing), and for the general plate treatment and fitting for " forming only " processes. Owing to good design, and oil quenching, fitting is at a minimum with the latter methods. On the high production machinery, which quenches the plates in the dies, comparatively little after-rectification is required, in some plants not more than 5 per cent. of springs going into the fitter's hands, and the maximum rarely exceeding 10 per cent.

There is always a tendency for a certain number of points to be " off " with one-radius springs. This is a matter of no importance with machine work, particularly when it has been through controlled furnaces, as all these points will come on with a relatively light load—depending upon the strength of the spring. In light springs, points that may appear off when the spring is lying flat on the fitting plate, come on merely by the general spring weight when it is placed in running position. Nevertheless, the practice of employing a sufficiency of rebound clips considerably assists in bringing any such points properly " on " when the spring is unloaded, and apart from other virtues of such clips, their value in this direction is not to be overlooked.

In machine fitting which includes immersion of the plate whilst held, it is obvious that the design of the dies occupies a position of the first importance. Solid dies, cast iron, with teeth, are used to a certain extent—but are relatively costly to make and fit up. For quantity production, however, this aspect is negligible—but such dies would clearly be of no value for the European trade. One disadvantage of all dies of this pattern—which are equivalent to the finger dies of Figs. 216 and 217—is the liability of obtaining soft spots under the die contacts. Continuous dies are better if they can be arranged, but no little difficulty presents itself with obtaining these of satisfactory design. Channelled dies are a good form, and obviate soft spots, if proper circulation of the quenching fluid can be obtained through them. With the water quenching process, it is possible to obtain hard plates up to a certain section by permitting the water to contact only with the edges of the steel. This has the disadvantage

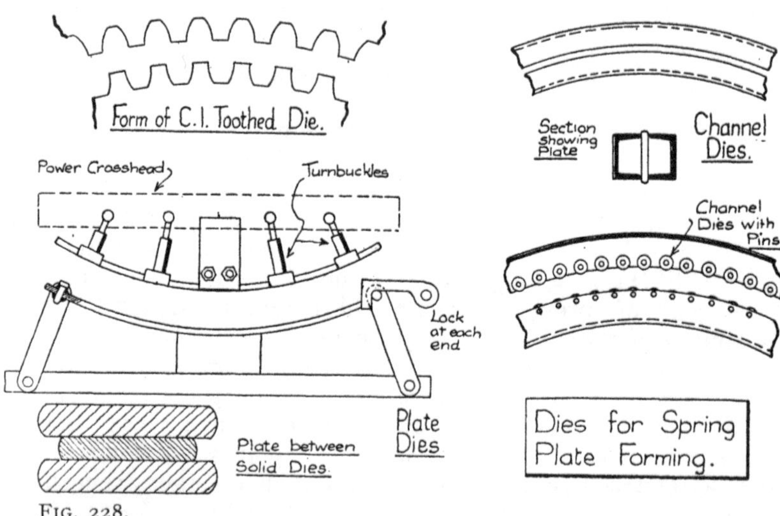

FIG. 228.

that the scale is not swept off, but the results obtained are satisfactory. One difficulty with water quenching on plates held between dies is the fact that the rapid contraction of the steel tends to produce cracks, and careful study of the die design is necessary to obviate this. No trouble with this is experienced, however, if side contact only is resorted to. Some used forms of dies are shown in Fig. 228, and include the toothed pattern, the quite solid die for edge quenching in water, the channelled type, etc. A very necessary item to be borne in mind if machine work is intended to be resorted to by manufacturers on this side, is the necessity for the employment of dies cheap to make, and easy to fix. The fixing device for the edge-contact dies referred to is shown in Fig.

228. The perfect die form is that which would give a continuous contact of the fluid over both surfaces of the plate, the plate meanwhile being firmly held. Obviously, however,

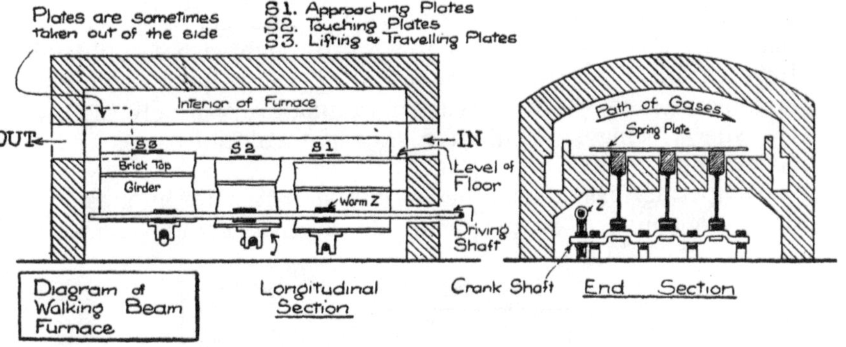

FIG. 229.

such form is abstract—and probably a large amount of experimenting will be found necessary to determine the design which satisfactorily fulfils all the conditions necessary, of obtaining a perfect quench, a simple form, and an easy fixing.

If the desired output of springs justifies the expenditure, the best form of heating furnace to employ is shown in Fig. 229. This is known as a " walking beam " furnace, and it will be seen from the diagram that inserted as part of the floor surface are three continuous beams (firebricked on top)

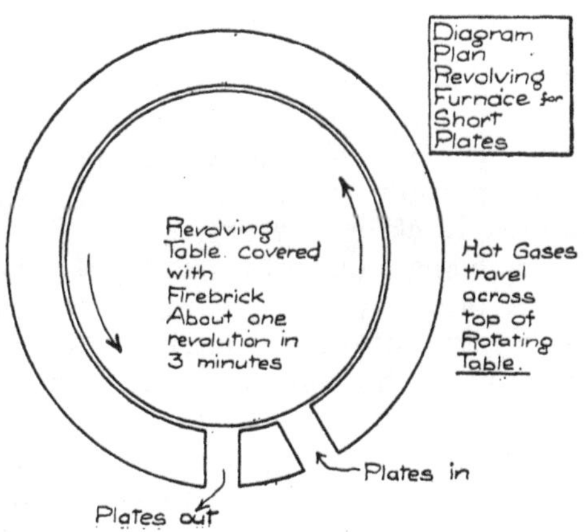

FIG. 230.

rising, falling and travelling longitudinally according to the movement of crank shafts, of small throw, underneath. These are generally driven by gearing from a motor-worked shaft running the length of the furnace. This furnace is an expensive unit, but capable of very high production. Short plates are obviously not easy to work through with certainty by such an arrangement, and for these a revolving furnace is sometimes used, as shown diagrammatically in Fig. 230, the speed being so graduated that a revolution gives the correct heat.

Tempering furnaces can be worked mechanically by means of known conveyor arrangements, as these will stand the heat

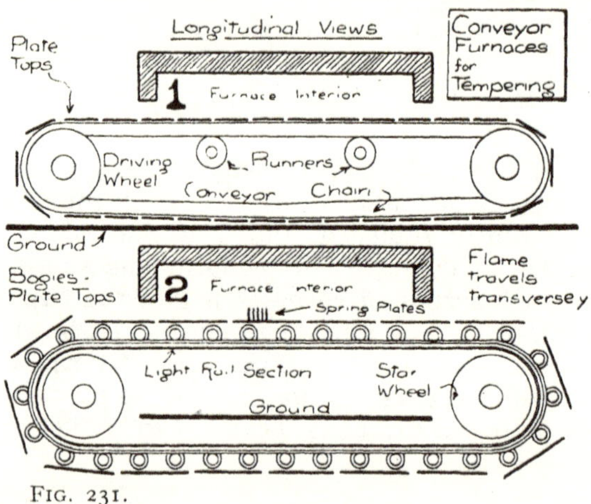

FIG. 231.

of 350° to 550° usually required for this operation. Fig. 231 (1 and 2) shows two furnaces of this type. No. 1 is provided with a link conveyor, which requires intermediate runners to keep horizontal the top portion of the chain. No. 2 is really an endless railway track in a vertical plane, the moving floor running thereon being numerous small four-wheel flat top bogies. This is probably a more expensive pattern than No. 1, but is a very substantial working unit. Tempering furnaces worked as continuous, by such arrangements, have naturally to be of considerable length, as both ends are more or less open, and the hot zone, of desired temperature, has therefore to be in the middle of the furnace length.

For all these furnaces with mechanically travelled floors, gas or oil forms the ideal fuel, owing to its ease of application with numerous burners, along the sides of the furnace. As would therefore be expected the development of these furnaces has received more attention in America than here or on the Continent.

It is more important accurately to control electrically the tempering heat than the hardening heat, and on large furnaces for tempering purposes, several pyrometers should be fitted at different parts of the hot zone to ensure uniformity and limitation of temperature. All these mechanical furnaces, electric controls, etc., entail largely increased capital and maintenance costs, and their existence is therefore only justified by correspondingly high outputs. On the other side, the cost of production per ton is decreased, and the question as to their installation or otherwise resolves itself into one of economics, as the resulting products are not apparently superior to the Sheffield hand-made spring.

Most American high-production plants in the automobile trade include control stations, with various coloured electric lights over each furnace. The control operator has a set schedule of timing for the working furnaces, and plugs in pyrometers at fixed intervals. Generally, one lamp glows continuously, and is white for " alright," otherwise, when the furnace is within the fixed limits of operation of 20° or thereabouts ; red for " too hot " ; and blue for " too cold." Other plants have an automatic control throughout, a rise in temperature cutting off heat, and a fall increasing it—but heat controls of this nature are obviously expensive items. With ordinary methods of control as employed in the States it is claimed that temperatures can be maintained within plus or minus 2 degrees (Fahrenheit) which is certainly a very creditable performance. It cannot be pretended that anything approaching this is so far worked over on this side, and from all practical points of view, probably plus or minus 10 dergees (Centigrade) is by no means an unsatisfactory standard to aim at and achieve—on the spring plates, and not only on the pyrometers. Various machine and hand processes now in use could be arranged to have the plates tempered in bulk, in a controlled furnace, were it not for the difficulties of the (economic) working methods involved.

A plan diagram of one of the latest U.S.A. plants for automobile spring production is shown in Fig. 232. Four units have been installed, and will give a total production of 500 to 600 tons per week. A brief description, keyed to the drawing, is as follows :—

Automatic Furnaces.
Main (Cold) Air Line
Preheater
Hot Air Line
Hot Air Line
Blower
End View
Oil Supply & Burner
Plan
6 ft. × 16 ft.
Hot Air Line
Hot Air Line
Main (Cold) Air Line
Motor Driven Air Blower
Preheaters for Air Blast, working with Waste Furnace Heat
Down Pipes to Burners
1
2
Holes
Plate
Plate
3
3
4
4
5
5
6
6
7
8
Squares thus are fixed positions of workers
155 feet overall
10
10
9
25'-0" between expanded metal fences
A
B
C
D
11
12
12
E
F
Diagram Plan of One Complete Unit.
Furnace 2 6'-0" × 16'-0'
8,
Conveyor 11, 7'-0' × 55'-0'
Machines near Conveyor 11
A Press inserting Bushes
B Side Grinder for Eyes
C Rivetter for Clips
D Reamer for Bushes
E Testing Machine
F Testing Jig
High Production Plant
Automobile Springs
American Autoparts Company, DETROIT.

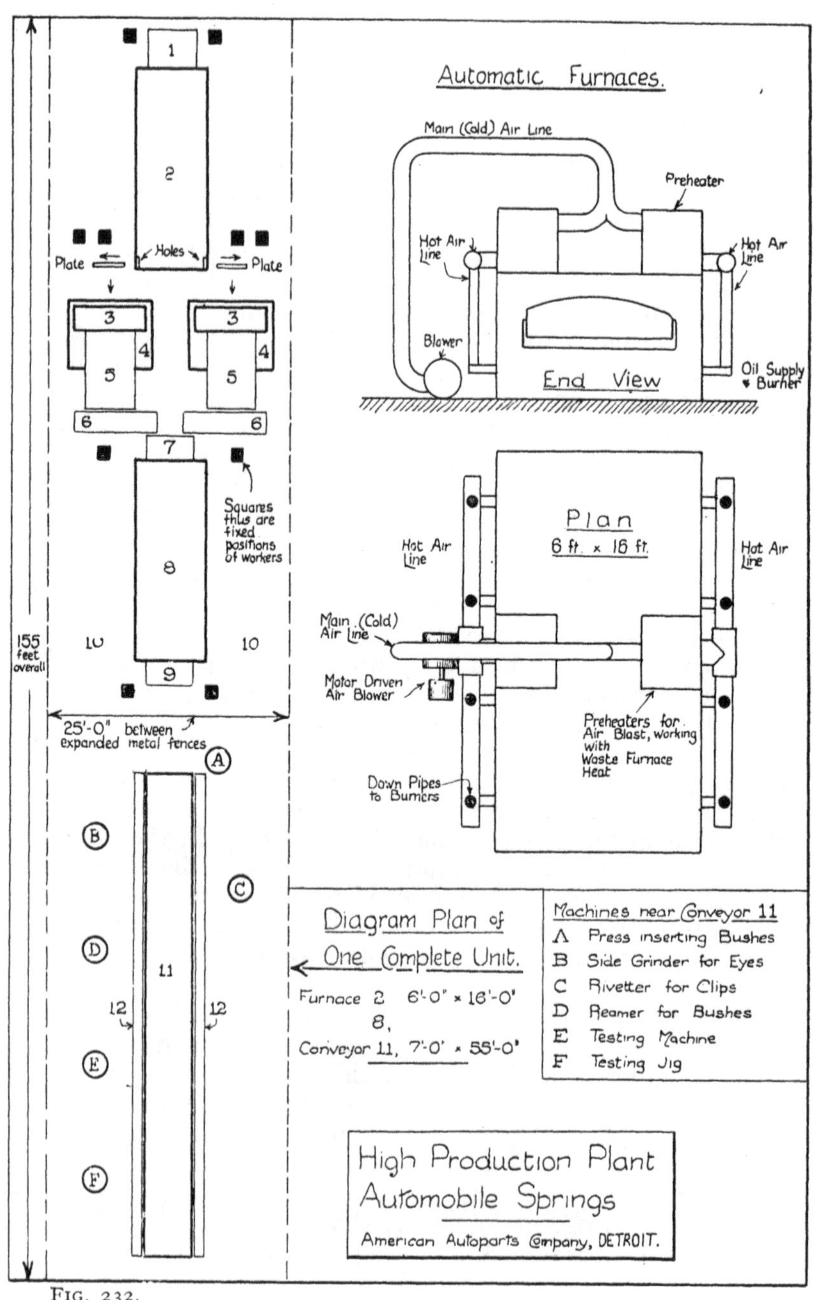

FIG. 232.

1. Is the charging plate, being an extension outside the furnace of the moving floor.
2. Is the hardening furnace, of the walking-beam type, inside dimensions 6 ft. 0 ins. × 16 ft. 0 ins. A plate takes 25 minutes to travel through, and the heat is so arranged that the first 15 to 20 minutes are employed in heating it up, and the remaining time for soaking. The hot end is closed, and the plates withdrawn from the sides.
3. Is the multiple die forming machine, of the revolving pattern (Fig. 223).
4. Is the oil tank in which the forming machine rotates.
5. Is the conveyor from the tank to
6. Which is the top landing of the conveyor.
7. Men stationed here take the plates from the tank conveyor landing and place them on the tempering furnace conveyor, this point being the conveyor prolongation external to the furnace.
8. Is the tempering furnace, 6 ft. 0 ins. × 16 ft. 0 ins. inside.
9. Is the finished landing from this furnace, from where the plates are stacked in the spaces
10. Which are the general stacking grounds adjacent to the final assembly.
11. Is the assembling conveyor, 7 ft. 0 ins. wide = 55 ft. 0 ins. long, running at just such a satisfactory speed as will keep the workmen busy.
12. Are the two stationary plates, guarding the men from the moving platform.

The various finishing machines pertaining to this assembling unit are noted on the drawing.

A special view is given of the furnaces, both hardening and tempering units being of the same general furnace construction but with the different moving floorings noted. Oil is used as a fuel. Necessarily, to keep running such large units as this entails very large orders being obtained, with a minimum of types; and, this condition fulfilled, the plant illustrated represents certainly the last word in modern spring making.

Automatic Spring Forming Machine.
Elevation
Plan View.
Plates travel always
Ground Level.

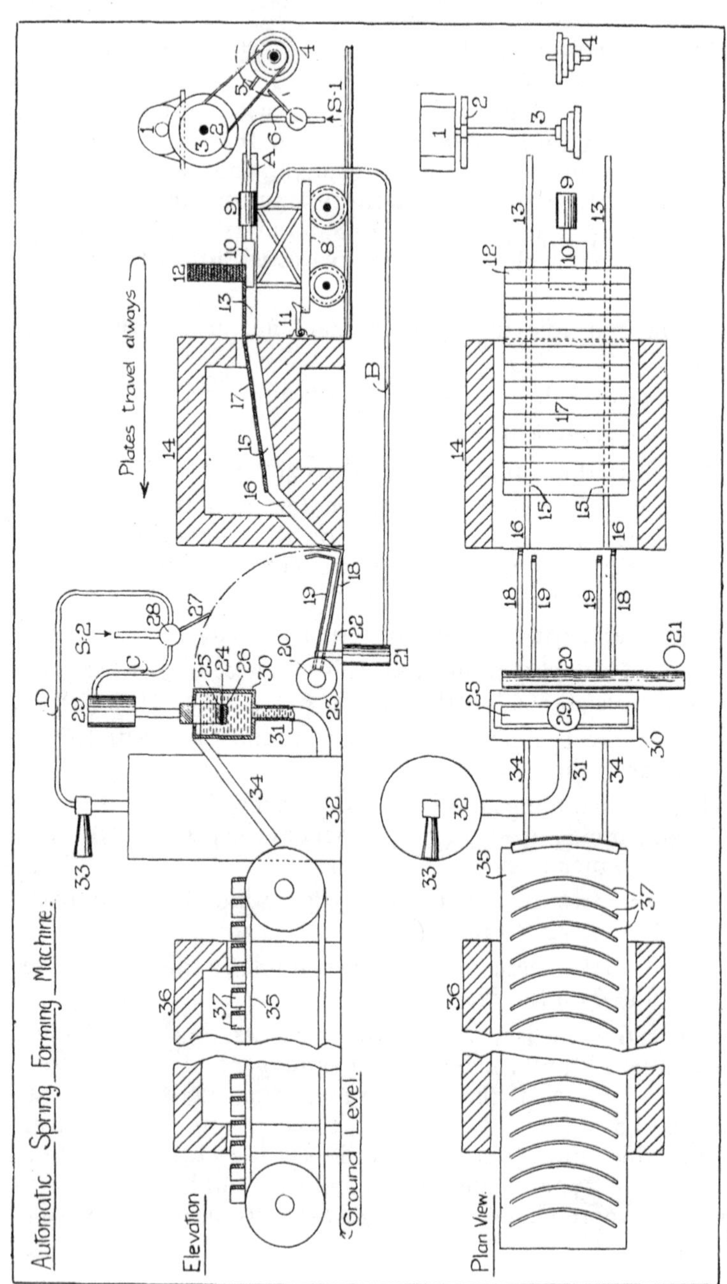

Fig. 233.

A very different continuous process is shown in Fig. 233, which is the subject of a recent British patent, and is sufficiently ambitious to endeavour to entirely replace intermediate hand labour. Briefly, the plates required are placed on to a railed bogie, fitted with a charging device, which pushes them, plate by plate, according to the fixed periods of the timing gear, into a continuous furnace, of short length, and with sloped flooring—thereby doing away with the need for mechanical means of movement through the furnace. The insertion of a plate at the cold end, causes one to be thrust out at the hot end, and it falls into the forks of the "transferrer" (18 and 19). This raises, and places the plate on the dies of the forming machine, which then operates, forms and quenches the plate, and the chain of operations from the charging bogie again being put into movement, causes a further hot plate to be brought up by the transferrer, which displaces the formed and quenched plate down the skids (34) on to the conveyor (35) of the controlled tempering furnace. Schemes of this sort are, however, in common with the American high-production units, valueless over here unless some standardization of springs is carried out. Railway springs will probably show a tendency to become of fewer designs—but the automobile trade seems as far off as ever coming to any rational arrangement scheme. No manufacturer would trouble about fixed designs if the orders came along in thousands, but when a firm of automobile builders, desiring about 2000 springs per annum, has 50 to 100 current designs, the difficulty of pretending to carry out machine fitting and furnace controls on commercial lines, is obvious.

CHAPTER XL

HOOP ATTACHMENTS

THE majority of railway and tramway springs have the plates centrally secured by means of a hoop or buckle, which either rests upon the vehicle frame, or is positively attached thereto. This arrangement of the attachment of the spring hoop or buckle to the spring is of importance, as certain methods of centre fastenings are seriously inferior to others, and are in themselves responsible for numerous plate breakages. There are a number of designs in general use, and also various patented types not in general use, and before dealing with hoops, it will be advisable to illustrate and describe some of these hoop attachments. (The relative efficiencies of the various patterns of centre fastenings, discussed from the point of view of the spring, have been dealt with in Chapter XXVIII.)

In Fig. 234—A shows the plain rivet fastening, with rivet countersunk each end. Sometimes the top end is left as an upstanding cup-head. This is the most common used form of fastening in this country, and is adopted for all classes of springs. From a manufacturing point of view, it is indubitably the best, as the spring fitter has a hole in his plates by which to handle them, and the final attachment of the hoop to the spring is very simple. For locomotive springs it is not ideal, as if used for springs on wheels which are liable to receive end shocks, there is a tendency for the rivet to shear, after which the spring will probably shift its position in the hoop. Iron or soft steel rivets are invariably used, so this liability to shear can obviously be lessened by the use of, say, 40-ton steel rivets. The efficiency of this form of fastening as against the solid plate is 70 per cent., on a basis of a hole $\frac{1}{8}$th the width of the plate.

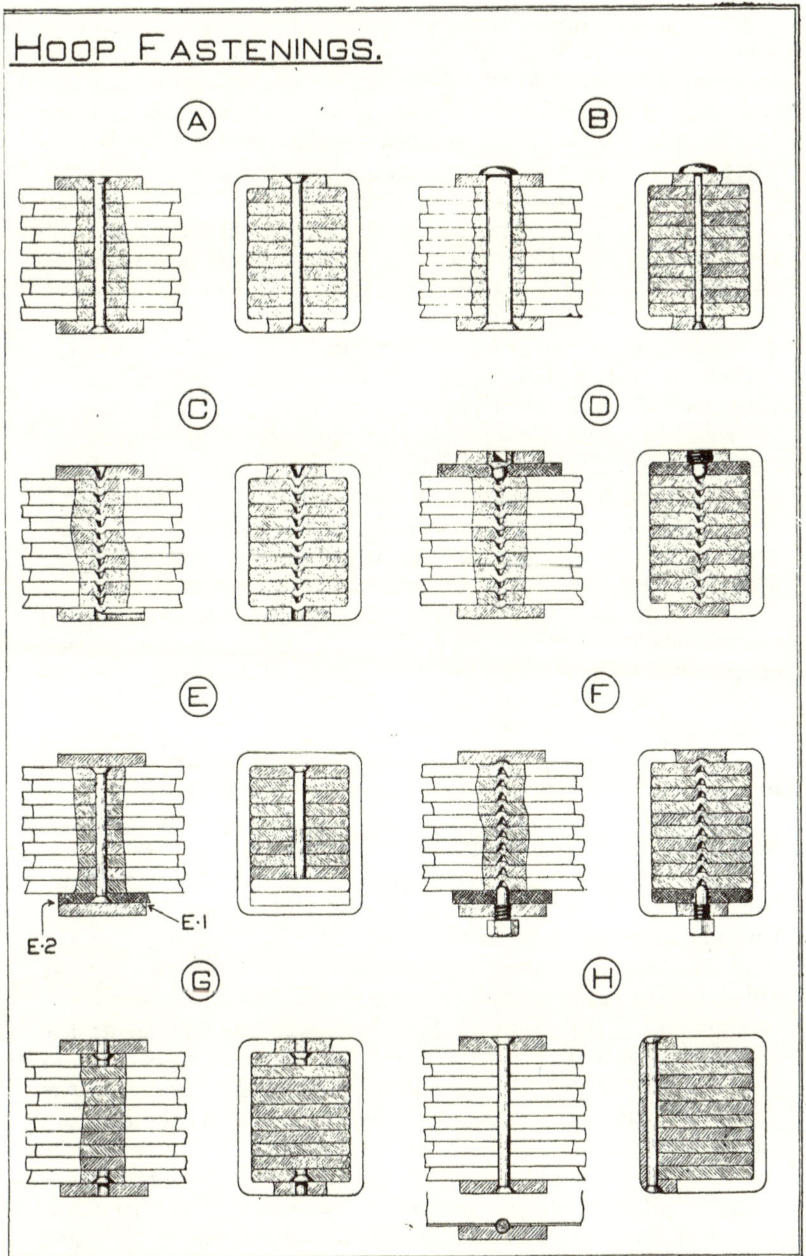

FIG. 234.

Fig. 234—B shows a cotter fastening. This is frequently adopted for carriage and wagon springs, but is not much used in locomotive work. The cotter hole weakens the plate less than the round hole, but it is not quite as easy to work as the round hole in manufacture. The cotter hole is also more difficult to make in the hoop, particularly if the latter has thick ends, and cannot be cold punched. The efficiency of a plate with a $\frac{5}{16}$-in. cotter compared with a 4-in. wide solid plate is 75 per cent., and on this basis it is hardly worth employing against the plain $\frac{1}{2}$-in. dia. hole with an efficiency of 70 per cent.

The foregoing two types of attachment are used to a limited degree in automobile work, chiefly on heavy commercial vehicles. Trouble is frequently experienced with the springs shifting on such vehicles, owing to soft bolts, rivets, or cotters, and it is advisable to insist upon the use of heat-treated high-tensile steels for this item. The centres of front springs are particularly liable to cause trouble, as, if the hoop is not firmly placed, the whole of the traction effort of the engine on the front axle is taken by means of the rivet or cotter in each spring. The rear springs are safeguarded against such troubles when radius rods are used. The hooped cotter-hole form for automobile springs of heavy pattern, is however, a sound design.

No. 234—C shows a form of " downward nib " attachment. It will be observed that the top of the hoop is forced into the depression in the top plate, and the nib on the bottom plate rests in a small hole in the bottom of the hoop, a narrow slot being cut along the bottom of the hoop to this hole, thus allowing the hoop to be driven over the nib. This slot makes the job look possible on a drawing, but as a matter of fact, there is no need for it, as the hoop will readily drive over the nib. For carriage and wagon springs, this is the highest efficiency fastening, the downward nib giving 85 per cent. of the solid plate strength, and the fastening itself being the acme of simplicity. If the hoop has to be stripped off (a consideration which always has to be studied), the nib on the underside will shear, so that if the same hoop is replaced on the spring, it should be drilled and tapped and a set screw inserted, as a substitute for the previous nib. This attachment is not very useful for locomotive springs, as the hoops are generally too thick to allow of a nib being effectively formed on them.

No. 234—D shows the downward nib form with a screw and packing plate. The hoop is placed in position, the pack then driven in, and when cold, the screw is inserted. This is a very usual type of hoop attachment in " British practice " countries, for carriage and wagon springs, but is not much used here for this purpose. The advantage it possesses over 234—C is the facility of re-hooping with the same hoop unamended. Generally, a standard set screw is used, but in the illustration, a special type of flush top screw is shown, rendering the fastening suitable for underhung locomotive springs. With such springs, it frequently happens that there is not sufficient clearance between the bottom of the tee hanger and the top of the hoop bridge to allow for the ordinary set

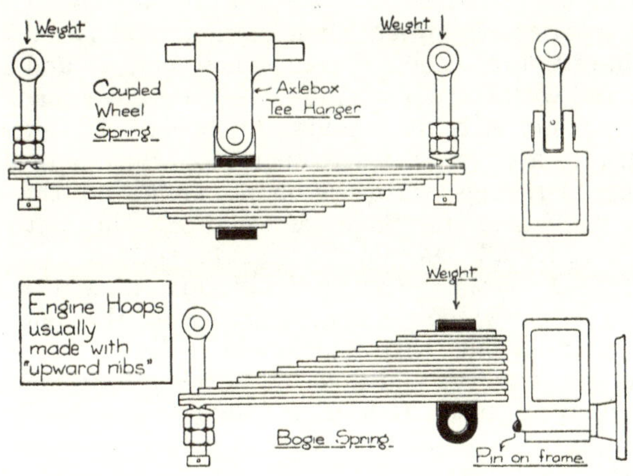

FIG. 235.

screw (see Fig. 235), therefore the designer argues that the set screw must go in at the bottom of the hoop, and the nib must consequently be reversed. A similar case occurs in many designs of inverted bogie springs, where the hoop is made of the " solid lug " type to take the weight-carrying pin which is fixed directly on the bogie frame. Fig. 235 illustrates this also. For this latter spring, it is clear that the downward nib type of attachment cannot be employed, and the pattern recommended would be 236—P (following), of the side cotter type.

As has been previously shown, the upward nib as usually made is the worst possible type of fastening, and this is reflected in the breakages caused through this pattern of nib,

By using a small screw, as in 234—D, the downward nib can be employed for coupled wheel springs. The use or disuse of a packing plate is largely a matter of taste as regards carriage and wagon work. With the set screw, and slotted hoop as in 234—C, there is no particular need for the pack, particularly if buckle stripping machines are installed to strip hoops. Without such machines, a packing plate facilitates removal of the hoop.

No. 234—E illustrates a very usual Continental form of attachment. A soft steel pack is riveted to the spring plates, and the hoop slightly tapered at the bottom inside edges (see E—1), is put on, after which the plate is caulked over into the hoop, as shown at E—2. There would not appear to be much to recommend this over the ordinary rivet pattern, 234—A, particularly when plain section steel is employed. The bulk of Continental springs are, however, rib and groove section, and with these plates, the use of a grooved pack plate obviates the necessity for grooving the hoop. It also retains the spring, as a separate entity, apart from the hoop, whereas with most fastenings, when the hoop, is removed, the spring resolves itself into its component plates. This pattern of hoop is used for all classes of Continental springs.

No. 234—F shows the upward nib pattern, with packing plate and set screw, 60 per cent. efficiency of the solid plate. As just pointed out, this attachment is limited to use on underhung locomotive and inverted bogie springs, and is the worst type that can be used, unless the nib is specially made as an " upward nib " and not as it is invariably made, as a downward nib, reversed.

No. 234—G illustrates a pattern introduced to avoid holes in the plates as far as possible. It will be observed that holes are only in the top and bottom plates, and small rivets hammered hot through the ends of the hoop, upset inside and secure themselves and the plates to the hoop. The intermediate plates are held practically free from end movement by the studs and slits which must be provided with this type of attachment. The advantages are not specially obvious, except theoretically. Clearly, the number of potential breaking points is reduced by only holing the top and short plates, but as, with nip-fitted springs and solid-end backs, the top and bottom plates are the most liable to breakage of all, it would seem that this fastening has no real advantage to offer over 234—A, and with 3-ft. springs or less, studs and slits are needed, which could otherwise be dispensed with.

No. 234—H shows a side rivet hoop attachment. A flat cotter is sometimes used instead of a rivet, in which case, no recess is made in the hoop, the plates only being side slitted to fit the cotter. When the rivet as illustrated is employed, it is sometimes insisted upon that the hole should be drilled through after the spring is hooped. From a machine shop aspect, this is not one of the happiest of jobs, particularly with a deep spring, as the drill is running in soft steel on the hoop side, and hardened and tempered spring steel on the other side. When plain hoops are used, made from the welded flat bar, this is not the only practicable method, as such hoops can be forged with a side groove, and the holes then drilled into them, the spring plates being punched separately, or shaped together, to remove the registering semi-circle required. With hoops cut from the solid block, the side hole can be drilled as a complete hole, and the material then machined out to the half-hole, the spring plates being dealt with as before. There are certain distinct advantages in favour of this side fastening, as for the same size cotter or rivet, less active material is removed than with the centre hole. The efficiency is 80 per cent., midway between the downward nib (85) and the centre cotter hole (75). This side attachment is not often used here, but is quite common on the Continent, with a cotter in place of the rivet.

No. 236—J shows a pattern used on rare occasions for locomotive springs—a few top plates, including the back, being side notched and a hole in the hoop slotted to correspond. After the hoop has been driven on, a red-hot locking piece is dropped in, and hammered, causing it to upset into the side notch in the plates. The remainder of the plates are held from end movement by the studs and slits. On the whole, this is preferable to the similar device in 234—G, but is slightly more expensive.

No. 236—K has nothing to recommend it. The upward nib throughout is a bad feature, and whilst the use of the triple nib on the back plate undoubtedly prevents it shifting in the hoop, there are other arrangements which have been illustrated which accomplish this quite as well.

No. 236—L is the "double downward-nib" pattern, used to a limited degree on the Continent for railway work, and to a certain extent in the U.S.A. for automobile work. In the case of the ordinary railway spring the use of this double nib removes the objection which can always be made to the true

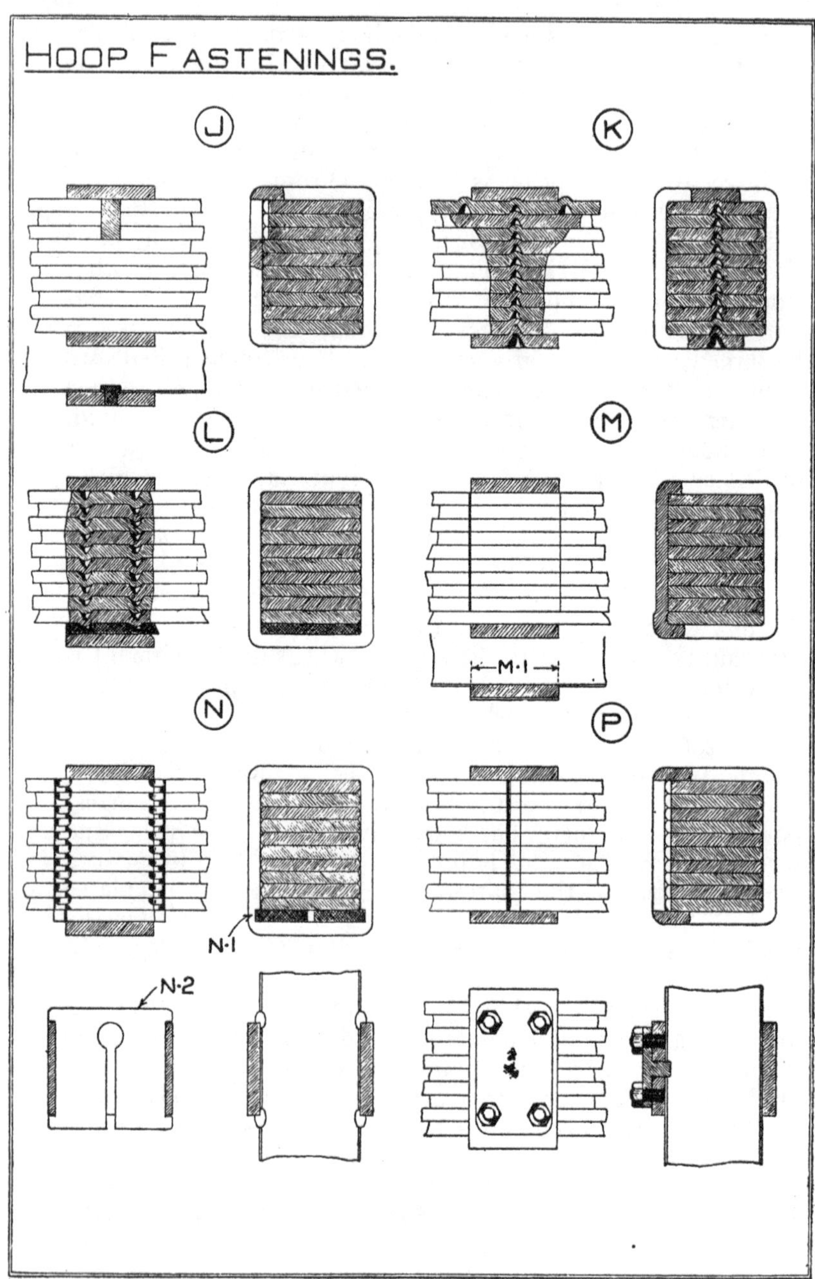

Fig. 236.

central single nib, of its being in the plane of maximum stress, but nevertheless, the double nibs may be in planes of " equal to maximum stress " depending on the plate layout of the spring. The width of the normal hoop limits the distance apart of the nibs to a very small figure, so that by the use of this attachment two potential breaking planes are actually obtained instead of one. In the case of automobile springs, this double nib is even more objectionable, as the working lines of the spring are bound to traverse these nibs, whereas, with normally wide spring seats and general spring design, so long as the clamping device holds firm, it is probable that the maximum stress lines remain clear of any " centre fastening." Fig. 237 illustrates this aspect.

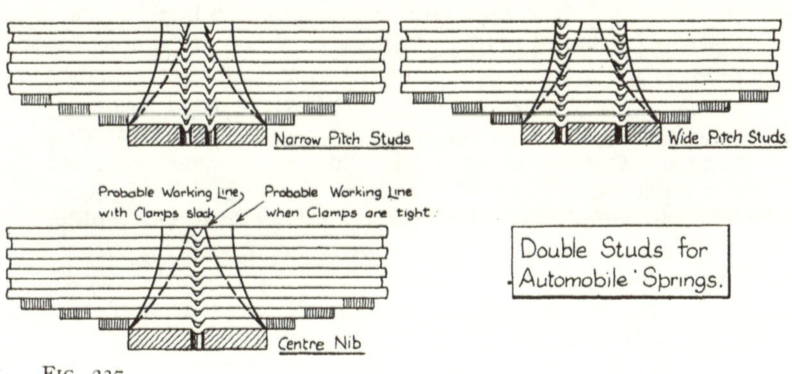

FIG. 237.

No. 236—M shows an attempt to produce a centre fastening without any holes or nibs being employed, the width at one side of the plate being reduced, say, $\frac{1}{16}$ in. for a distance of the width of the hoop, see M—1. All the plates are treated like this (by milling or shaping) except the short plate, which is left full width. The hoop is put on and pressed into the side flats. No fastening at all is provided on the short plate, as no difficulty is found with this shifting, if the hoop is reasonably tight. This type of hoop attachment has a very high theoretical efficiency, being practically equal to the solid plate. There is always a tendency, however, for the plates to cut through the hoop, owing to the very small bearing they have on it. For any springs likely to be affected by shocks, this pattern is not recommended, but otherwise, the side flat is quite a practical type, and could be usefully employed far more than it is. In this type M, to remove the hoop, the short plate is knocked out, and then other plates dropped into the place of the short plate and knocked out, until sufficient have been removed to enable the balance to come out by turning them in the hoop. From a manufacturing point of view, the side flat is not very popular, as there is no very distinct point on the hot spring plate to which the fitter can work, and such springs require more time for assembling for hooping than the hole or nibbed types.

No. 236—N is another attempt to eliminate a centre fastening, by the stamping of four small lugs on the spring plates, these being afterwards fitted to embrace the hoop. It is clear that these have to be introduced one by one into the hoop, through the wide groove, N—1, which latter is then filled with a packing plate. This type of fastening involves hooping with the hoop cold, and therefore demands a higher standard of workmanship than is usually regarded as necessary for spring work. The side lugs present the same disadvantages as the two nibs in 236—L, as two potential breaking planes are presented close to the longitudinal centre of the spring, and they also involve hot working the middle with machine stamping tools. A special packing plate, adaptable for use with types of attachments additionally to N, is shown at N—2. It will be noticed that it is split, and has lugs embracing the hoop. Driven in from one side it is opened out with a wedge, and is then secure until the spring needs dismounting, when, being made of soft steel, the lugs can be driven in to narrow the pack so as to clear the bottom groove. For hooped springs on automobiles, this type has certain advantages.

No. 236—P is a variation of 234—H, all the plates of the spring being notched, side cotter shape, and the hoop correspondingly slotted. A " T " plate is then screwed on to the side of the hoop, and introduces itself into the plates, therefore effectively securing all. This is a pattern frequently met with in Continental locomotive designs, and has an advantage over 234—H, inasmuch as by the removal of the " T " plate, it can be observed whether any spring plates have broken in the middle. A duplicate function is therefore performed by this plate, as Continental engine builders frequently slot the hoops through the side the full depth of the plates, thus providing a permanent inspection slot. This is a very good fastening, but relatively expensive.

Fig. 238—Q shows a variation of the side flat pattern illustrated in 236—M (ante). The distinctive feature of the one under notice lies in the top packing plate, which is grooved each side to correspond with the spring plates. To remove the spring from the hoop, this packing plate has the lugs Q—1, at one end, chipped off, when it can be driven through the hoop. The hoop is then knocked slightly out of square which allows the other plates to be withdrawn, bottom plates first, as solid-end or rolled-eye backs, unless very small, will not go through the top space. This fastening is not infrequently used for carriage and wagon springs, and is a good solution of the attempt to obtain the perfect type. It might be improved by the provision of the split pack plate shown in 236—N as this is readily removable without damage, but as shown, it is the most used of any special fastening.

Fig. 238—R shows a downward nib form, the pack plate and corresponding hoop surface being arranged with inclined faces, to ensure the pack wedging up tight all the plates. The pack is then drilled in position and the set screw inserted. From a designing point of view, this is perhaps nearer perfection than its original, 234—D, but in practice, no advantages present themselves sufficiently valuable to entail this taper top hoop and wedge-shaped pack.

No. 238—S is another variation of the side flat pattern, adaptable for springs subjected to abnormal running shocks. The material is removed from the spring sides in the shape of an arc, the hoop is similarly formed to match and then pressed on. As shown, such a hoop would be difficult to strip, but there is no reason why a bottom pack should not be introduced, which, when knocked out, will allow the other plates to be driven out through the pack space. Hoops for this attachment are not easy to make, as the curved side has to be forged to shape.

No. 238—T is a side-notch pattern, with a portion of the hoop pressed into the notch. The objection to this form is that the hoop must be cut off any broken spring, thus involving a new hoop for replacement.

No. 238—U is a Continental pattern of the downward nib and packing plate type, similar to 234—D, but with a wide "cotter," formed of bent steel strip, in place of the set screw, the former being a somewhat cheaper fastening.

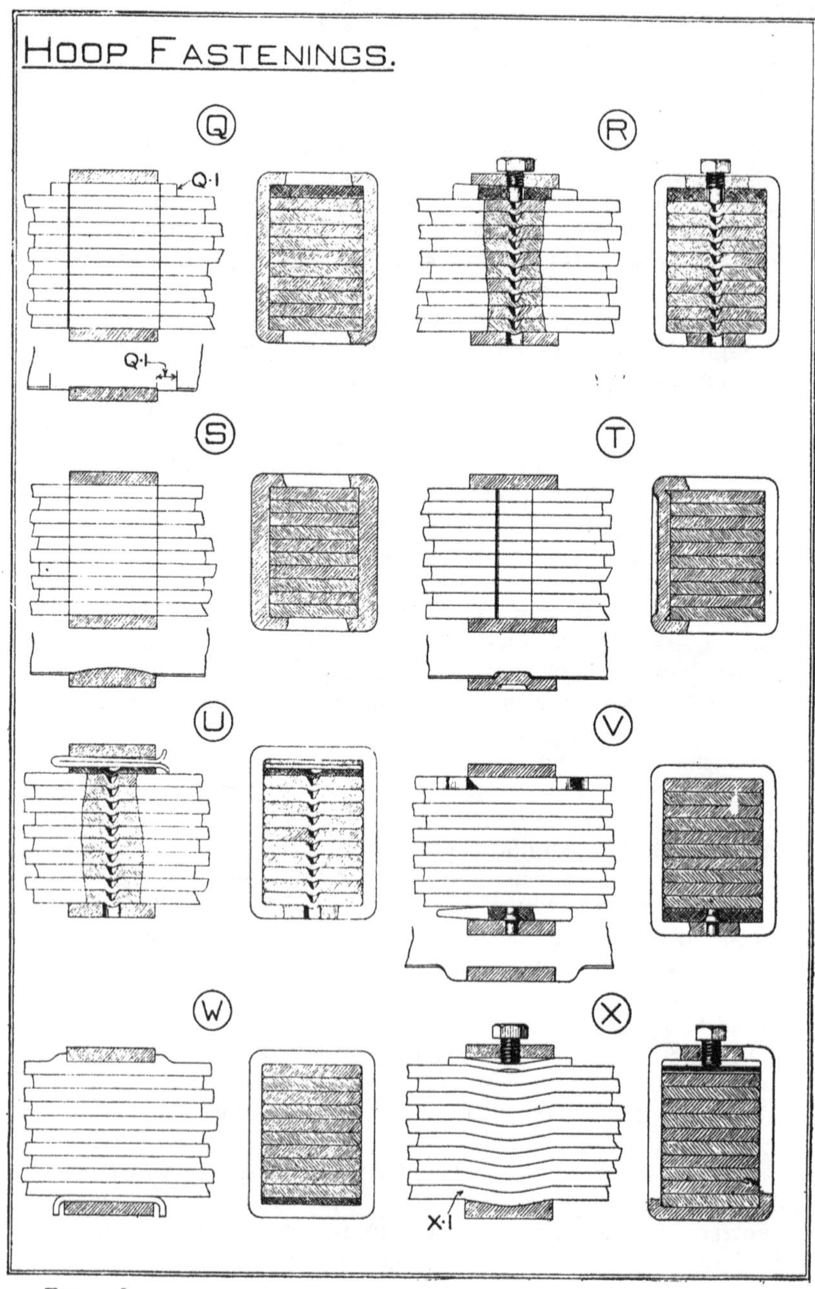

FIG. 238.

No. 238—V shows a back plate with lugs formed from the solid, the plate being rolled to the overall width of the lugs, and then machined down as required. By this means the spring is very securely centred and held with the hoop. Intermediate plates are kept from endways movement by studs and slits, and the short plate is secured with a jumped rivet through a wedge pack plate.

No. 238—W is a variation of this, with the lugs formed across the top of the back plate. These can be made by drawing the back from the solid billet and then stamping the lugs to shape. This also holds the spring securely and centrally, but, like 238—V, is an expensive form, with the back plate none too safe. The bent-over pack at the bottom fills the excess depth of hoop which it is necessary to allow for its passing over the lugs on the top plate. Intermediate and short plates are held by studs and slits.

No. 238—X shows a form with all plates "kinked" in the middle to register into one another, the extra space in the hoop necessary to allow for this kinking being filled with a soft steel pack, bent when in position by the set screw. This general form of fastening is capable of variation, the plates sometimes being curved the opposite way to those shown. Whilst theoretically this would appear a good form, the distortion of the material due to the bends at X—1, coupled with the fact that the short plate works from the edge of the hoop, results in practice in the type being no more immune from breakages than any ordinary attachment. It has all the manufacturing disadvantages of the side flat patterns, without their efficiency.

On occasion, ordinary fastenings are desired with patterns of hoops to which they cannot be applied in the usual way. The "tail" hoop is one example, in which the tail is frequently much longer than the hoop and a through rivet a non-practical job. Such cases can be met by the arrangements shown in Fig. 239. No. 239—A shows a rivet fastening, a blind hole, perhaps $\frac{3}{4}$ in. deep being drilled in the bottom of the hoop, and the short plate countersunk on the underside. The point of the rivet is then heated and on being hammered in, "jumps" itself into the counter-sink of the short plate. No. 239—B shows a long tailed set screw, the end of which enters a blind hole in the hoop bottom. No. 239—C shows a downward nib fastening, which only requires the hoop countersinking to take the nib, and for hoops of this pattern, this type of attachment is indubitably the best.

From the foregoing *résumé* of a varied assortment of hoop fastenings, it will be noted that none stand out as meeting all practical requirements, which are—

(*a*) Equal efficiency of spring plates compared with the solid plate.

(*b*) Ease of manufacture in the forging shop.

(*c*) Appreciation by the spring fitters.

(*d*) Capable of being rapidly and correctly hooped.

(*e*) Permitting ready removal and replacement of the hoop.

(*f*) Entailing minimum of cost in the manufacture of the hoop and accessories.

Generally speaking, if the first requirement (*a*) is met, (*c*), (*d*) and (*f*) have a tendency to suffer, whereas if the last five requirements are met, the first one is neglected. The perfect fastening has not yet been evolved, and will not be until the whole of the above requirements have been carefully embodied in the design. Balancing all things, the author would recommend the following patterns :—

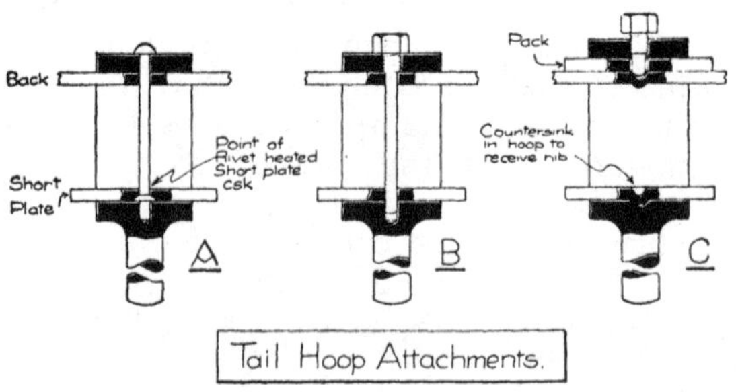

FIG. 239.

Railway Carriage and Wagon Work, Tramway Vehicles—

No. 234—A, plain rivet type. This is easy to forge, fit and hoop ; the hoop is easily removed and replaced at the expense of shearing a rivet, and no cost is entailed in hoop accessories.

No. 234—C, downward nib type. This is easy to forge, not quite as easy for the fitter to handle as 234—A, but the hooping is facilitated owing to the registering action of the plates. (With the plain rivet hole, plates have to be knocked endwise until the holes are in alignment before the rivet can be driven.) The plate centre section is much more efficient than 234—A, and no extra cost is needed for the hoop.

Locomotive Work—

No. 234—D, downward nib type. The same general remarks apply as above, but a packing plate and set screw are necessary owing to the greater thickness of engine hoops.

No. 236—P. This is easy to forge and reasonably easy for the fitter and hooper, as the latter can knock the plates into longitudinal register when the hoop is in position. The expense of the hoop is slightly more, owing to the side slot and " T " plate involved (an " L " plate is frequently employed), but the fastening is very efficient, and the small extra cost is negligible in the case of locomotives.

Automobile Work—

No. 234—D downwards nib type. As most automobile springs are not supplied with separate hoops, but are held to the axle by divers clamping devices, no packing plate is necessary, the top clamp plate being nibbed to register into the spring back plate. If double nibs are used, the top clamp must be treated accordingly. A small hole, either in the axle, bottom axle-pad, or equivalent position, will hold the nib of the short plate. A suggested arrangement, which, it is considered, is far superior to the too general centre bolt, is shown in the next chapter.

The one fastening of all to avoid is the " upward nib," 234—F, and it is to be hoped that this pattern will entirely disappear at no distant date.

Universal practice favours the through rivet for all work except locomotive work. For the latter, this country employs generally upward nibs, downward nibs or rivets, the Continent uses rivets or side cotters (236—P type), and America prefers the downward nib, without pack plates or set screws.

CHAPTER XLI

TYPES OF HOOPS

THE types of hoops in use (as distinct from types of hoop attachments) is legion. A few will be illustrated and described in detail, after reference has been made to qualities of the materials used.

Hoops can generally be sub-divided for this purpose into " band " and " special " hoops, the former including only types which embrace the spring in the simplest possible way, and the latter all those which include external bearings or attachments—generally as Fig. 240. Sketch A shows a usual carriage or wagon spring, with plain band hoop, dropping into lugs on the top of the axlebox. It should be noted that the short plate should clear these lugs. This it does not always do, owing generally to the lugs being somewhat deeper than they should be, with the result that the short plate bears thereon, until it wears and breaks. Sketch B shows a form of " pap " hoop, seen often in use for locomotive tenders. Sketch C shows a Continental form of " special " hoop, of the " double cross lug " type, and D shows an inverted bogie spring, with a tail hoop, arranged with adjustable (screwed) end. World practice then goes along the following lines :—

British.—Carriage and wagon hoops for bearing springs are nearly always of the " band " type, together with a small proportion of engine hoops. Such hoops are generally made from puddled iron bar, worked as required by the pattern of hoop and then welded. The minority are cut from solid steel blocks. The quality of iron for use in hoops, British standard specification requirements, demands a tensile strength of 20 to 24 tons per square inch, with 20 per cent. elongation in 8 ins., and bending tests according to thickness. Steel has to be 24 to 30 tons, with 20 per cent. stretch. Fig. 241 illustrates these test requirements. The use of steel is not prohibited for welded hoops, but, as long as good puddled iron can be obtained at a reasonable price, this material is preferred, as it is generally considered easier to work and weld.

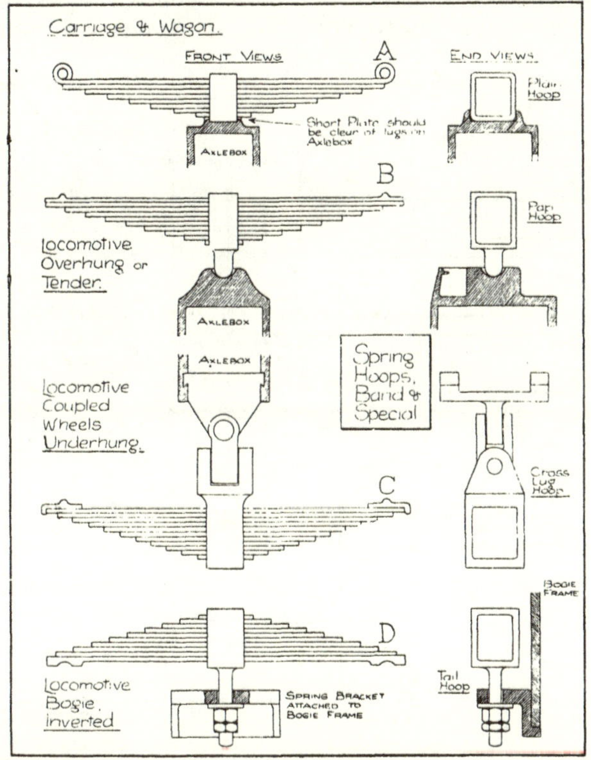

FIG. 240.

Locomotive hoops are mostly of the "special" type, and in these days, are usually made by machining from the solid steel billet, quality as noted above. Prior to the introduction of the heavy puncher-slotter machine (to be later described), very few engine hoops were made from the solid, nearly all being forged from Best Yorkshire Iron billets, in divers manners. Best Yorkshire Iron is well known as the most reliable material obtainable for intricate forging and sound welding, and whilst it was (and is) high in price, there was little economy in saving a penny per pound on iron (pre-war Best Yorkshire Iron was about twopence per pound) and then spending a shilling per pound on forging, with ultimate uncertainty as regards soundness. However, very few hoops to-day are made of Best Yorkshire Iron, as steel hoops made from the solid have proved quite as satisfactory in service, and are substantially more economical as regards production costs. Best Yorkshire Iron test requirements

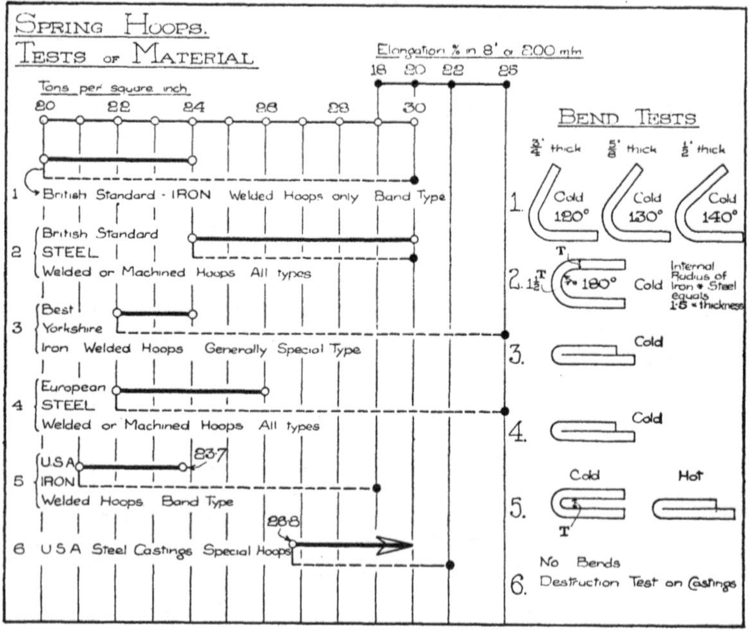

FIG. 241.

were 22 to 24 tons per sq. inch, with 25 per cent. elongation in 8 ins. and 45 per cent. reduction of area. It was the latter feature on which Best Yorkshire Iron always indicated its superiority to other puddled iron, and its approximation to the homogeneity of steel. All Best Best (B.B.) and Best Best Best (B.B.B.) wrought iron would give 22 to 24 tons, and the B.B.B. in the neighbourhood of 25 per cent. elongation, but the 45 per cent. contraction remained a feature of the Best Yorkshire Iron on hoop billet sizes.

Buffing spring hoops are quite a feature of the railway trade in Great Britain, and to-day, nearly all these are made from the solid. The proportion of welded hoops is becoming steadily less, and the only attempt which has been made to displace the "solid" hoop has been by the drop-stamping process. This can be made to produce a good hoop, but it is doubtful whether it can compete commercially with the machined hoop.

Continental.—Carriage and wagon hoops are generally of the "band" type, welded up from dead soft steel (flusseisen) of about 22 to 26 tons tensile strength, with 25 per cent. stretch in 8 inches. A certain proportion are "special" hoops, and these are frequently drop stampings from the same quality of steel. In Germany, weldless band hoops are made by two very different processes to be afterwards described.

Locomotive hoops are of the "special" variety and are chiefly machined from the solid. As a rule, the engine building plants make their own hoops, and buy the spring only, and consequently they do not run a specialized hoop plant, ordinary heavy locomotive machinery being sufficient to deal with such relatively small incidents as spring hoops.

Transverse buffing springs are largely used, and the hoops are now invariably made from the solid block, replacing the welded steel process.

American.—Car springs have always simple "band" hoops, welded up from either soft steel, maximum carbon content 0·15 per cent., or wrought iron having a tensile strength of 47,000 to 53,000 lb. per sq. inch (21·0 to 23·7 tons) with 18 per cent. elongation in 8 inches. Cold and hot bends, and fibre fracture tests are also specified.

Locomotive spring hoops are nearly always of the "band" type, and are made of the same material as the car spring hoops. When "special" hoops are necessary, they are generally made as steel castings, tenacity 60,000 lb. minimum (26·8 tons) with 22 per cent. stretch, and 30 per cent. contraction, all carefully annealed.

Transverse buffing springs of the European type are, of course, unknown, owing to the central buffer coupling being employed, and the plate spring arrangements used in the draft gears of these buffer couplings, are invariably enclosed completely in the draft gear casings.

Band hoops are all of the one general form, but have distinguishing details as indicated in Fig. 242. Herein, No. 1 shows the most ordinary type, the " plain hoop," sides and ends being of the same thickness, and the hoop being made direct from a flat iron or steel bar, which should be rolled about $\frac{1}{32}$ in. thicker than the dimensioned hoop thickness, to allow for scaling losses in the heating necessitated by the manufacture and subsequent spring hooping. European practice finishes these plain hoops to shape, as shown, but American practice simply bends the bar into a rough ellipse and welds it, the size being such that it will go over the spring. The final shape is obtained by the pressure of the spring hooping press when the hoop is in position. This type of hoop is

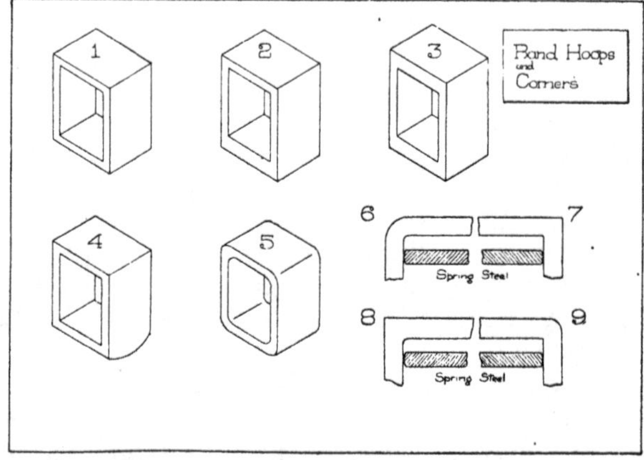

FIG. 242.

invariably used in tramway practice—and generally also when hoops are fitted to automobile springs.

No. 2 shows a " one thick end hoop " and No. 3 shows a " two thick end hoop," the terms being self-explanatory. A top thick end is generally used where set-screws are employed as the centre fastening medium, and a bottom thick end is to permit a thicker wearing (or rusting) surface on the bottom where the axlebox bearing may be taken. It is not infrequent to find thick end hoops designed with $\frac{1}{2}$ in. sides and $\frac{9}{16}$ in. or $\frac{5}{8}$ in. top and bottom. Such hoops are much more expensive to make then a plain hoop of the maximum thickness, as the iron stock should be ordered of the greater end thickness and then drawn down to the dimension of the sides. In fact, with the increase of machining in this trade, it is now cheaper to make a " two thick end hoop " as a " plain hoop " and then machine the sides down to the desired thickness. Owing to the extra forging required on this type of hoop, there is no difficulty in including square outside corners (as shown in the sketches), but considerable hammer work is required to bring up such square corners on a " plain hoop." No. 4 shows a round-ended " one thick end hoop," of a type which can be finished off under the hammer. No. 5 shows a plain hoop with round corners, inside and outside, which type is the most practical from all points of view. Nos. 6, 7, 8 and 9 show corner finishes, of which Nos. 6 and 8 are recommended, the former for plain hoops and the latter for thick end hoops. Square inside corners are by no means satisfactory, particularly when the springs are hooped under a power press. One thick end hoops, with a difference of not more than $\frac{1}{8}$ in. between the end and side thicknesses, can usually be made by arranging the weld on the thick end, but this is not customary unless the thick end happens also to be the bottom end.

All this class of hoop work is carried out under the usual pattern of smithy hammers, generally about 5 cwt. capacity. The tools used are of simple design and the production of plain hoops, of good finish, is very rapidly effected.

" Special " hoops are a very numerous class and a few types will be illustrated and described. Amongst locomotive hoops are—

Fig. 243—A. Double Lug Hoop. Probably the most-used form for underhung locomotive springs. This sketch shows all the refinements of finish in the way of chamfers.

Fig. 243—B. Double Lug Hoop. For the same purpose as 243—A, but simpler in design. It will be noticed that the suspension hole is counterbored for the pin from the T-hanger on the axlebox. The hoop is also shown with a slot in one side, acting as an inspection slot for the detection of plates broken in the centre. The adoption of this slot is almost confined to Central European practice.

Fig. 243—C. Double Lug Hoop. For the same purpose as 143—A, but counterbored under the pin, owing to restricted distance between axlebox and rail head.

Fig. 243—D. Solid Lug Hoop. Generally employed for inverted springs on locomotive bogies. Also used on automobiles, when the springing is of the half-elliptic cantilever form.

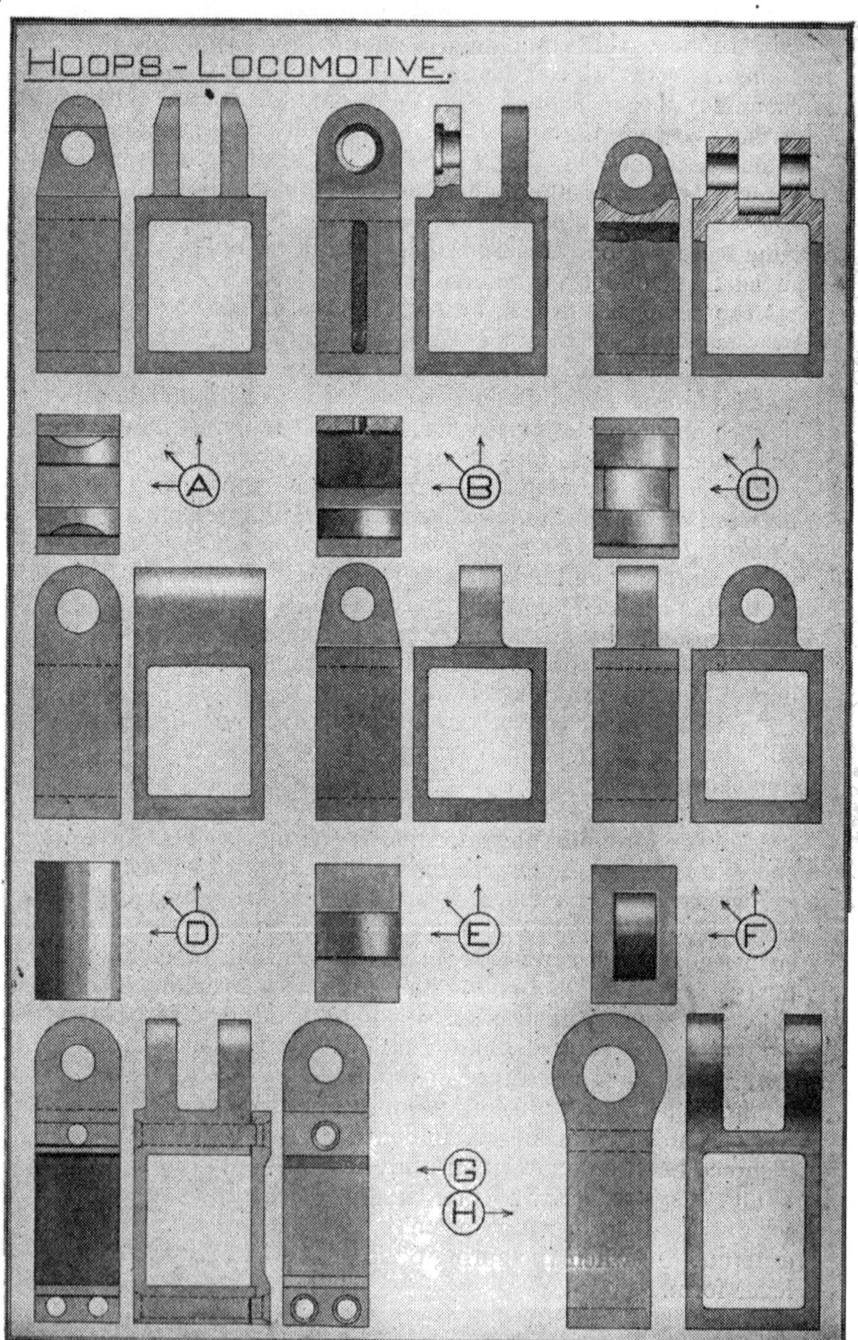

FIG. 243.

Fig. 243—E. Single Lug Hoop. A locomotive pattern not frequently used. Adapted for a bent bar axlebox hanger instead of the more usual T-hanger form.

Fig. 243—F. Single Cross Lug Hoop. A locomotive pattern used with an intermediate link between the axlebox and the hoop.

Fig. 240—C (*ante*) Double Cross Lug Hoop. Similar to 243—F, but stronger design.

Fig. 243—G. Double Lug Hoop. This is a locomotive hoop, with one side open, either secured to the main hoop with turned bolts or turned rivets. A great deal of machining is necessary, and it would doubtless be a good pattern to use with some of the special fastenings which have been described, but usually some very ordinary type has been employed. The hooping of a heavy spring is facilitated, without danger of damage to the hoop, but it is obvious that to be of any value in this direction, the machining must be extremely accurate and each hoop must be fitted to its own spring.

Fig. 243—H. Battledore Hoop. So called owing to the enlarged eye end. Sometimes used for inverted bogie springs, but more often for tender buffing springs, for the engine and tender draft and buffing gear.

Fig. 244—A. One thick end hoop. Pattern for wheels of locomotives having overhung springs. Round-ended flat bars (sometimes in guides) act as compression members between the slots in the hoop bottom and the top of the axlebox.

Fig. 244—B. One thick end hoop. Similar to 244—A, but with square bottomed slots. This is not a very practical type, as a certain amount of hand work is required to finish the slots.

Fig. 244—C. One thick end hoop. Similar to 244—A, but arranged for one solid bar between the hoop and axlebox top.

Fig. 244—D. One thick end hoop. Similar to 244—A, but spherical ended and bearing direct on axlebox.

Fig. 244—E. One thick end hoop. Pin rocker type. Used largely in American practice for all classes of semi-elliptic engine springs.

Fig. 244—F. One thick end hoop. Spring pillar pattern —a round bar acting as the compression member between the axlebox and the hoop bottom.

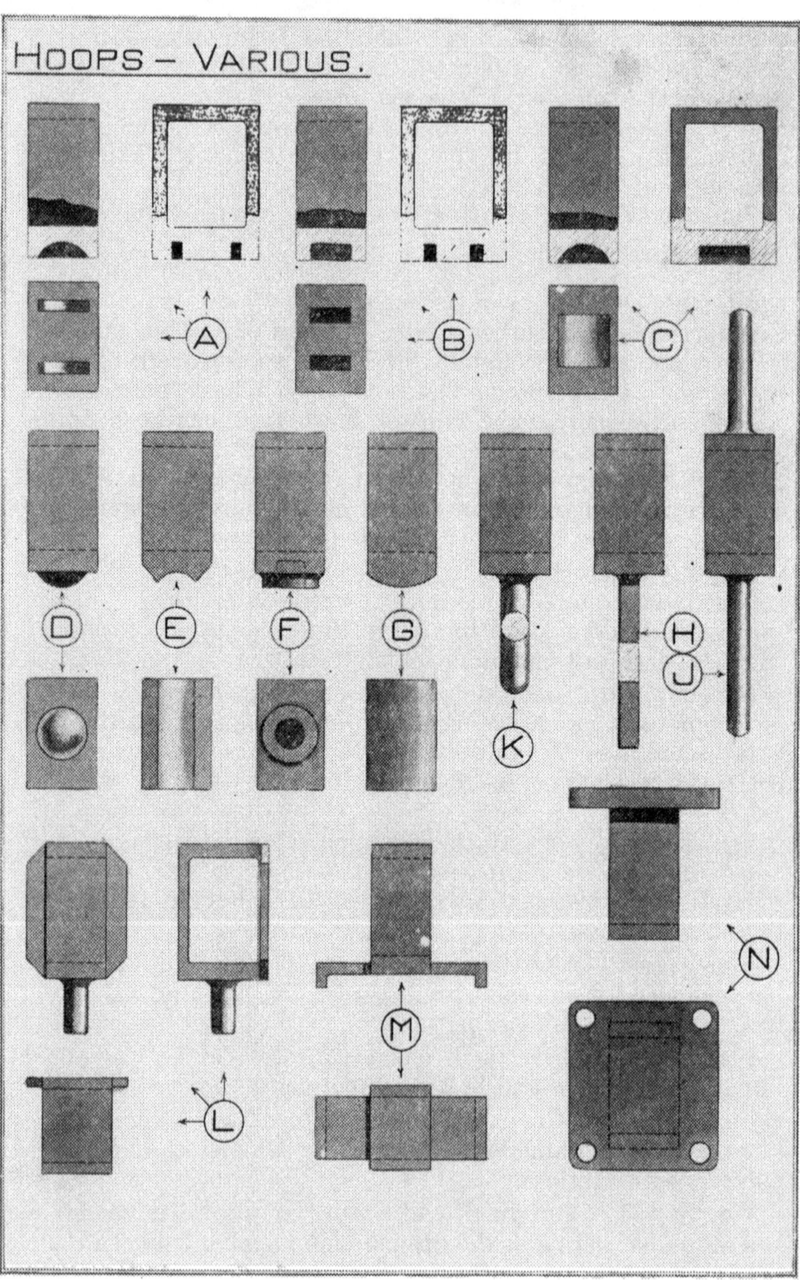

FIG. 244.

Fig. 244—G. One thickend hoop. Rocker type. Another usual feature in American practice.

Fig. 244—K. Tail hoop. In this pattern the round tail is solid with the hoop and bears on the axlebox top. All lengths of tail are included in the designs, hoops with small tails, say 1 in. long, for locomotive bogie springs, etc., being known as "pap hoops." (240—B *ante.*)

Fig. 244—H. Tail hoop. This type has a rectangular tail and is frequently used on heavy crane or gun trucks, where it is necessary to place packing blocks between the hoop top and the truck frame, so as to relieve the springs of the weight when the truck is in action. The spring hoop is then required a distance off the axlebox, which is arranged by the length of the tail. Fig. 245 shows two methods of arranging the relief of springs on such trucks.

Fig. 240—D (*ante*). Tail hoop. The hoop in this case is arranged on an inverted bogie spring and the tail is placed accordingly. Provision for a certain amount of adjustment is shown.

Fig. 244—J. Double tail hoop. Sometimes used for locomotive tenders, the bottom tail on the axlebox and the top tail through a guide.

Fig. 244—L. Flange hoop. A locomotive tender pattern, not often employed. The rear flanges work behind flats attached to the tender frame, which act as guides.

Fig. 244—M. Flange hoop. More particularly used in carriage and wagon work and arranged to embrace the top of the axlebox.

Fig. 244—N. Flange hoop. This is a special type, for tender bogies in which the main arch bars are arranged on top of the hoop. A similar form is sometimes used on a special type of locomotive bogie, wherein the main frames of the engine bear directly on the inverted side springs of the bogie, in which case the hoop top has the additional function of a rubbing block.

The "Italian" truck employs a heavy hoop of this type, when the weight of the coupled axle is taken by a transverse laminated spring.

All the hoops so far described are in modern practice cut out of the solid steel block. Variations of the flange hoop, 244—M, are made as drop stampings, but this procedure is only warranted where large quantities (say 2,000 per order) are required. At times, a certain amount of forging is done to the steel block before the machining is commenced, but this depends chiefly upon the quantity of hoops required and the sizes of material in stock or readily obtainable. For instance, it might under certain circumstances, be advisable to drift out a small hole in the battledore hoop 243—H, so as to obtain the enlargement necessary round the eye. In the tail hoops, it might under certain conditions, be as cheap to swage down the tail nearly to size as to cut it from the solid—it depends largely on machining facilities at the moment, and in this case, also on the length of the tail. The flange hoop 244—N, would be, in many cases, machined from a block with the flange roughly formed thereon. Generally speaking, how-

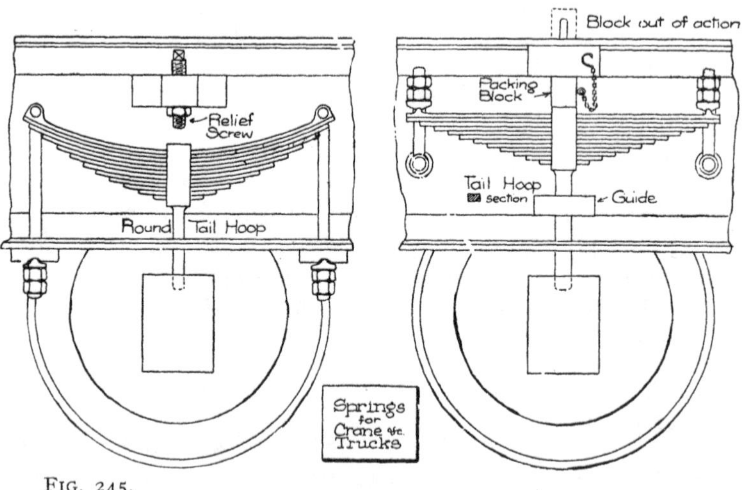

FIG. 245.

ever, with proper tools, the most economical and satisfactory method to-day is the machining throughout of these hoops.

A further assortment of special hoops is shown in Fig. 246, as follows :—

Fig. 246—A. Double Double Lug Hoop. This is the standard R.C.H. buffing spring hoop and, in slightly modified dimensions, is also used largely by the railway companies here. Outside this country it is little employed, but here some thousands are made weekly, far more than all other types of " solid " hoops combined. The double double lug is necessitated by the through draft gear arrangement.

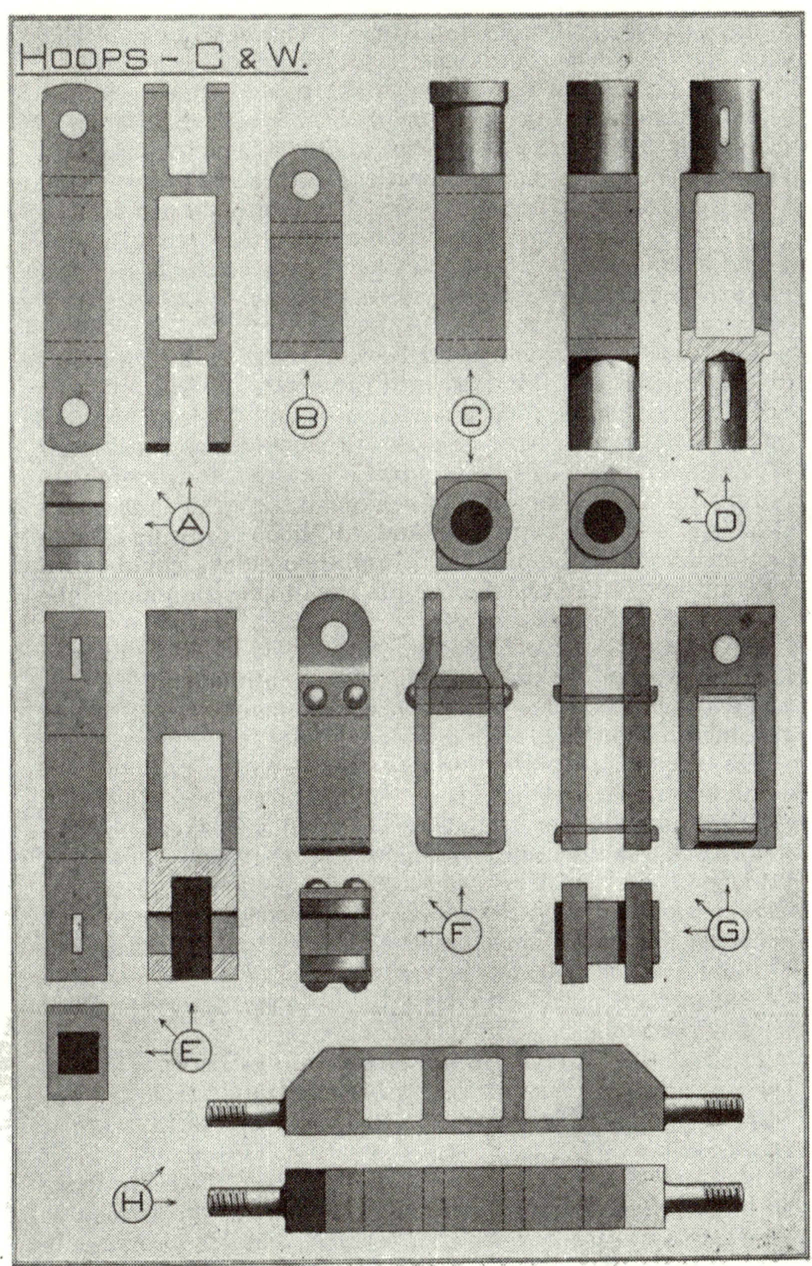

FIG. 246.

Fig. 246—B. Double Lug Hoop. This is a buffing spring hoop for use where continuous draw gear is not employed. Note the radius at the top, which is struck from the centre hole of the drawbar pin. It would be as well if this were now supplanted by the radius shown at the ends in 246—A. The 246—B radius was in order when these hoops were welded up, and the bar ends were speared in hammer tools, but now practically all of this hoop type are cut from the solid, the said radius becomes a milling cutter dimension and as such, on the heavy duty machines used, 246—A is far more practical.

Fig. 246—C. Single Round Socket Hoop. This takes a round drawbar end, cottered into the socket. The enlarged collar on the socket, when shown, invariably falls outside the rolled section of the bar (see plan) which renders it necessary for all such hoops to be "jumped" at this end, merely to bring up the collar size. Such details as these are anachronisms—and many a standard hoop drawing which dates back to the universal iron hoop days could, with advantage, be overhauled, with a view to revision along the lines of modern practice.

Fig. 246—D. Double Round Socket Hoop. Practically the same as 246—C, only arranged for continuous draw gear, and without the collar—the whole hoop thereby falling within the lines of the rolled section.

Fig. 246—E. Double Square Socket Hoop. Adopted as standard on certain lines here. As a practical iron hoop job it is a sound design, but with the gradual decrease of this method of manufacture it is perhaps the biggest problem that the solid steel hoop advocate has to tackle, as whilst these deep square holes can be machined out, there is not sufficient demand for machinery. When made in the solid, the end socket holes are generally drilled, and then drifted square in the smithy—a relatively expensive process. This hoop is also made as a Single Square Socket.

Fig. 246—F. Strap Hoop. For standard British buffing springs. The flat iron or steel is bent to the strap as shown, and the top packing block, of solid steel or cast iron, then riveted in.

Fig. 246—G. Washer Hoop. For standard British buffing springs. Large section flats ($4\frac{3}{4}$ ins. $\times$ $1\frac{1}{2}$ ins.) are generally used as the washers in modern practice, and the box is cut out of these flats. The top and bottom strips are put in the washers as the spring is ready to hoop, and the whole then driven over the spring to the centre, a through rivet then holding together the top and bottom strips and the spring.

Fig. 246—H. Triple Hoop. A very special hoop, used on rare occasions for the bottom halves of bogie bolster springs (coaching stock). The spring plank usually carrying this bottom half is eliminated, and the ends of the hoop have the swing links directly attached thereto.

No spring fastening means have been shown in these various hoop sketches, as all ordinary types can be used. Most of the engine hoops have upward or downward nibs, or rivets —the buffing hoops mostly rivets, and the carriage and wagon hoops downward nibs or rivets.

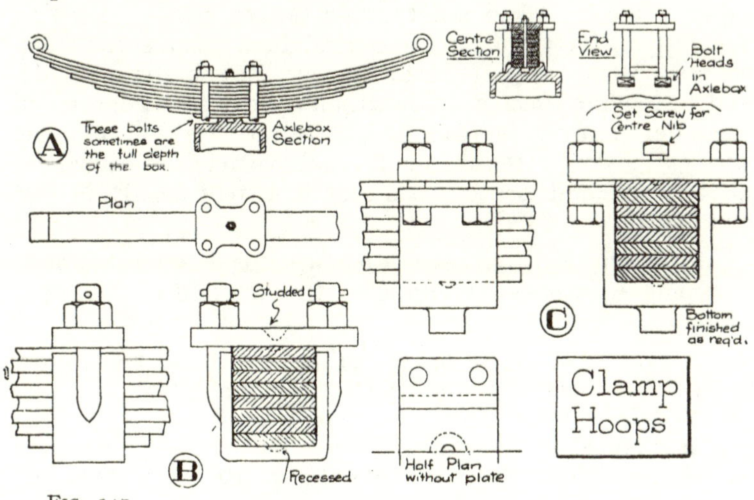

FIG. 247.

It is not infrequent to find railway springs fastened to axleboxes by devices which are not properly hoops. Fig. 247 illustrates two examples, 247—A showing a form employing four bolts and a top clamp plate. The (bottom) tee head of each of the four bolts is inserted into a recess cast in the axlebox, and the whole spring thereby is clamped firmly to the box. The clamp plate must be curved to the general sweep of the spring. Fig. 247—B shows a U-hoop type, in which the hoop is a forging, with the screwed ends solid therewith, or alternatively with them welded on. The bottom is sometimes plain, otherwise with a suitable pap (as at C), and a small top clamp secures the whole spring. It is clear that such form of U-hoop is not easy to make from the solid, as usually designed, but it could be re-designed as regards certain dimensions which would facilitate this. Another type is shown in 247—C, which can be fairly readily made as a "solid" job. Both B and C are, however, relatively expensive patterns compared with the plain hoop, which could substitute them without serious loss in efficiency. Theoretically, it is a very excellent thing to have a type such as B or C, from which the spring plates can be easily removed to repair breakages. Practically, such idea is hardly worth entertaining, as these U-hoops are generally used on wagon stock, and more wagon springs are taken out of service through old age and lack of maintenance having caused them to rust away, than are taken out through broken plates. Such as the latter, if returned to a central depôt, have the hoop stripped with a minimum of damage, new plates inserted, and re-hooping effected, for less than the extra cost of some of these devices over the plain hoop. In the case of automobile springs, however, the clamp type above referred to has its advantages, and is used to a certain extent mostly in this country and on the Continent.

In Fig. No. 248 are shown a few usual types of automobile clamping arrangements, and a brief description of each is as follows :—

A. A pattern in use for heavy springs, which has many good points. The top and bottom clamp plates are well rounded off each end, which has a great influence in the avoidance of fractures. Two bolts are in permanence through these plates, so that the spring can be removed or replaced as a complete entity. The type lends itself to either centre nibs or centre bolt ; if the former, the top clamp must be nibbed to suit the spring, and if the latter, a clearance hole for the nut must be drilled. The register to the axlepad is shaped on the bottom clamp.

B. A much used type, particularly in the U.S.A., for all classes of springs. The top clamp piece is a malleable casting, recessed to take the U-bolts, the nuts of which are put on under the axle flange. The pattern has the advantage of cheapness and simplicity, but is not as good as (A) for heavy trucks. A general disadvantage is that the spring plates are "loose" when the spring is removed from the vehicle.

C. This is a U-hoop with a top clamp plate only, which is drilled for the through bolts to the axle seating. The use of a U-hoop permits a short "short" plate, and by consequence, a well designed spring; and it has also the advantage that the whole of the spring plates are held together.

D. This shows a method of fastening shown in Mr. Remington's Paper, 1922, Institution of Automobile Engineers (London). It is stated that the idea is to reduce the

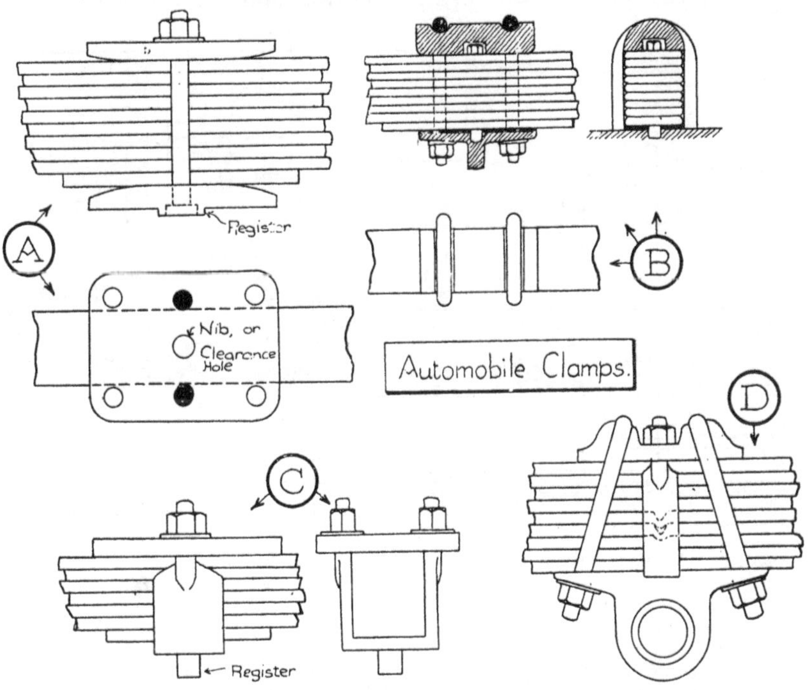

FIG. 248.

amount of dead material in the middle of the spring, but as this turns entirely upon the short plate bearing, the reduction of the top clamp bearing area would not produce the desired effect. Nevertheless, this is a good type, and has all possible practical advantages, the spring being held complete, and also nibbed in the middle.

Most designers rely on top and bottom clamp plates of simple form, which are, on the whole, unexceptionable;

the weakest point being the small size of the bolts generally employed. With the continual vibration, the spring plates work loose, and in tightening small bolts, they invariably stretch, and aggravate the trouble. Clamp bolts should be as large as possible, and of some quality certainly very much stronger than the too usual mild steel.

Overhung springs—that is—springs fixed above the axle-boxes to which they apply—have the bottom of the hoop in compression, the sides and top acting only as a means for holding the spring together, and taking none of the vehicle

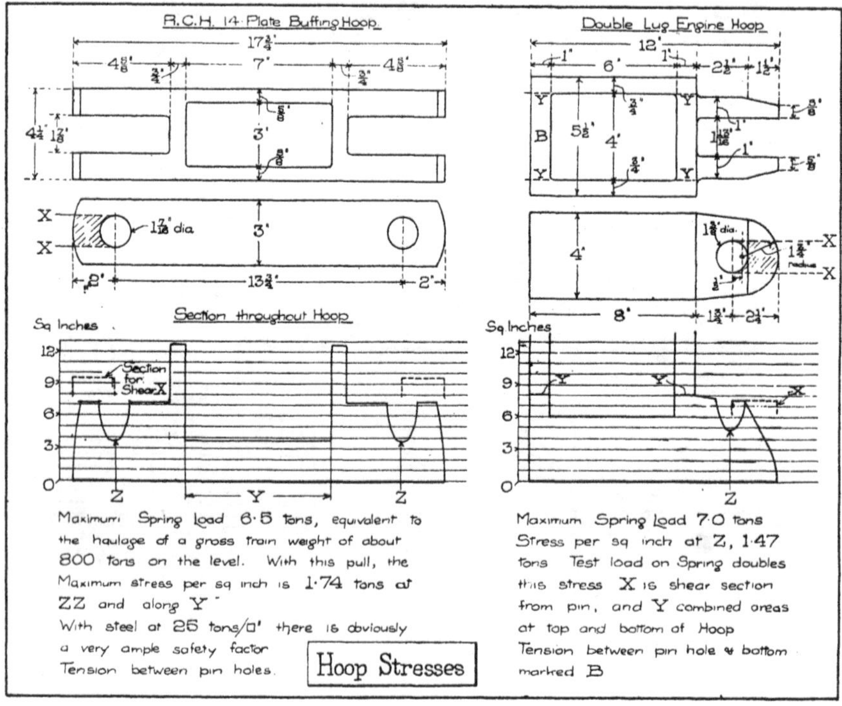

FIG. 249.

weight. With underhung springs, however (springs below the axleboxes), the hoop sides are always in tension due to the vehicle weight being transmitted through them to the T-hanger suspension pin. Inverted bogie springs may have hoops of either pattern (240—D shows a "tension" hoop, and 244—N what might be termed a "compression" hoop) according to whether the engine weight is taken directly on top of the hoop (spring in position) or whether it is taken by a pin or tail from the bottom of the hoop.

The point as to whether a hoop works as a " tension " or " compression " detail should be carefully borne in mind in the manufacture and ultimate inspection, as many slight imperfections can be safely passed for " compression " hoops that are very objectionable, and sometimes dangerous for " tension " hoops. All square corners (outside edges excepted) should be omitted from tension hoops, and they should be carefully treated when put on the spring. The designing dimensions are generally very liberal from the stress point of view. Fig. 249 gives dimensions of the R.C.H. standard buffing hoop, and also of a representative double-lug engine hoop, with particulars of the working stresses on various sections.

Some railways make it a practice to case-harden the bearing pin holes, or other bearing surfaces of hoops, such as underside paps, etc., and others put hard wearing bushes in all the pin holes. Opinion differs considerably as to the need of such refinements, and the bulk of hoops for British and American practice are not so treated—Continental designers tending chiefly to the case-hardening and bushing features.

CHAPTER XLII

MANUFACTURE OF WELDED HOOPS

HAVING taken a brief review of types of hoops, it will be advisable to refer in detail to the methods of manufacture of certain " special " hoops, and it will be of interest to touch first upon some typical hoops made as forgings.

Fig. 250 shows three methods of making, under the smithy hammer, of a locomotive " double-lug hoop," commencing as an iron billet. In older days, it was customary to weld separate lugs on to the welded box, but this is clearly an unsatisfactory process, and in later years, the methods sketched out have been generally used. In 250—A are shown the lugs left solid in the centre of the billet, and each end drawn down therefrom to form the sides and ends. These are then bent over suitable blocks and welded at the bottom end of the hoop. The lugs are finished by machining. No. 250—B works off a similar billet, probably somewhat less thick, and leaves the bottom of the hoop in the middle, working sides and lugs out to the ends. The forging is then shaped round blocks and welded between the lugs. Method 250—C comprises three pieces of material, the lugs being formed by bending into a " U " shape and stamping to form a bridge piece. A side and half-end is then welded at the bottom of each lug, and the whole then shaped round blocks and finally welded at the bottom. Of the three methods shown, 250—B is the best, as will be gathered by examination of the stress locations in the finished hoop. Only a first-class smith can make an entirely satisfactory job of 250—C, in view of the three welds required, the corner welds ultimately being nearly coincident with shear planes.

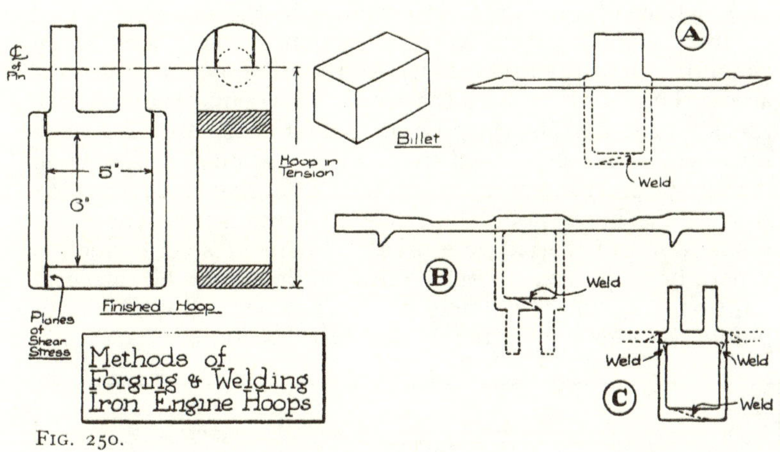

FIG. 250.

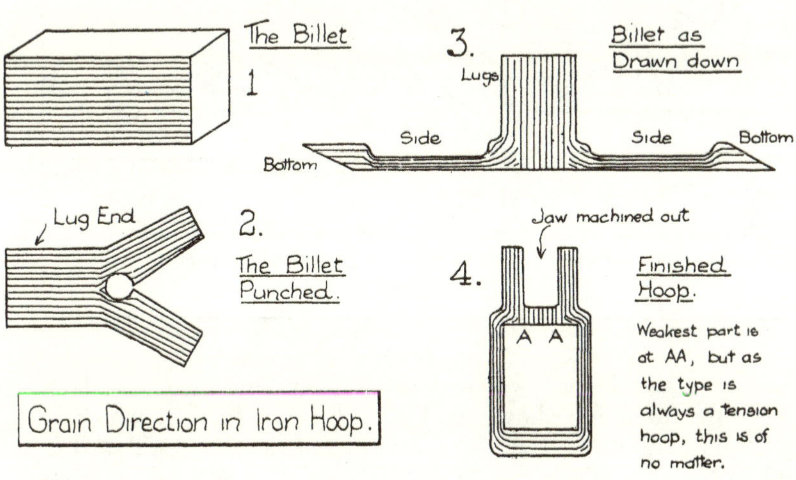

FIG. 251.

A matter of very great importance in the manufacture of an iron hoop is naturally the direction of the fibres—as in general, iron of the best quality is at least 30 per cent. weaker across the grain than with the grain. This point is given very special attention by leading manufacturers, and Fig. 251 illustrates some aspects thereof. For instance, in the hoop made by the process 250—A, if the forging were done as shown, the hoop would be liable to break across the main suspension pin if iron were used. With soft steel—which is generally employed on the Continent for work of this description—the question does not arise, as steel well-worked is of practically the same strength in both directions.

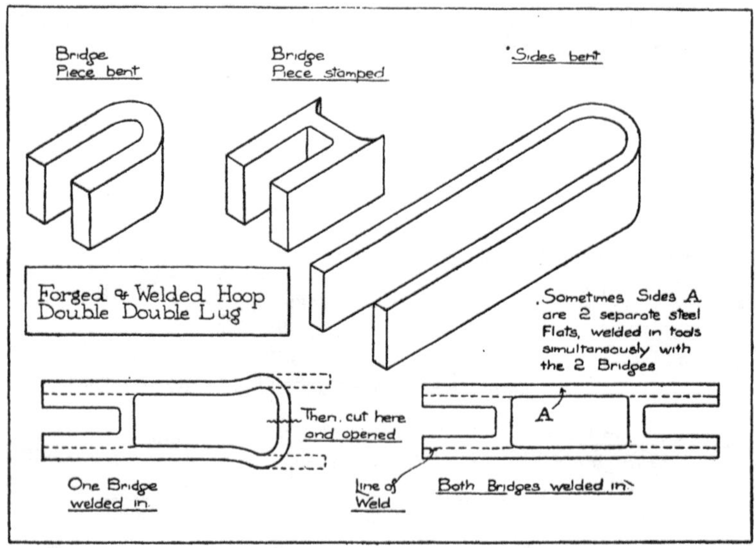

FIG. 252.

Fig. 252 shows two methods of making the standard "double-double" shown in Fig. 246—A. In one, which is still largely used, the sides A which take the pull of the drawgear, are of steel, and the two bridges are welded between with special tools and arrangements. As a welded hoop, the manufacture is sound, owing to the traction stress being transmitted through the two mild steel sides, and the hoop gives satisfaction in service. The other method shown involves first the bending of a flat of correct side section (actually slightly thicker). The first bridge is then inserted into the U and welded in. (For a double-lug hoop, this completes the manufacture, after the box has been squared up). The bottom end of the U is then cut and flattened out, and the second bridge then welded in. By thus handling the job, it is considerably facilitated for the smith, as he has a solid half completed hoop to work with when finishing the second bridge.

Fig. 253 shows a method of making the square socket hoop shown in 246—E. The first operations are on the same lines as for the double-double. When this stage has been reached, pieces are dabbed on, as shown at 253—B, and the whole end

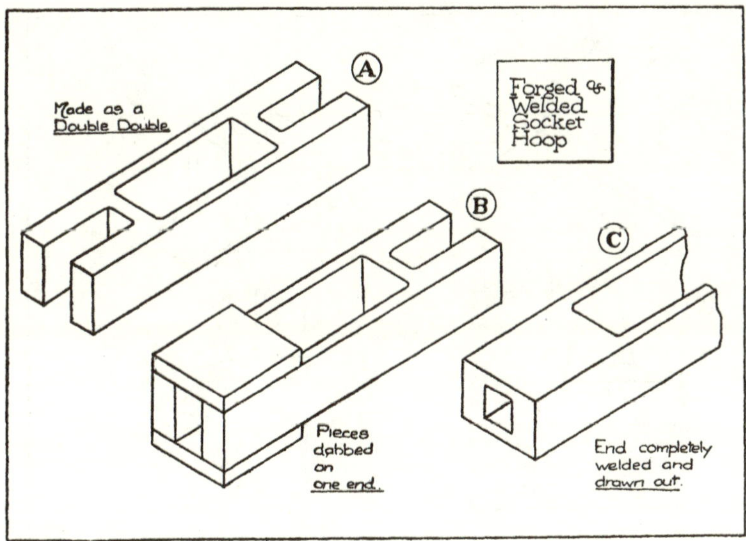

FIG. 253.

then raised to a welding heat and worked over a square mandrel—the socket lugs being drawn out in length as the closing pieces are gradually welded in.

Fig. 254 is the usual method of manufacture for flange hoops—involving as shown, the crossing of two strips of iron, welding them and then bending to the box shape and further welding. The lips on the flange are formed by a further heating and bending under special tools, the finished hoop being as shown at 254—B. No. 254—C shows a Continental carriage and wagon flange hoop, with a thick bottom end and pap. This is generally made from soft steel—a piece of rolled material being drop-stamped into the cross shape shown at 254 D, the stamping forming the pap. The sides are then bent up and the hoop is welded at the top in the usual way. These hoops have not always the small flanges at each end of the main flange, frequently two slots or holes being arranged in the ends, and securing bolts passed through these into axlebox recesses.

The manufacture of welded " band " hoops, particularly when round corners are shown, and the hoop is one thickness on all four sides, is very simple, and these can be turned out at a high rate of speed, exact to dimensions, and of very good

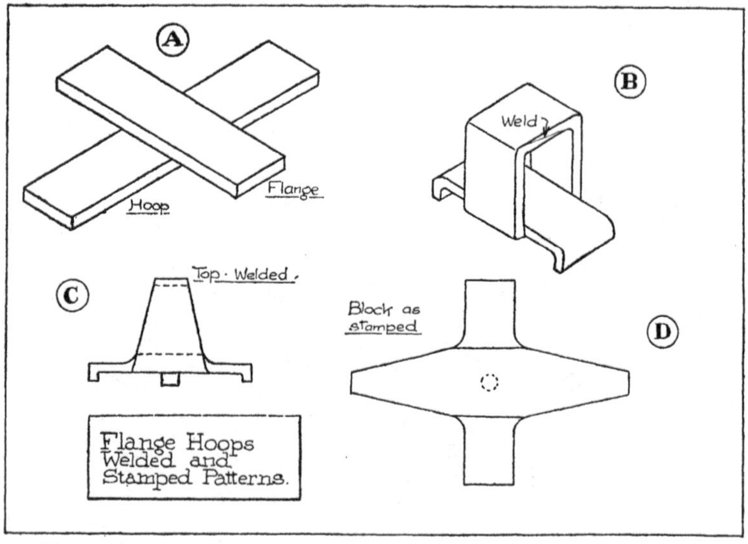

FIG. 254.

finish, by a smith working with a few cheap tools under a light steam or air hammer. Hoops of the " one-thick-end " or " two-thick-end " types are more costly to make, and in many cases, there is no particular reason for their employment. On occasion, " one-thick-end " hoops can be manufactured by causing the weld to come on the thick-end. If this is, however, too large in dimension to be thus accomplished, the bar stock must be ordered of greater thickness, and drawn down to suit. In American practice, all such hoops are made as plain hoops (one thickness all round) and a seating to match is " welded " in to bring the bottom to the required dimension.

Rounded ends on welded hoops are generally produced under smith tools by stamping. Another method is to weld up the bar, and then mill the sides down as required. These three manufactures are illustrated by Fig. 255.

On occasion, forging machines of the bulldozer pattern are used in America for the manufacture of band hoops, but these are relatively expensive tools, and a simple machine for the purpose is the hydraulic press, Fig. 256. Suitable dies, not shown in the picture, are affixed to the top and bottom vertical rams, and the sizing block is attached to the horizontal ram. The stock piece, being placed on the bottom die is formed into the necessary "U" shape, by the movement upwards of the bottom ram against the sizing block. The top ram

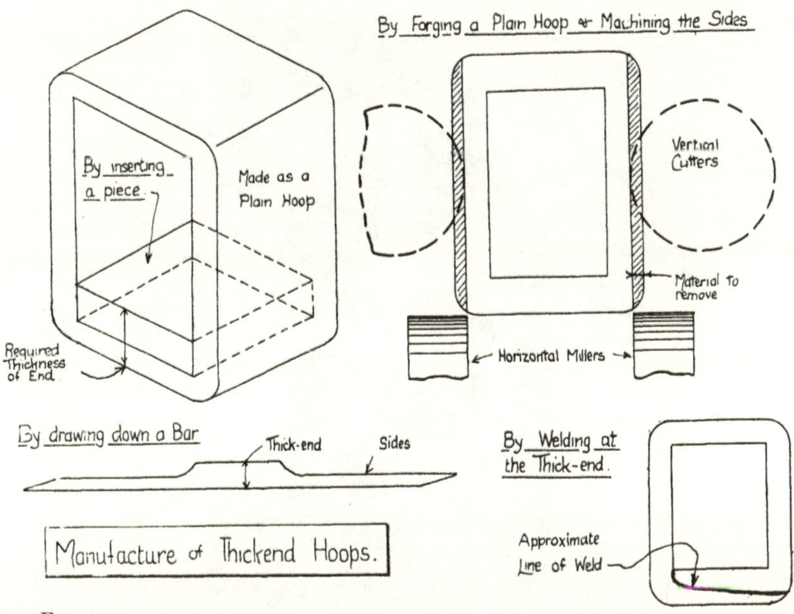

FIG. 255.

then advances and throws over the outstanding legs of the "U" to an angle of about 60°. The top ram then returns, and a loose tool is introduced which flattens down these bent-over legs and completes the hoop—the bottom ram meanwhile having held it up to the block. The horizontal ram is then returned to its "in" position, and thereby strips the hoop off the block.

Such a machine is of value where semi-skilled labour only is available—but the manufacture of the plain hoop has been so specialized in this country and the Continent, that it is doubtful whether sufficient advantages present themselves to justify any change-over from existing methods. In such manufacture, European methods are easily ahead of American practice, when all things are taken into consideration. Also, without any doubt whatever, the Sheffield-made railway spring hoop of this type is from all points of view—accuracy, soundness of welding, and finish—better than any contemporary production.

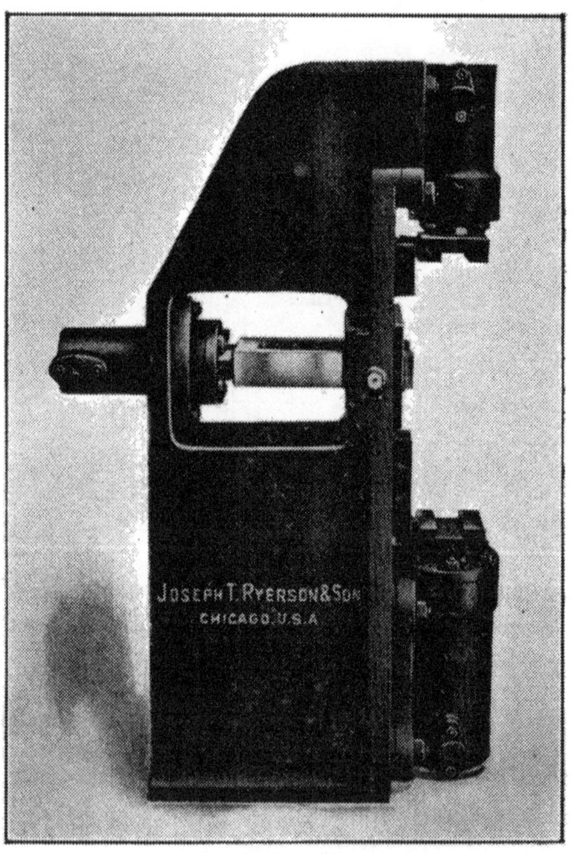

FIG. 256. PRESS FOR PLAIN HOOP MAKING. J. T. RYERSON & SON, CHICAGO.

CHAPTER XLIII

MANUFACTURE OF SOLID HOOPS

THE great diversities of practice between Europe and America as regards rolling stock construction are nowhere more clearly indicated than in the design of spring hoops. American practice is ignorant of almost every pattern of hoop except the "band" variety, and when this is used, it is always of simple design. On the other hand, it is no uncommon thing to have twenty well-recognized patterns of "special" hoops, with perhaps five designs off each pattern, on the floor of a Sheffield shop at the same moment. The result of this is that manufacturers on this side have not been slow to seize upon any process which would produce such hoops with greater satisfaction and lower cost than the forged hoop, and this has been found by the assistance of the machine tool firms, and tool steel makers, by which the method of "machining from the solid" has become recognized practice, to the almost entire exclusion of the forged "special" hoop.

The machining of large quantities of hoops from the solid has been entirely developed in this country, and has been made a commercial proposition chiefly by the introduction and continued improvement of the heavy "puncher-slotter." Clearly, hoops could always be machined out of the block, and frequently were by European locomotive builders, with the ordinary machinery of engineering shops, but as the heaviest operations involved are the slotting of the box and lugs, such hoops were something of a luxury from the point of view of cost. The heavy slotter, has, however, by general design, solidity of build and ease of operation, brought into being here the now general procedure of cutting "special" hoops from the block, instead of forging and welding them.

The standard R.C.H. double-double hoop (246—A) is now produced in very large quantities, and a *résumé* of the detail manufacture of this will probably be interesting, as it is the only type of hoop employing specialized machinery. Fig. 257 gives views of the successive stages in the progress of this hoop. The bar is rolled to $4\frac{1}{4}$ in. by 3 in. section (257—A), with practically square edges, from dead soft steel blooms. It is sometimes hot-sawn to 18 ins. long (one block), but more frequently supplied in multiples of this length, generally 2, 3 or 4, as, in high production mills, the hot saw cannot deal

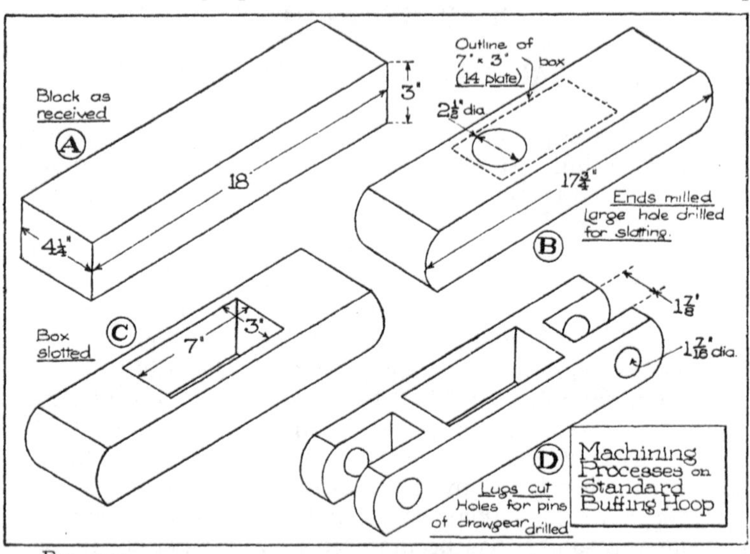

FIG. 257.

with 18-in. lengths. Such multiple lengths are left to be cold-sawn in the spring machine shop. It is very essential that the section should be truly rectangular and not more than $\frac{1}{16}$ in. over the nominal sizes, otherwise great trouble is caused with the ultimate machining. The first operation consists in milling to length, with curved ends—the block after this being as shown in 257—B. This operation is invariably performed on a double vertical spindle milling machine, carrying cutters of the necessary form—the jig employed being arranged to hold six or twelve blocks. Fig. 258 illustrates a machine of this pattern, which is admirably suited for heavy milling work. The ended blocks are then drilled with a large hole, from 2 ins. to $2\frac{3}{4}$ ins., according to the size of the slotting tool to be used. This is done on a heavy pattern machine, the time for a 2¾ in. hole being about three minutes. Fig. 259 is an illustration of such heavy drilling machine, which requires no refinements as regards feeds and speeds, as the material to be operated on is always the same and the hole always the same. In any shop with

FIG. 258. DUPLEX MILLING MACHINE. KENDALL & GENT, LTD., MANCHESTER.

a reasonable production, one heavy machine of this description should be allocated for this work only. The block is then marked out for the slotting of the box, in the instance shown, 7 ins. by 3 ins. Two patterns of puncher slotter are employed, the double and single. Fig. 260 shows the duplex type, in which each head carries two tools and each table carries two hoop blocks, rendering it possible for four blocks to be machined simultaneously. Each head is independently, driven, having its own motor, so that either can be stopped for re-setting as required. The provision of the two heads is particularly useful on heavy jobs where two tools cannot be employed on each head, such as cutting out lugs, as in a case of this sort, one head can be working two tools on boxing hoops, and the other one tool on lugging hoops.

For continuous work on double-doubles, the single type machine illustrated in Fig. 261 has certain advantages, as it

FIG. 259.—HEAVY DRILLING MACHINE. W. MUIR & CO., LTD., MANCHESTER.

is cheaper than the double machine and has about two-thirds of its capacity, owing to the provision of the revolving jig, shown in position on the table, which enables the idle time of the machine to be at a minimum, one pair of hoops being set whilst the other pair are being machined. The slotting tools used are of square section, generally jumped to $1\frac{1}{4}$ ins. or $1\frac{1}{2}$ ins. square at the cutting edges. The corners should be rounded with about $\frac{1}{8}$ in. radius. Tools up to $2\frac{1}{2}$ ins. wide

FIG. 260. DUPLEX PUNCHER SLOTTER. W. MUIR & CO., LTD., MANCHESTER.

FIG. 261. PUNCHER SLOTTER WITH SPECIAL JIG. W. MUIR & CO., LTD. MANCHESTER.

Fig. 262. Special Milling Machine. W. Muir & Co., Ltd., Manchester.

can be used for cutting lugs, etc., of this width The boxed hoop is as shown in 257—C and the next thing to be considered is the making of the jaw openings, taking the drawgear. These can, of course, be slotted on the machines illustrated, but a special milling machine has been evolved for this work, which gives good service. Fig. 262 illustrates this machine, generally known as the "twelve-cutter," and as will be seen, it is arranged with two shafts, each separately motor-driven and carrying each six milling cutters. Both shafts are fed simultaneously into the hoop blocks mounted in the central jig. With a shop having sufficient trade in the standard buffing hoop to keep one of these machines going, the provision of such is justified. but it will be clear that it is limited to this

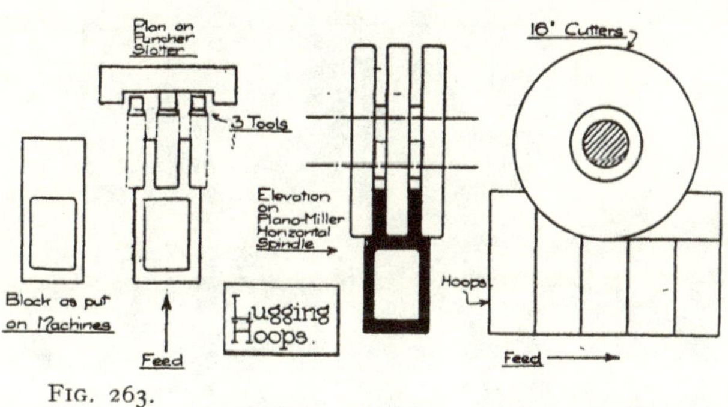

FIG. 263.

one type of hoop (meaning thereby the double-double 6 in. and 7 in. long boxes, or 12-plate and 14-plate) whereas the puncher slotter, which will perform almost equally well this lug forming, is an extremely elastic machine for general hoop types and can be diverted also to spring back-plate machining. With the introduction of the "solid" hoop, it would doubtless have been a good policy for the British spring makers to have contributed to a central plant for all the hoop and spring machine work, in which case specialized machines, like the "twelve cutter" would have justified their existence, but under present circumstances, only about two firms in the trade could really keep such a tool running all its time.

With the formation of the lugs completed, the hoop has then only to be drilled for the draw-bar pins (257—D)—best effected on a two-spindle drill similar to that shown in Fig. 183—and finally with the small rivet holes through each bridge to take the centre fastening.

The sequence just described in detail is fairly general for most hoop types, the greatest diversity occurring in the forming of the lugs, which are sometimes milled and sometimes slotted. Fig. 263 shows the tool arrangement for the

Fig. 264. Horizontal Milling Machine. Kendall & Gent, Ltd., Manchester.

two methods, and of the two, the milling is probably the better, as a run of hoops can be set up and finished throughout. A heavy machine for this class of work capable of using 16-in. cutters, is shown in Fig. 264, which is not only useful for the lugs, but also for machining blocks to width. The standard rolled sizes obtainable are the $4\frac{1}{4}$ in. by 3 in. and $4\frac{1}{2}$ in. by $3\frac{1}{2}$ in., most other hoop sizes having to be made as forged blocks. Frequently, however, by judicious arrangement, the standard rolled sizes can be employed, and generally speaking, it pays to machine such where required rather than to have forged blocks for special sizes.

Apart from the special machine tools described, it is obvious that it is necessary to have milling machinery of the ordinary pattern, and a good shaping machine is always an adjunct of value. For cotter holes, such as are needed in buffing hoops of the socket type, or for hoop slots (Fig. 117—B), it is possible to drill two small holes and slot to them with a small tool, or otherwise, use a key-waying machine. The latter is the better method, and a machine of this type which will slot each side simultaneously, with fully self-acting reverse and feeds, is one of the best for this work.

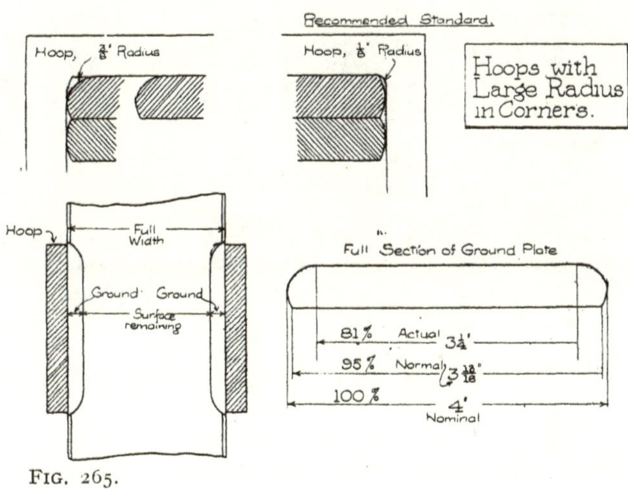

FIG. 265.

In some designs of hoops the inside radius of the corners is shown as ¼ in. and sometimes ⅜ in. Such patterns necessitate the edges of the spring plate being ground to allow for the hoop fitting. It cannot be called good practice as such grinding involves the removal of a fair percentage of very valuable material, particularly in a back plate, with a centre hole or similar fastening. Generally speaking ⅛ in. radius has been found quite satisfactory for inside corners, and if trouble has been experienced owing to fractures here, the whole question of design and material should be gone into before increasing the corner radius. Fig. 265 illustrates the point in question.

In addition to the testing of the material used in the manufacture of hoops, it is frequently specified that a certain percentage of finished hoops should be tested to destruction. In the case of " compression " (overhung) hoops, there is not much point in doing a great deal of finished hoop destruction. If the hoop is rejectably unsound, the probability is that it will crack when being driven on to the spring, or when finally flattened or pressed. Lugged (tension) hoops when welded, give information with destruction tests, but as every hoop cannot be so tested, there is always an element of uncertainty as to the absolute soundness of a batch. With steel hoops

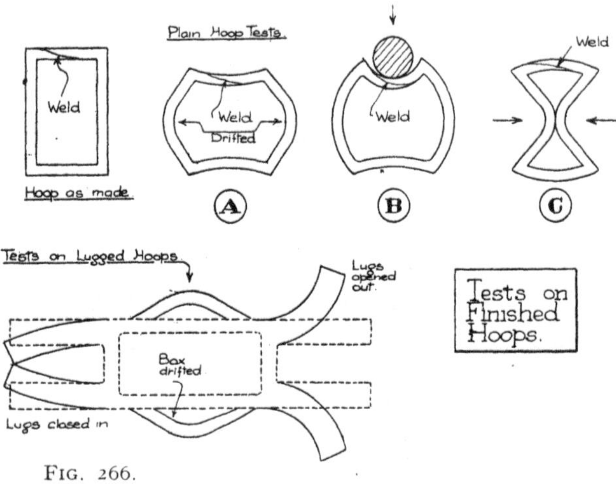

FIG. 266.

made from the solid, if the steel quality is right and the machining reasonably smooth, there is no need, except as regards general interest, to take any destruction tests. At times, a complete hoop of this type is taken for a tensile test, which will be found to confirm the calculated results based on the hoop design. Fig. 266 illustrates some tested hoops. The best test on the plain hoop for weld soundness is the one shown in 137B. It is not as spectacular as some others, but more severe on the weld.

Continental practice frequently insists on every hoop of the tension type being subjected to a specified load before it is put on the spring. This is usually done on a standard tensile testing machine, which is a relatively slow job, and a special arrangement was designed and made in this country to deal with quantities of hoops ordered under such test. This was such that it could be used on any ordinary load-testing machine, such as are essential in spring shops, and an illustration is given in Fig. 267. A general specification for this hoop test demands 20 tons pull on each hoop, and the machine was designed with this in view. It has, however, worked up to much higher loads. As will be seen from the small key drawing, two cantilevers on a common pivot are employed, and a cambered spring spans these and bears on the free end of each, the hoop bearing being one-quarter of the total lever distance. Accordingly, a pressure of 10 tons on the spring converts into a pressure of 20 tons on each hoop—the machine having been designed of the balanced form to take two hoops. Owing to the high stresses involved, and the limitation of the lever sizes by necessity of the hoops having to pass over them, careful calculations were necessary for these levers, which were made of special alloy steel, carefully heat-treated. The operation of the machine is clear from the drawing—it can be used either under a load-testing machine, or under a scrag, the loading indication for the latter being the deflection of the spring, as when this is straight, 20 tons is applied to each hoop. No difficulty is experienced in testing 120 hoops per hour with this arrangement.

Other processes have been more or less experimented with, or employed, for the manufacture of hoops from the solid, without machining. One of the weak points of the machining method is the large waste of material involved. This is not such a matter of high importance if the machine shop is in the possession of a steel plant, or if there are possibilities of closely adjacent works using the scrap cuttings. If, however, the steel blooms or bars have to be transported some distance, and then the scrap has to have a further transport, the economics of the job have to be studied. With "special" hoops this is not such a consideration as with "band" hoops, owing to the very high forging labour costs on the former compared with the latter—which cause it to be on the whole, much cheaper to cut 50 per cent. of the bought material to waste by making the hoop from the solid. When, however, "band" hoops are specified to be cut from the solid, it becomes a matter of serious study—as, made in the ordinary machining way, they become a very expensive hoop. Generally, a large block is taken of multiple size,

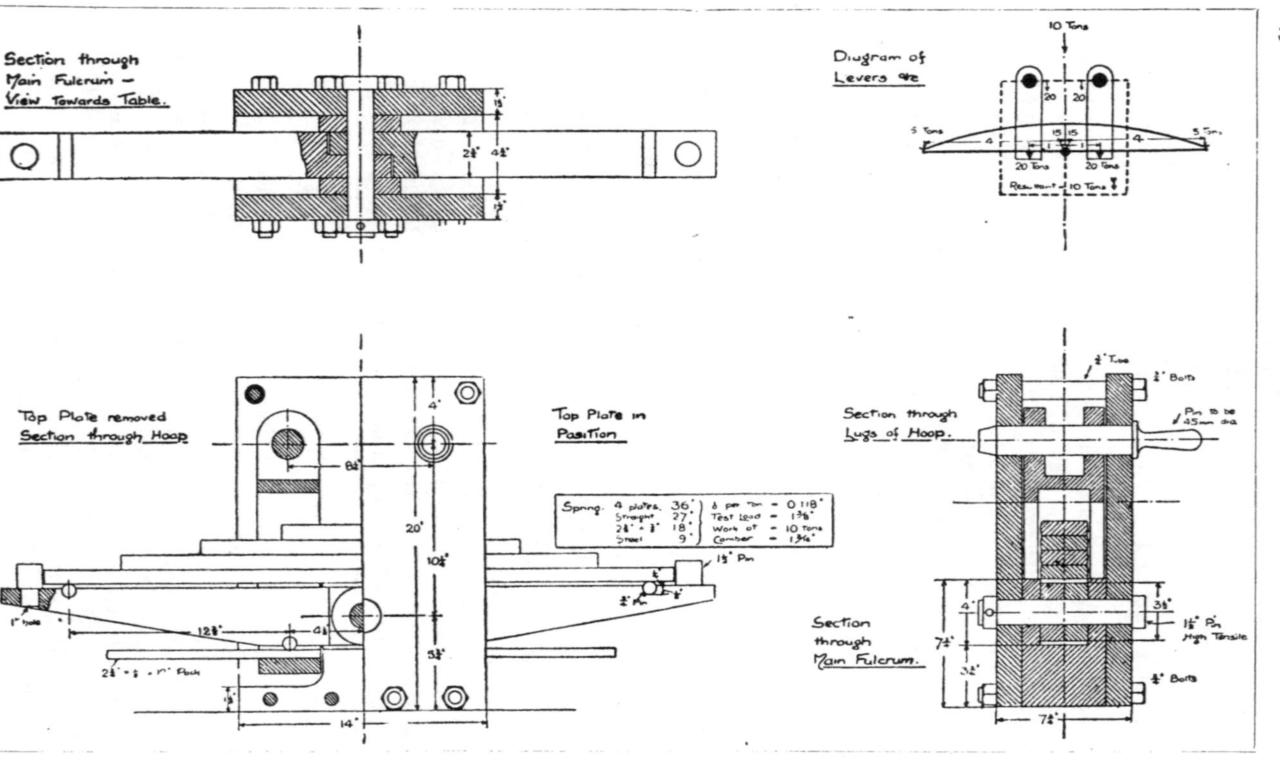

FIG. 267. MACHINE FOR TESTING SPRING BUCKLES.

the boxes of perhaps four hoops slotted in, and the block then parted up, either on a saw of some form, or on a parting machine. Two alternative methods to this are illustrated in Fig. 268. In the first, the basis of the hoop is scrap ends off round bars. Any size available is taken, and the necessary calculations made to determine the size of the drilled hole necessary. This is put in to suit, and the ring then drifted out to a specified size—afterwards being squared up on a block of the correct dimensions. Thick-end hoops can be made this way by machining down the two or three sides as required. The drifting operation can be replaced by a

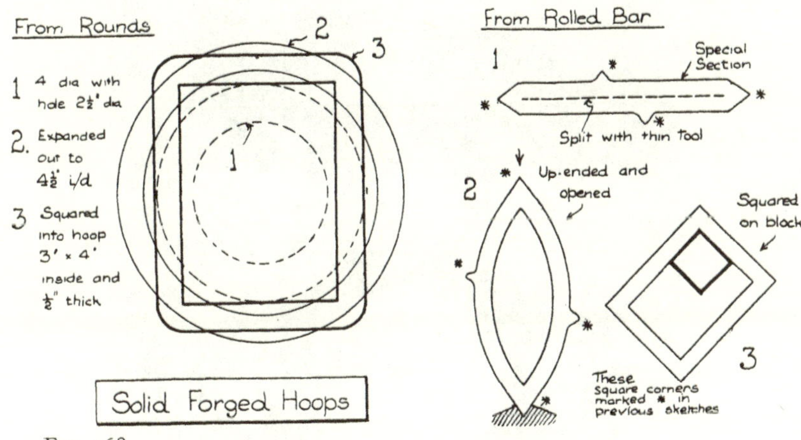

FIG. 268.

miniature rolling mill, and this is the case in certain Continental production. In any case, the demand of one standard sort has to be high, and furthermore, the basis material must be extremely cheap, otherwise hoops made by this process cannot compete in price with hoops made from the welded bar.

The second process shown depends upon a special rolled section, and obviously considerable quantities would be required to justify this. The method is being used in Germany, and the illustrations make clear the sequence of operations, all of which are done hot. A " hoop from the solid " is undoubtedly obtained by this method, but it would not appear to be as low-priced as the hoop welded from the bar. Attempts have been also made to electrically weld up two " U " pieces, but these will not stand the shrinkage which comes when they are cooling on the spring.

Drop-stamped lugged hoops have been attempted with a certain degree of success, but here again, it is necessary that large orders should require fulfilment before the cost of the stamping tools can be justified. Fig. 269 shows a double-double hoop of this kind " as made " and " as tested." The saving in weight is not as great as might be imagined, owing to the loss involved in the high heats to which it is necessary

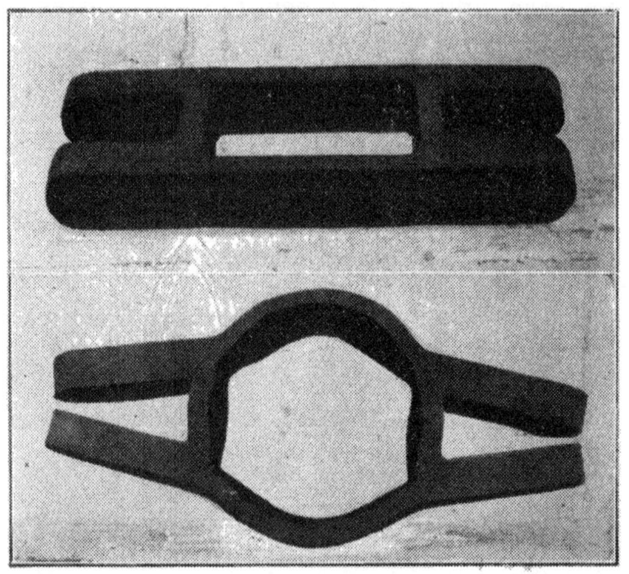

FIG. 269. DROP STAMPED HOOPS.

to subject the steel. Additionally, they are somewhat heavier than the welded or " solid " hoops, owing to the " draw " necessary to enable the stampings to be removed from the tools. The double sloping sides caused by the need of this " draw " result in making the hoop very unpleasant to drill—this cannot properly be done unless the centre position of the holes is first flatted by grinding or other means. Doubtless this objection would be overcome were there sufficient demand for the drop-stamped hoop, as centres could be pressed in as the hoop came hot from the stamps. However, preference is always given to the machined-from-the-solid hoop, and there are such a number of specialized plants now engaged in this country on this particular work, that the prices are at a low level, and it is difficult for any other method of hoop manufacture to successfully compete.

One of the most difficult hoops to make from the solid is the " flange " pattern (Fig. 254). Certain definite requirements, however, continually arrive, and at least one firm now buys these as-rolled bars, available for several thicknesses of spring. This considerably relieves the amount of machine

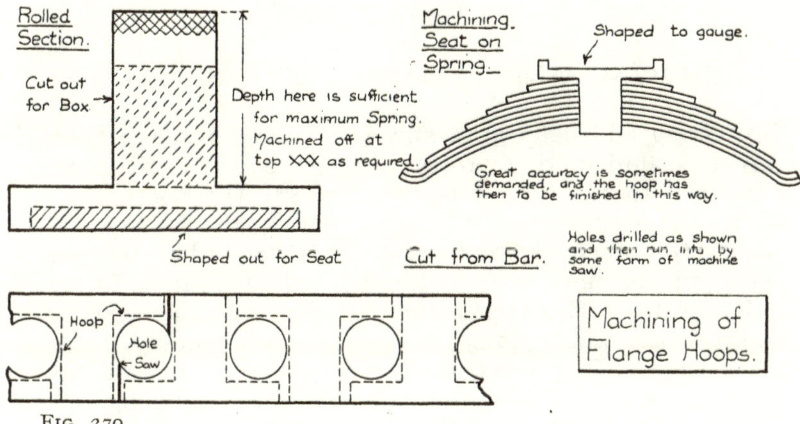

FIG. 270.

work. The section supplied is shown in Fig. 270 together with the usual method of manufacture when the hoop is entirely machined out.

Within recent months there has been considerable development with a flame cutting arrangement which is claimed to burn the hoop shape out of the solid at a cheap rate and of such perfection as to require no after machining. So far, this new method has not been adopted by the trade here, but its possibilities are great, and the developments should be watched with interest.

CHAPTER XLIV

HOOPING AND FINISHING THE SPRING

The spring and hoop have now been traced through their various manufacturing processes and the two have to be assembled to form the complete article. The hoop is invariably heated for putting on the spring, the heating expanding it sufficiently to allow of its being readily dropped over, and the subsequent contraction tightening it on to the spring plates. This contraction is sometimes hastened by dropping the whole spring in a water bosh. Badly maintained hydraulic presses, are however, quite an efficient substitute for the bosh, and save additional handling. It is also generally specified that hoops are to be pressed on with some form of pressure. Hand screw vices have been used for this purpose, but they are now obsolete and all manufacturers employ some form of power press, designed to exert a pressure of about 40 tons. The hoops are heated to about 800° (Centigrade) in either a gas or coal-fired furnace, and whilst heating, the spring is screwed up tightly in a vice. The hoop should be of such size as to require lightly driving over the plates. This is a very important matter, as if eth hoop is too small in the box, heavy driving is required, which may distort the lugs and damage the surfaces, and on the other hand, if the hoop is too large in the box, the subsequent pressing on will cause the corners to be lapped, owing to the surplus metal forced into them. Owing to the curvature of the spring, a gap is left at the edges of the hoop, which is frequently caulked up with a misguided impression that it improves the appearance, as it makes the hoop look tight. It is clear that this caulking is of no value, as it disappears as soon as the spring takes the load and flattens out. It is still worse practice to " pan " the hoop to the spring curvature, as in this case, the short plate works from the edge and has its effective length reduced accordingly, tending to assist in an early fracture, as railway springs are invariably designed, and tested, on the basis of the whole working length being effective. Springs with numerous plates of different thicknesses, with high camber present a difficult problem to the

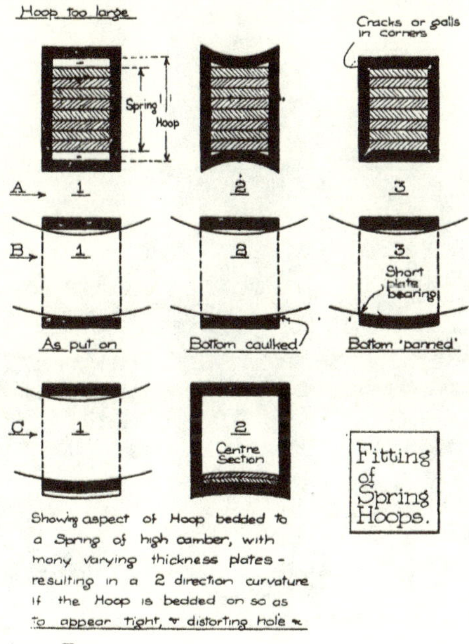

FIG. 271.

hooper, as to make the hoop to appear anything of a fit, the bottom has to be curved in two directions. Fig. 271 illustrates the foregoing points.

A hooping press of English pattern is shown in Fig. 272. Such presses have one open side and vertical and horizontal cylinders, arranged with either hydraulic power or balance weights for the return (idle) stroke. The press shown has the latter arrangement. The open side has its advantages, particularly in the case of the long springs which are made for buffing work, but it is also an element of weakness in the design, and it will be noted that the American pattern, Fig.

273, is a four-sided press, with very heavy tiebolts in addition. The spring drawback to the main rams is a good feature, as it simplifies the valve gear in comparison with the hydraulic drawback. The press shown in Fig. 272 has combined with it a buckle-stripping device and this is shown in operation in the illustration. The spring is gripped between the heavy screws shown and the cross lever, with suitable packs, bears on the buckle. Application of pressure to the horizontal

FIG. 272. HOOPING AND STRIPPING PRESS. CRAVEN BROS., LTD., MANCHESTER.

ram moves forward the cross lever and strips off the hoop. When required for hooping only, the whole of the stripping gear swings out of position.

In Fig. 274 is illustrated a hooping press of Continental design and manufacture, which is more complex in construction than the usual British pattern, but is more economical in the use of pressure water, owing to the idle distances worked being operated by the small cylinders only, the pressure only coming on to the large cylinders when the tool faces are up to the hoop. Whilst the economy of hydraulic power is undoubted, it is probable that this would be to a certain extent outweighed by the higher capital cost and maintenance needed. The practices of both this country and America are to employ the simplest possible devices for spring work, which is regarded as a rough trade, and most firms here prefer to pump more water for simple presses, than to operate and maintain more valves and attachments for the benefit of water-saving devices.

FIG. 273. HOOPING PRESS. TINIUS OLSEN, PHILADELPHIA.

As previously hinted, hydraulic presses in spring shops are not always the happiest of tools, as furnaces and dirt are very prevalent, and on presses of this description, quick operation is required—with good organization and simple springs, 60 and more per hour can be hooped. It is not a matter of surprise, therefore, that press leathers become leaky, and frequently no time can be spared for their replacement, resulting in constant streams of water along the rams. Accordingly, pneumatic presses are in common use in America,

FIG. 274. HOOPING PRESS. ENGEL & BIERMEYER, HAGEN, GERMANY.

where the use of air, as is well-known, is more highly developed than on this side. Air can leak, but with less unpleasant effects than water. Fig. 275 is a drawing of a pneumatic type, intended for a pressure of 60 tons on each ram. The valve control is, of course, simple compared with that of the hydraulic press, although the latter can be made simple enough if manufacturers would try and dispense with controls requiring numerous leather packings, which are generally a constant course of trouble. Another improvement that

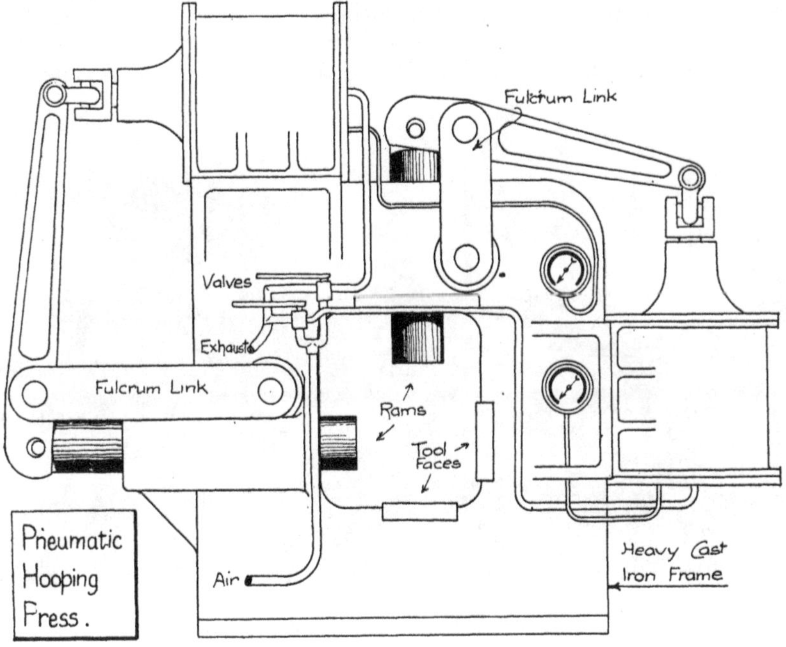

FIG. 275.

could be made with advantage in the direction of simplicity, and which can be recommended to European makers, is shown in the American design, Fig. 273, namely, the substitution of springs for the normal drawback rams. Such arrangement reduces the number of valves and pressure leathers, in addition to ensuring that there are no drawback pistons to fly into the shop roof. It is generally considered by designers that brass must be used for the drawback rods, and this is a very doubtful material for working in tension under high pressure.

A separate hoop-stripping device is shown in Fig. 276, working on the principle as that on the press, Fig. 272, the action of which will be readily understood from the above description. Such machines are useful accessories to the railway shops, but are not particularly needed in spring manufacture. The American hoop-stripper is shown in Fig. 277. This works along very different lines, as indicated, and it would not strip the long springs which are common to British practice. It is quite suitable, however, for the shorter and lower cambered American spring.

A simple method for the heat-stripping of hoops, with no damage to the hoop, is shown in the sketches, Fig. 278. The

FIG. 276. STRIPPING MACHINE. CRAVEN BROS., LTD., MANCHESTER.

spring is placed on a gantry between two small blowers, with guard plates each side of the hoop, which is rapidly heated to redness, and drops off the spring, or can be tapped off with a hammer. This particular device is Continental, and would not be as easy of application to the usual British spring which is rivetted through. Many thousands of Continental springs, particularly those of the Latin countries, have the spring plates only rivetted together, the rivet not going through the hoop. These plates cannot therefore, get apart through any "nip" being present, and spring on to the hoop—which renders it the easy matter just described to perform the stripping.

A device sometimes used in U.S.A. practice is the hydraulic clamping machine. It is a very heavy job to handle a large engine spring for hooping in the way it is usually done, namely, to stand the spring vertically in a heavy vice, tighten it up, then remove any temporary hoop or fastening which

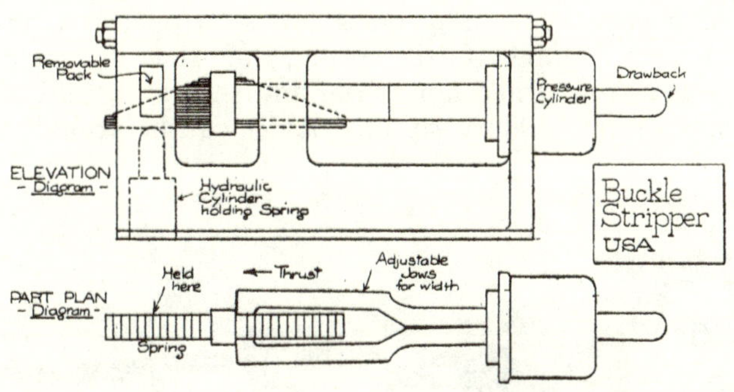

FIG. 277.

has clipped it together, drive on the hoop, then carry the spring to the press and lay it there for final pressing. In the U.S.A. it is still more of a heavy job, owing to the great weights taken per axle and the necessarily heavier springs needed to carry such weights. Owing to this, the clamping machine has been developed and Fig. 279 shows one pattern. The spring is placed plate by plate on the table and the hydraulic cylinder then operated, which forces all the plates

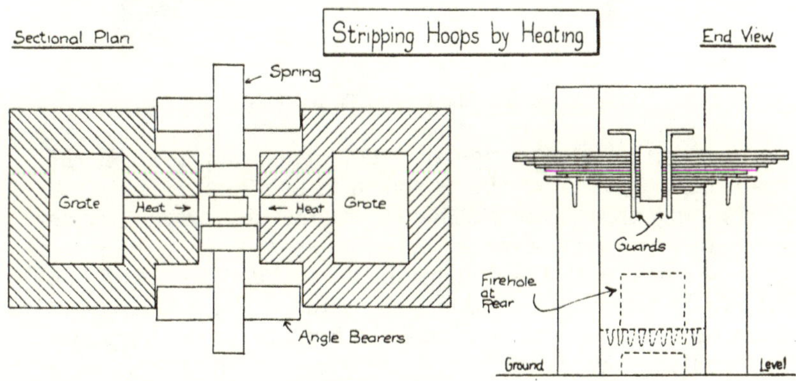

FIG. 278.

into close contact. By withdrawing the support marked on the left, one half the table tips over into a vertical position, when it is possible to drop on the hoop and release the pressure.

The last word in hooping devices is probably the American press, which combines in the one machine (see Fig. 280), a clamping press, a hooping press and a stripper. After the hoop has been dropped on the spring, the table of the clamping press is again turned horizontal and the spring slid under the hooping rams. The stripping device can be brought into

FIG. 279. SPRING CLAMPING MACHINE. TINIUS OLSEN, PHILADELPHIA.

action by the introduction of a supplementary ram on the main horizontal ram; this is shown in position.

Hooping presses are excellent machines as regards saving hard and heavy hand work, but it is essential that the same care be taken with the dimensions of the hoops as was taken when all springs were hooped by hand. The tendency of such presses is to lead to hoops being made on the large size to facilitate "dropping over" the spring, and then leaving the press to force the superfluous material into sides and corners.

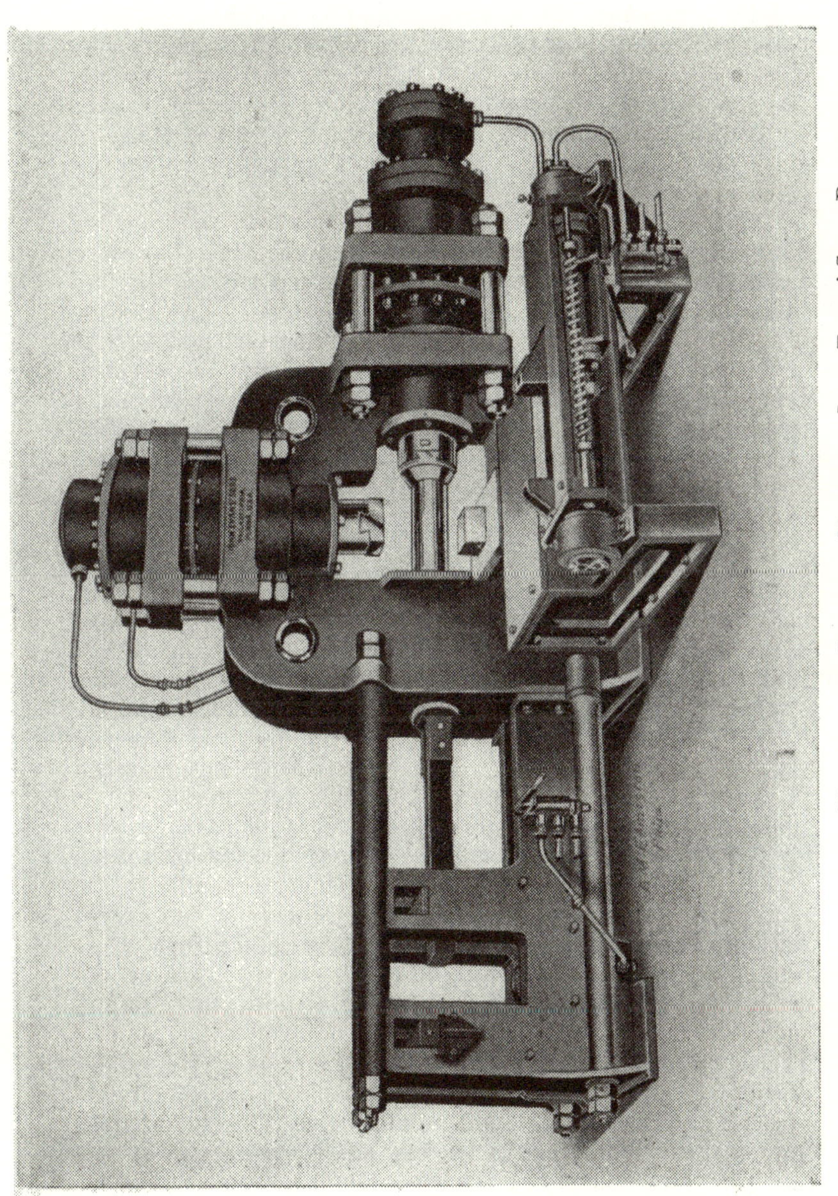

FIG. 280. CLAMPING, HOOPING AND STRIPPING MACHINE. JOHN EVANS' SONS, PHILADELPHIA.

Before hooping, all springs should be painted, plate by plate, with either oil, or a mixture of oil and graphite. An incidental advantage to the handling of each plate for this painting is that on occasion the men engaged find a cracked or broken plate which has not previously been detected. A good graphite coating will remain as a rust-preventer for considerable periods.

Automobile springs are frequently specified to have the plates ground all over. On rare occasions this has been called for in the railway trade, but except in one or two Continental instances, is now entirely obsolete. The grinding of automobile spring plates on their surfaces should also be obsolete. Sometimes when this bright finish is specified, every plate is ground all over, and at other times, and more frequently, only surfaces visible when the spring is bolted up are ground. The claims made for grinding reduce the three :

1. It reduces interplate friction.
2. It removes all skin cracks.
3. It improves the finish.

The last claim is obvious, and quite non-technical, the only comment necessary being that as the spring is painted over the bright grinding, it cannot then be distinguished from the humbler unground spring. As regards the first claim, interplate friction is very much less than is generally supposed, Advocates of laminated springs, however, frequently justify their arguments on the basis of plate friction, and then probably include grinding in their specification. It is considered that the friction damps down oscillation, but this idea is doubtful, the sluggishness of the laminated spring, with its apparent dampening effect, being rather due to the large amount of stressed material therein, which under tension and compression has to move from, and return to, its original position. Generally, however, if the idea of the reduction of friction is at the bottom of the desire for grinding, such can be entirely eliminated, as it is not worth the expense. The second aspect—removal of skin cracks—is very specially unsound. Any spring steel will probably show, under fine wheel grinding, slight longitudinal seams, due to inherent ingot defects, but these in no way affect the spring service, and can be completely neglected. The transverse cracks are the danger, and these will not reach the grinding wheels if a proper scragging test has been applied to the complete spring.

On the whole, grinding provides more definite troubles than it eliminates imaginary defects, as, owing to the awkward shape and section of curved spring plates, the only practicable method is by hand. It is generally performed on a large grit-stone, and when finished, it is doubtful whether the

FIG. 281. GRINDING MACHINE.

thickness of the plate is as uniform as it was originally. Variations in thickness, along the same plate, obviously set up varying stresses, with potential breaking points additional to those already in the spring in the form of centre holes and other breaks in the regularity of the plate. Various machinery has been evolved for automatic grinding, but the ordinary stone still performs the bulk of the work. The practice is almost entirely confined to medium and high priced pleasure cars, made in this country and the Continent. It is interesting to note that the automobile springs produced in the U.S.A. off the wonderful high production plants, with their controlled furnaces, are not ground—whereas the "coach springs" which are water sprinkled whilst held between forming dies, and thus "hardened and tempered" are ground bright every time. It is merely a matter, one assumes, of trading in competitive markets.

A certain amount of finishing is sometimes required with railway springs, which is generally confined to trimming up with a file, and reamering holes. A swing grinder is of use on occasion if hoops are desired polished bright. Automobile springs require bushing, various types of bush being employed, including open-seam steel tube, cast iron, and different yellow metals. The job of importance to be performed on this class of spring after completion is the side-grinding of the eyes. This can be done with single or parallel emery wheels—the latter being obviously the better method as it reduces handling. If parallel wheels are arranged to run at a fixed distance, the introduction of the spring eyes casues severe wear on the stones, and a multiple machine intended to obviate this is illustrated in Fig. 281. The spring end is fixed in a jig, readily adjustable for various types, and the wheels are set about ½ in. wider apart than the spring width. Both wheels are fixed, and the saddle carrying the spring is traversed each way until the eye width is ground down true to gauge. Provision is made also for swinging off alternate heads if required, for grinding ends of cantilever or quarter elliptic springs, and a special traversing slide is arranged for these.

The lubrication of railway springs is a matter which does not trouble users—it is true that "private owners" wagon springs frequently are seen rusted up in service, but they have probably been on the vehicle for 30 years, without repainting. On the other hand, the lubrication of automobile springs receive a great deal of attention, the favourite device in this country (apart from the application of spring gaiters and similar fitments) being to interleave the springs with brass or yellow-metal sheets, machine perforated, and the perforations filled with grease of some description. An illustration of this arrangement is given in Fig. 282, which includes also views of the " dope-cups " stamped on occasion in the plate ends of American springs for the same object. Whether the points aimed at by the provision of either " interleaves " or " dope-cups " are fully attained, remains a matter for argument, but it would certainly seem that, automobile springs being comparatively easy to dismount, pull to pieces, and re-assemble, the soundest way of ensuring freedom from rust would be to take the spring down say, every six months and clean them through.

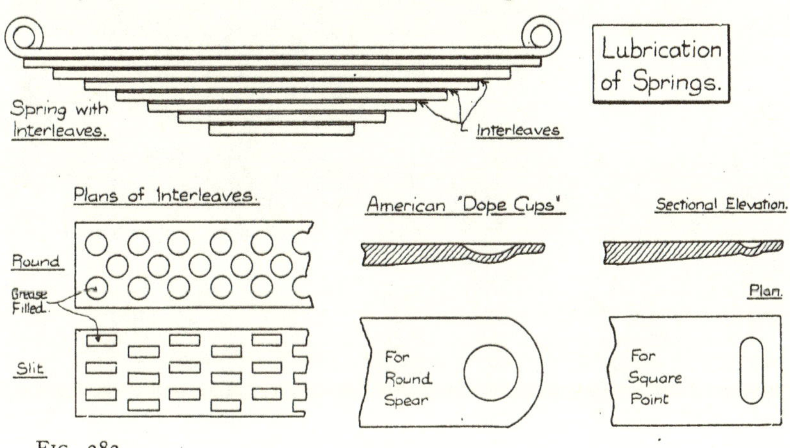

FIG. 282.

INDEX

C

G

H

www.ingramcontent.com/pod-product-compliance
Lightning Source LLC
LaVergne TN
LVHW091632100826
845152LV00001B/12

* 9 7 8 1 4 2 7 6 1 9 4 5 7 *